COMMUNICATIONS

SATELLITES

Third Edition

by Larry Van Horn

Grove Enterprises, P.O. Box 98, Brasstown, NC 28902

Printed in the United States of America

Cover Photo

The Apollo 11 moonship sat on the launch pad like a queen on a throne, illuminated by spotlights that crisscrossed and fanned outward in a rainbow of colors against a blue-black sky. White vapors of liquid gasses snaked outward and upward from the side of the ship as pre-flight crews readied her for launch. Her stately image reflected with magnificence from the calm waters of an inland waterway that separated the launch pad and press site.

At dawn the moonship rose slowly on an 800-ft-tail of red-orange fire; the earth shook beneath our feet as the rocket, three miles from where we stood, yawed and then straightened out in vertical flight. Hundreds of news cameras, including mine, clicked and whirred as the rocket ascended. But somehow, those action pictures lacked the elegance and the majesty that had radiated from the moonship on the launching pad the night before.

Wayne Mishler
October 1987

ISBN 0-944543-01-4
Library of Congress Catalog Card No. 87-082311

Introduction

For months I had been reading *Popular Electronics* and *Electronics Illustrated.* I would send in reader service cards and send off for catalogs to check out the latest in receiving equipment. I devoured every ad and daydreamed in school about owning a big, multi-knob shortwave receiver.

Finally, Christmas 1965, my dream was fulfilled. To a youngster of 13 even a used Hallicrafters S-120 receiver was a machine of wonder and amazement. After the antenna went up and the rig was turned on for the first time, the warm glow of the Hallicrafters' five tubes convinced me that I was going to hear what I wanted to hear -- satellites.

Back in those days satellite signals on shortwave were plentiful. Columns in the major electronic magazines gave numerous HF radio frequencies for satellites, both U.S. and Russian. There were even voice channels on shortwave to listen to.

Since those early days I have spent a lot of time in front of many different radios seeking out the weak telemetry of a geostationary satellite or voices from manned spacecraft.

The introductory chapter will examine the different aspects of satellite monitoring for the beginner -- frequencies, receivers, antennas and accessories, and general information. We will provide computer programs for geostationary satellite pointing, reception techniques, hardware, and sources for additional information (also to be found in the Appendix). The necessary terms, conversions and formulas that might be needed to get you started are covered in the Glossary section.

We would appreciate your contributions as well; they may be sent to the author at the following address:

Communications Satellites
c/o Larry Van Horn
160 Lester Drive
Orange Park, Florida 32073

Author Larry Van Horn

Credits

Since the first edition of *Communications Satellites* a great deal has happened in the space program. January 1986 saw the success of Voyager 2's photo mission as it passed Uranus, and the tragedy of the Space Shuttle explosion on liftoff.

A whole series of international satellites visited Halley's Comet in March of 1986 and the launch of the world's first high power space telescope from the Space Shuttle will aid scientists to see farther than ever before from beyond our atmosphere.

Our future in space will require listeners to keep pace. The advent of new receivers will make the challenge a little easier. New home computers help turn data telemetry into useful information for the monitor.

This book has not been a one man show; many aerospace industry public relation shops have provided me with valuable information. I am deeply grateful to the following companies and government agencies for their support in making this edition of *Communication Satellites* possible: AEG-Telefunken; Aerospatiale Inc.; American Satellite Corp.; Ball Aerospace; Boeing Aerospace; Bonneville Satellite Corp.; British Aerospace Dynamics Group; COMSAT General, Inc.; Department of Commerce NOAA/NESS; Dornier GMBH; European Space Agency (ESA); Fairchild Industries, Inc.; Ford Aerospace; Hughes Space Divisions; Indian Space Agency; Lockheed Missiles and Space Co.; SA Matra; Messerschmitt-Bolkow-Blohm (MBB); National Aeronautics and Space Administration (NASA); National Space Development Agency-Japan; RCA American Communications; RCA Astro Electronics; Rockwell International; SPAR Aerospace Ltd.; TRW, Inc.; United States Air Force-NORAD; United States Air Force Space Division; Western Union; and AMSAT-US.

Several individuals also have made personal contributions. I appreciate the assistance that these experts and monitors have given me in the preparation of this book. These people include: Bob Grove, Mark Severance, Brian Webb, Robert Horvitz, Nicholas Johnson, Mike Pruden, Lynn Kennedy, David Coursey, Fred Varian, Robert Popham, Grant Zehr, Tom Van Kuiken, Paula Murphy, members of the Kettering Group, Stephen Padar, Richard Kramer, Jeffery Wallach, Michael Morley, Chris Rodgers, Dave Headland, Geoffrey Falworth, and John Biro.

I would also like to acknowledge all of the individuals who wish to remain anonymous. Words cannot begin to express the respect and pleasure of knowing each of you, and the value and quality of your information.

I would like to dedicate this book to three people who have been very influential in preparing this third edition. To Wayne Mishler, thanks for being a true friend, a great boss, and a fine journalist and teacher. To my wife Gayle, I cannot begin to thank you enough; your love and devotion has made this third edition possible. And finally to Bob Grove who believed in my dream and has given me the chance to let others share in the enjoyment of listening to communications satellites.

On behalf of my son Loyd, Gayle and myself, I am proud to present to monitors everywhere this third edition of *Communications Satellites*.

Larry Van Horn
August 1987

CONTENTS

Chapter One

MONITORING TIPS AND TECHNIQUES

"It is difficult to say what is impossible for the dream of yesterday is the hope of today and the reality of tomorrow."

Robert H. Goddard
The Father of Modern Rocketry

The Satellite Spectrum

One of the first questions the new satellite monitor must answer is, "What type of satellites do I want to listen to?" This will then determine the frequency range and receiving equipment needed to hear the target satellites.

To give you a better understanding of what satellite frequencies are available, what follows is a very simple synopsis of satellite downlink frequencies that can be heard with easily obtainable equipment. Appendix B in this book gives a more detailed, specific list of frequencies.

HF: 3 - 30 MHz

While most activity on shortwave has long ago moved higher in frequency, two distinct areas of the spectrum remain. These areas are used primarily by amateur radio satellites and Russian spacecraft.

The frequency spectrum from 29.3 to 29.5 MHz is used by Soviet amateur spacecraft of the RS series for downlinks. The listener will hear CW, RTTY and SSB voice transmissions. These satellites represent the best opportunity for the beginner to monitor satellites. Orbits are predictable and information on these satellites is widely available.

The other range in the HF spectrum also contains Russian satellites. Listeners will find the majority of the activity in the 19.0 to 20.1 MHz range. The signals will consist of data telemetry from a variety of Soviet orbiting spacecraft. No voice will be heard.

VHF Low Band: 30 - 50 MHz

Currently, the only known activity here centers around 40 and 41 MHz from the Soviet Interkosmos satellites. Listeners might want to watch for data signals with doppler shift and report what you hear. I know of no voice activity in this range.

6 Meter Ham Band

The Russians have invaded the 6 meter amateur band; their satellite CW beacons in the 51.0 to 52.0 MHz range are coming from some of their most sensitive intelligence satellites and there could be more yet to be discovered in this range of frequencies.

VHF High Band: 135.555 - 144.000 MHz

This range of frequencies includes the 136-138 MHz satellite band. For years satellites have filled that spectrum with all types of telemetry downlinks. One of the few known voice channels in this range comes from the American ATS-3 downlink (135.555 to 135.650 MHz). Russian voice from the MIR and Salyut 7 spacecraft are also located in this band.

Transmissions can be monitored via narrow band FM. Weather satellite FAX signals are widely heard in this range and can readily be demodulated. Most of the rest of the satellite signals consist of telemetry data signals.

2 Meter Ham Band

The frequency spectrum from 145.8 to 146.0 MHz supports numerous amateur radio satellite downlinks. CW, RTTY, Packet, and SSB voice transmissions are the major modes. The Uosat satellites stand out as their 145.825 MHz can be widely heard on the simplest of equipment using narrow band FM. These satellites even carry digital voice downlinks.

149.9 - 150.1 MHz

This frequency range consists mainly of U.S. and Russian navigation satellites sending out telemetry. I have one report of possible voice from the ATS series of satellites on 149.999 MHz.

150.1 - 225.0 MHz

I know of only sporadic activity in this range mainly from Soviet spacecraft using wideband data (FM-300 kHz wide).

Military Aircraft: 225.0 - 400.0 MHz

This is probably the richest region for satellite signals today. Most of the satellites that operate here are military, both Russian and U.S. The Space Shuttle also has voice downlinks in this area. Just about any mode can be heard in this range with satellite a mix of geostationary and orbiting.

400.0 - 1300 MHz

Activity in this area is sporadic, consisting mostly of wideband and narrowband telemetry. Some Soviet spacecraft activity can best be heard on SSB receivers.

1300 - 3000 MHz

There are some very rich satellite signals in areas of this spectrum for the non-voice satellite experimenter. The easiest to decode are from geostationary weather satellites sending FAX signals to ground stations. Beginners should ignore this range, definitely for experienced monitors.

3000 - 5000 MHz

This is the domain of the TVRO satellites. All forms of voice, video and data channels will be present. Equipment costs have come down over the last few years and all of the satellites will be geostationary. This makes the tracking equation very easy. There is a lot of activity here and the satellite monitor equipped to monitor this range will find it rewarding.

5000 MHz and Above

With the exception of the Ku TVRO band (equipment is readily available) all of the activity is for the experimenter only. The move over the last few years has been toward higher frequencies; I see that trend continuing as the available spectrum becomes more crowded.

Receivers

To help the beginning satellite sleuth, this section will discuss receiver philosophy as it pertains to satellite monitoring.

By now you should have an idea what frequency ranges you want to monitor. Now you must decide what your level of involvement in the hobby will be and how much you are going to spend on your equipment.

Opinions on equipment are like opinions in general: Everybody has one. In this article I will give you my opinions on the best equipment to use in satellite monitoring based on 21 years' experience in the hobby. I am sure there will be more than a few that will argue my selections.

HF Receivers

To hear the bulk of HF satellites, you only have to concern yourself with three frequency ranges: the 10- and 15-meter amateur bands and 18.0 to 20.5 MHz. There are a multitude of HF receivers and the smart monitor will purchase a receiver for general monitoring as well as satellite sleuthing. Get the best general coverage receiver you can buy.

Keep the following attributes in mind: A stable receiver, sharp selectivity, selectable modes, digital frequency readout, and high sensitivity. These characteristics are a must for a top-notch receiving station.

Those Specs:

Receiver stability is very important for accurate Doppler determination and telemetry analysis. If the receiver drifts these important tools used in satellite monitoring cannot be used.

The HF bands are very crowded and hearing a weak satellite signal on a busy frequency can be difficult. The receiver needs to be selective in order to separate satellite signals from earth-bound signals.

Proper mode selection is important to hear the satellite. Monitoring an SSB satellite signal in FM will not yield any intelligence for the monitor. I have found that USB/LSB/CW selectable type rigs perform better overall than BFO product detecting receivers.

Digital readout is very important for getting you on the right frequency to hear a satellite's signal. Without an accurate means of determining your frequency, you will have to perform band sweeps. This may work for finding new satellite frequencies, but not to monitor known channels.

Finally, the single most important attribute to look for in a good HF satellite receiver is sensitivity. You have to hear 'em to work 'em. Bottom line is get a sensitive receiver.

To make an intelligent choice on an HF rig, check back issues of *Monitoring Times* and previous editions of the *World Radio TV Handbook* for receiver reviews. Another excellent source of information is Larry Magne's *White Paper* receiver reviews (*Radio Database International,* Box 300, Penn's Park, PA 18943).

VHF/UHF Receivers-Scanners

Most scanner buffs are usually pretty satisfied with their programmable scanners. They get the usual police, fire and ambulance calls. However, I repeatedly hear, "I can't hear any satellites on my Bearcat 210 or Radio Shack Pro-30 scanner. How come?"

Frankly, these scanners just don't cut the mustard as satellite receivers. The bulk of satellite activity today starts at 136 MHz and up, and the receiver is a critical element in hearing weak satellite signals.

Sensitivity is a very important consideration in VHF/UHF monitoring; the monitor must strive for every bit of noise-free signal your receiver can muster. Noise in the VHF/UHF region can be your biggest enemy. Self-generated receiver noise is the most predominant source of noise and if your scanner is eaten up with microprocessor noise, weak satellite signals will not make it to the speaker.

Selectable modes are a must. If your scanner microprocessor controls mode selection based on frequency, you will not be a successful satellite monitor. I recommend also that you try to obtain (if finances permit) one of the newer tunable VHF/UHF receivers. The single sideband modes they offer can be crucial in detecting many satellite downlinks.

Get the widest frequency coverage your budget will allow. For instance, if milsat monitoring is your aim, 225-400 MHz coverage is essential. The more frequencies you can cover, the more satellites you can monitor.

Additional considerations are the rig's search and scan speeds and memory channel storage. Slow speeds and small memory storage make satellite listening more difficult.

Pay close attention to reviews and talk with friends who own radios you might be interested in using. Try to get some hands-on monitoring time on the rig you're interested in, using known satellite signals.

An easy way to measure a radio's satellite-monitoring potential is by listening for the Russian NAVSAT channels on 149.910 149.940 149.970 150.000 150.030 MHz. If you haven't heard RTTY-sounding signals ("diddly-diddly") on any of these channels after a couple of hours, pass the receiver by.

The best VHF/UHF radios today for satellite listening are the ICOM R7000 and the Yaesu FRG-9600. The 7000 seems to have the better track record in reviews but is quite expensive. The FRG-9600 has had mixed reviews; some obtain good results while others achieve no results at all.

Here is an interesting satellite listening tip for Yaesu FRG-9600 and ICOM R-7000 owners. The digital channels on the Fleetsatcom satellites are weak and it is easy to tune past them. Use the SSB mode on your receiver to find the carrier, then switch to the proper demodulation mode.

Surplus Gear

As always, you have a surplus option. Some of the stuff is highly suited for satellite work because it was made for satellite telemetry reception in the first place. Take your time -- know what you are buying and the condition of the gear. Nothing impresses your wife more than that 300 pound boat anchor you just brought home that doesn't work!

Most ham magazines list ads by surplus dealers. This is always a great place to start. Hamfests can be another great place for bargains. But remember, know what you want to listen to and what frequency and modes the satellite(s) operate on, then make your purchase accordingly.

TVRO Equipment

Needless to say, there is a lot out there. My only recommendation is that you buy from a reputable

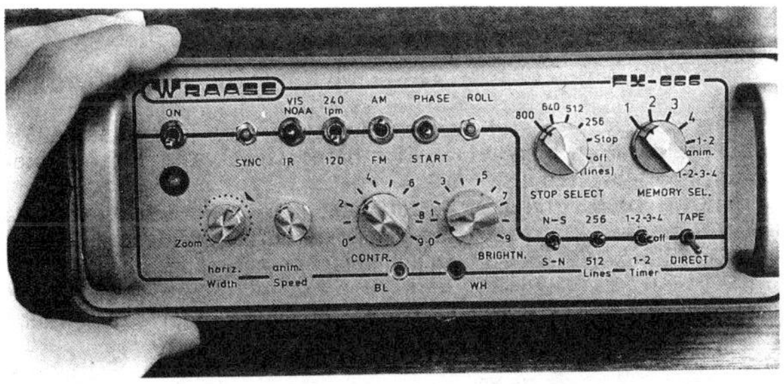

Right: An example of the high resolution reception possible with top-notch weather satellite receiving equipment like the WRAASE FX-666 pictured above.

dealer. Make sure that he guarantees not only the equipment, but the installation as well.

Weather Satellite Reception

An example of commercially available weather satellite communications reception equipment is the WRAASE weather satellite receiver ($800) and weather picture (APT/WEFAX) imager ($1535). Viewing this device has been likened to looking out the window of the Space Shuttle while in oribt.

The receiver scans the VHF weather satellite frequencies, automatically selecting the satellite as it comes into range, and turns on a stereo deck to record data for playback. It has 40 kHz bandwidth and an extremely good audio product which faciliatates greater detail when fed into the FX666 video imager.

The FX666 imager produces breath-taking panoramas as viewed from the advanced very-high-resolution radiometer on board the Tiros weather satellites in low earth orbit. It displays, in video form, images with 512 x 512 pixel resolution in both composite and black and white to drive a video recorder (VHS or BETA) and a monitor. It also has an R.G.B. port for false color viewing.

The unit sports four channels of video memory as well as animation of GOES frames as well as Meteosat and also has four power "zoom" capabilities as well as replay features off audio tape.

The imager and receiver are capable of receiving, in the correct formats, pictures from the following generations of weather FAX satellites:

NOAA	(USA)
METEOR	(USSR)
METEO	(ESA)
GOES	(USA)
[with GOES converter-$274]	
METEOR ADV	(USSR)

The receiver has two channels for receiving data from other satellites that are not as well known. These quality devices are currently marketed in the U.S. by G.P. Mengell of A.P.T. Associates, 2685 Ellenbrook Drive, Rancho Cordova, CA 95670.

Best Bets

After much study, analysis and use, these are our recommendations for utility satellite monitoring. If properly installed and with a suitable antenna these receivers and scanners should perform well in any utility satellite monitoring station.

HF

Kenwood	ICOM
R5000	ICR-71A
R2000	ICR-70

Yaesu	
FRG-8800	FRG-7700

JRC	
NRD-525	NRD-515
NRD-505	

VHF/UHF

Regency	ICOM
MX-7000*	R7000
MX-5000*	

Radio Shack	Yaesu
Pro-2004*	FRG-9600

* *Selectable AM, FM wide and narrow modes only*

Antennas

There are two ways to go with antennas: build your own or buy commercially made. There are two basic antenna designs one should consider when putting your station together: omnidirectional and directional antennas.

Omnidirectional Antennas

These antennas are satisfactory for general satellite work. Chances are your first reception will be on an omni-type antenna. The omni antenna will allow the beginner to concentrate on finding the satellite signal not having to worry about accurate tracking and pointing of the antenna.

The main drawback to the omni is the lack of gain needed to hear weaker satellite signals. This situation can be improved through the use of preamplifiers. One must keep in mind that preamplifiers are almost a must at the higher frequencies and the main characteristic to look for is a low noise figure.

Since the omnidirectional antenna we are primarily talking about is the discone, you will need to get a wide frequency coverage preamplifier. There are trade-offs and the individual can judge best his own situation.

One word of caution here is the use of indoor antennas such as table-top and back-of-the-set whips. For satellite use these will give poor results at best. You might be able to hear the more powerful signals such as the Russian navigation satellites on 150 MHz, but don't expect to hear much else.

For a commercially available discone antenna, the ICOM AH-7000 is probably your best bet. It consists of 16 rugged, stainless steel elements arranged on a 24-inch support pipe equipped with two mounting brackets. The frequency coverage is from 25 to 1300 MHz. Also include is a 40-foot length of low loss 50 ohm coaxial cable with N connectors. This antenna should serve the beginning satellite monitor very well, especially if a preamp is added to the setup. Similar discones include the Grove ANT-4, Diamond D-130, Heath HA2513, and Radio Shack 20-013.

A preamp that I have found works well is the Radio Shack inline coax preamp (Cat #15-1117). The $14.95 cost is very reasonable. Another preamp that performs well is the Grove Power Ant III ($39 plus $1.50 UPS shipping from Grove Enterprises, P.O. Box 98, Brasstown, NC 28902).

Directional Antennas

This is probably the best way to go for satellite antennas. Yes, there are some additional headaches and more expense, but the results will be far better.

Probably the best design compromise in the directional antenna area for satellite work is the log periodic design. You will get a wider frequency coverage with the log periodic and this will result in an overall lower cost for your station.

If, however, you want to listen to one area, say 240-270 MHz, then you might want to consider a Yagi style antenna. Several readers report excellent results with commercial 220 MHz amateur radio radio antennas that have been cut down and modified for this frequency range. Similar discones include the Grove ANT-4, Diamond D 130,

By concentrating on a more limited frequency range for your antenna you will also find it easier to find or make preamplifiers that are more sensitive and have lower noise figures.

For those with no spare time and two left thumbs, the Grove Scanner Beam ($49.00 plus $3 UPS shipping), coupled with a rotator, will give the listener up to 8 dB gain over other types of antennas. This antenna is of the log periodic design and covers 25-1300 MHz. I use one on my roof and it really drags in those weak Fleetsatcom channels that are hard to hear when the wideband channel gets saturated with a lot of users. For a full catalog contact Grove Enterprises at the address given above.

Some Final Thoughts

Patience is the name of the game. You will not be able to turn on your receiver and have it fill your speaker with satellite signals. There are no schedules, no guarantees and only those with patience will be rewarded.

Doppler Shift

The single most important indicator of an orbiting satellite (as opposed to geostationary): Doppler shift.

If you were standing alongside a train track when a train passed blowing its whistle, the pitch of the whistle would appear to descend. However, if you were on the train that pitch would appear to remain constant. This is because the frequency of the sound you hear depends on the relative motion between the source and you.

As a passenger on the train, you move along with the source so the frequency will remain constant; however, if you are on the side of the tracks the distance between you and the source is constantly changing so you hear a changing audio frequency. The phenomenon is known as Doppler effect (named after Johann Doppler, 1803-1853).

A satellite approaching you will exhibit an "up-Doppler" (frequency appears to be increasing) while a satellite moving away from your location will exhibit a frequency decrease of "down-Doppler."

Geostationary satellites also orbit the earth, but because they are placed in an orbit that matches the earth's rotational speed, they appear to remain stationary over one spot. As a result there is no apparent difference in motion between the observer and the satellite and no Doppler shift of the downlink frequency.

Gain of a Dish

When determining the size of an antenna, it is advantageous also to know the theoretical gain of a given antenna, relating directly to the expected transponder signal strength to be delivered to the Low Noise Amplifier (LNA).

3,700-4,200 MHz (C Band)

Antenna Size Feet	Antenna Size Meters	Approximate Gain (dB)
4	1.2	31.0
5	1.5	34.0
6	1.8	35.0
7	2.1	36.0
8	2.5	37.5
9	2.8	38.5
10	3.0	39.5
11	3.4	40.5
12	3.7	41.0
13	4.0	42.0
14	4.3	42.5
15	4.6	43.0
16	4.9	43.5
17	5.3	44.0
18	5.6	44.5
19c	5.9	45.0
20	6.1	45.5
23	7.0	47.0
26	8.0	48.0
29	9.0	50.0
33	10.0	51.5
36	11.0	52.0

* = *55% rated efficiency @4,000 MHz with greater than 30 degree look angle*

Tuning in on Satellite Communications

by Bob Grove

TV satellites do double duty--not only do they relay entertainment video to earthbound viewers, but they also carry two-way communications on the transponders on subcarriers. It takes a general coverage short wave receiver to monitor the "utilities"--the two-way users operating single sideband, facsimile, radioteletype, and data.

But where do you connect the receiver? Look for an output jack on the TVRO satellite receiver marked "video," "6.8," "baseband video," "6.8 sound" or a similar designation indicating output for a special tuner.

Connect the antenna jack of your general coverage receiver to that output jack on the satellite receiver, set the short wave radio to receive upper sideband, and start searching. While most of the subcarrier services will be heard in the 5-8 MHz range, users have been reported anywhere from 1 to 14 MHz! The nominal center frequency of this subcarrier band is 6.8 MHz.

Let us know what you hear!

Monitoring FM-SCPC

by John Wilson W4UVV

This article will discuss briefly three popular methods of FM-Single Channel Per Carrier (SCPC) reception which is rapidly becoming an increasingly popular, cost effective and efficient method of satellite audio and data transmission.

For transponder usage efficiency purposes, FM-SCPC signals are usually transmitted on dedicated FM-SCPC transponders. Two of the more popular SCPC satellites are Westar III transponders 1, 3 & 7 and Westar IV transponders 1, 2, 3 & 4. Satellite hopping through the arc on other non-video and common carrier transponders will disclose other transmissions.

Some Intelsats and the Russian Gorizont 7 satellite also have FM-SCPC. You may wish to keep your own reception log since there currently is no "TV Guide" listing of what's up there; however, at least two industry publications have recently begun listing programming services on an occasional basis for some satellites.

Let's Tune In

As with RTTY, FM-SCPC reception requires a special signal demodulation process for proper reception. FM-SCPC signals differ from the 5.5-8.0 MHz subcarrier audio "piggybacked" signals heard on video transponders: Each FM-SCPC signal is a "stand-alone" carrier typically 60 kHz wide.

When tuning your present satellite TVRO receiver across an FM-SCPC transponder, you will probably see and hear only "white noise" or a slight darkening of the video; you would not know these signals were there.

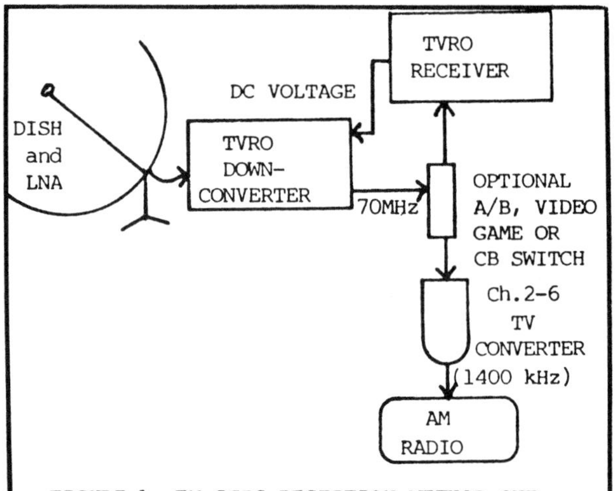

FIGURE 1: FM SCPC RECEPTION METHOD ONE

There are three popular methods of receiving FM-SCPC transmissions; all require at least a four-foot antenna (preferably six-foot or larger on the coasts) and a 100°K LNA (low noise amplifier) correctly installed and aimed at the desired satellite with the proper polarization set for the transponder to be searched.

The first method (see figure 1) involves the use of a TVRO down-converter and a VHF-TV (Ch.2-6) converter interfaced with an AM radio set at approximately 1400 kHz. The TVRO separately-powered down-converter's IF output at 70 MHz is connected to the input of the VHF-TV converter which, in turn, is fed to the AM radio. Tuning from channel 2 to 6 on the converter corresponds to the 40 MHz bandwidth of the transponder being searched.

Separating the received signals may be a problem; signal-to-noise ratio may be poor because the converter's frequency bandpass is about 150 kHz wide and the FM-SCPC signals are only 60 kHz wide (or less) so some noise will also be passed. This receive method assumes that you TVRO down-converter is separately powered and can lock onto the FM-SCPC signals.

I tried this first approach and had no success because my TVRO receiver/ down-converter was designed to lock onto 4.5 MHz video transmissions, not these narrow-bandwidth transmissions. At least 90% of TVRO receiver/ downconverters probably cannot lock onto these transmissions although this method has been proven to work with certain receiver configurations such as Scientific Atlanta.

My FM-SCPC receive system is the second method (see figure 2). The system is not cheap, costing approximately the same as a small-dish

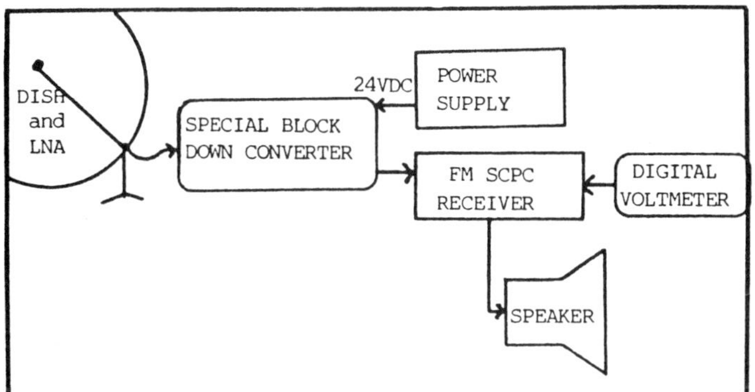

FIGURE 2: FM SCPC RECEPTION METHOD TWO

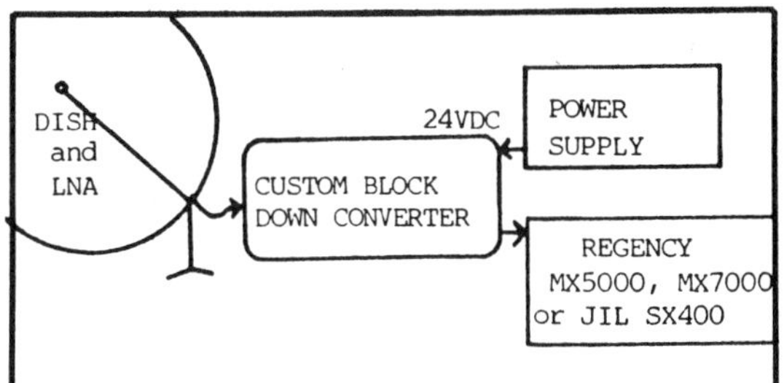

FIGURE 3: FM SCPC RECEPTION METHOD THREE

TVRO system. A special purpose FM-SCPC receiver designed for 60 kHz reception is used offering continuous tuning, triple conversion, good sensitivity, and selectivity. With a 100°K LNA and a six-foot antenna my reception is noise-free plus for over 90% of my monitoring. Of course, a larger antenna will improve any weaker signals such as some found on Westar IV.

The block down-converter is a special component interface to the receiver, powered separately from a 24VDC supply. Jacks are provided for 300 and 8 ohm audio output or you may connect the converter to a general coverage receiver such as the ICOM R-71A or NRD-515 which will tune to 10 kHz as partial interface for Frequency Division Multiplex (FDM) reception of UPI RTTY news (or to an FDM demodulator for reception of other "hidden" RTTY transmissions).

A digital voltmeter connected to the tuning control provides a rough frequency cross-reference when tuning a particular transponder. Slight variances in readings will occur with remote mounting of the downconverter due to thermal drift of the local oscillator.

The third method of reception requires the use of a customized block down-converter and a Regency MX-5000, MX-7000, Yaesu FRG 9600, Icom R7000 or JIL-SX400 scanner (see figure 3). A typical TVRO "off the shelf" block down-converter *will not* work (or, at best, will not work properly) and interfacing with any other scanner would yield inadequate tuning range. The cost for this method, if you presently own either scanner and have your own TVRO antenna and LNA, is comparable to the cost of a general coverage HF receiver.

Method three has been proven to work well with detailed frequency readout for cross-reference purposes using the scanner's search mode. Again, slight variances in readout will occur if the down-converter is remotely mounted.

The Menu

The following list represents recent typical FM-SCPC programming found on just one satellite--Westar III--in 1985. Some programming is occasional or seasonal such as football, baseball and basketball. Network and state news are usually on the hour or half hour and music services usually offer 24 hour programming.

TRANSPONDER 1

Approx Freq MHz	DC Digital Voltmeter	Service
3720.5	15.3	Tick-Tock Carrier/Sprts Line-National call-in talk program
3723.5	17.3	Carrier/Wall St.Journal Rpt/Dow Jones Talk Phone (LSB 19.9kHz subcarrier)
3724.3	18.3	Radio 650-WSM Nashville, TN-Country music format
3724.5	18.5	Tone Carrier/AP News (FAX ∓10kHz subcarrier AP News LSB 20kHz sub.)
3725.2	19.1	Unidentified Carrier
3726.8	20.2	Unidentified Carrier
3727.5	21.1	UPI Radio News (English & Spanish/FDM 50 state UPI RTTY (LSB 11.4kHz [E.Coast] subcarrier & LSB 15.4kHz [west coast] subcarrier)
3727.6	21.4	WPAT-Patterson, NJ-good music format

Transponder 3

Approx Freq MHz	DC Digital Voltmeter	Service
3742.6	1.5	Transworld Radio (L Ch Stereo)-Colorado Sprgs, CO-Current & past pop hits-commercial free
3743.0	1.8	Transworld Radio (R Ch Stereo)-Colorado Sprgs, CO-Current & past pop hits-commercial free/ LSB 17.5 kHz RTTY-Unid
3739.8	3.0	San Francisco Giants Baseball Network
3739.9	3.1	Chicago Cubs Baseball
3740.0	3.2	Tone Pulse Carrier-Unident.
3740.2	3.3	Detroit Tigers Baseball Radio Network
3740.4	3.4	Milwaukee Brewers Radio
3742.0	4.0	Carrier/Michigan Farm Radio Network/Spartian Baseball Network
3742.8	4.2	Tick-Tock Carrier/ RKO News/Fla Network News/ Florida Sports Network
3743.5	4.4	KKJO-Los Angeles, CA-Jazz music format/Motor Racing Network
3744.9	4.7	Unidentified Carrier
3747.0	5.4	Unidentified commercial free good music prog
3747.2	5.5	Unidentified commercial free pop music prog
3748.2	5.8	Minnesota Twins Baseball
3750.4	6.8	Unidentified Sports
3751.2	7.5	Carrier/ABC Radio Sprts /Oklahoma News Network/ ABC News
3751.8	8.2	Tick-Tock Car/RKO News
3752.4	8.4	FM100-Columbus, OH-country music format
3752.6	8.7	Carrier/ABC Contemporary News
3752.7	9.2	KAKZ News/Kansas Agriculture Network
3752.8	9.4	Tone Carrier/Oklahoma News Net/WYNY-NY-good music format
3752.9	9.5	Unidentified pop music
3753.4	9.8	Carrier/ Kansas Information Network
3755.4	10.5	Kansas Information Net
3755.6	10.6	Carrier/Southern States Network News
3755.7	11.0	Unidentified Carrier
3757.9	12.4	Carrier/Texas State Net English & Spanish News
3761.3	15.0	Carrier/ ABC Dimension Network/Missouri Net News/Sports (LSB 15.9 kHz subcarrier)
3761.8	15.6	Carrier/ ABC News
3761.9	15.7	Carrier/UPI Radio News/ Midwest Info Network
3762.1	15.9	Carrier/ ABC Direction Network/N.C. News Net
3762.2	16.0	Carrier/Kansas Info Net /ABC News/MO Net Sports
3762.9	16.4	Carrier/ ABC Dimension Net/Cyclone Baseball
3763.6	16.9	Tick-Tock Car/RKO News/ ABC Dimension Network
3764.0	17.2	Tick-Tock Car/RKO News/ Louisiana Network
3765.2	17.8	Carrier/Miss. News Net RKO News/Capital Sports Network(North Carolina)
3765.4	18.0	Unidentified pop music
3767.4	19.0	Carrier/N.C. News Net
3767.5	19.2	Carrier/ UPI Radio News
3767.7	19.6	Va.News Network/ UPI Radio News
3771.4	21.8	Tick-Tock Carrier/ Ark. Radio Network
3781.9	22.2	Carrier/Arkansas Radio Network/ RKO News
3872.5	22.6	Unid Religious Music
3373.6	23.6	Unidentified Carrier
3374.0	23.9	TalkNet/ WTMJ Mil, WI
3374.1	24.0	Unidentified Carrier
3775.0	24.6	Carrier/Okla. Sports
3775.4	24.9	Z100-Portland, OR/ Blazer SportsTalk/ Brownfield Agri Net
3775.8	25.1	TalkNet/FM100-Columbus, OH-Popular music format
3770.0	25.7	KMPC-Los Angeles, CA-'40s & '50s music

Transponder 7

Approx Freq MHz	DC Digital Voltmeter	Service
3832.0	2.0	Bible Broadcasting Net (L Ch stereo)-Relig
3832.5	2.2	Bible Broadcasting Net (R Ch stereo)-Relig
3841.0	14.6	Unidentified FDM

A list of FM-SCPC equipment sources is available from the author. Send an SASE to: John Wilson (W4UVV), 6413 Bull Hill Road, Prince George, VA 23875.

Brushfire Fighters use Milsat

Several listeners have reported hearing the California brushfire fighters communicating via the U.S. Navy's FLEETSATCOM communications satellite utilizing 12-watt, hand-held, FM transceivers.

The frequency reported, 269.95 MHz, was referred to as a secondary channel; the primary was not found. The HT's were on loan from the Navy.

It would seem that the wide area affected by the fires required reliable communications coverage, and a repeater at an altitude of 22,500 miles is hard to beat!

Additional information from our readers on voice modes used on satellites would be greatly appreciated.

Computer Tracking

Computer information and tracking programs are available for a wide variety of computers from:

AMSAT Software Exchange
P.O. Box 27
Washington, D.C. 20044 USA

The following versions of the W3IWI orbit prediction program--in Basic--are available:

1.* TRS-80 Model I and II Basic. Some have had problems running this version on minimum machine configurations. You should have at least 32K or plan to do some customizing.

2. Radio Shack TRS-80 Model III. The program will run on either 32K or 48K machines, but not 16K. At least one disk drive is required.

3. North Star Basic under North Star DOS for 5-1/4 inch hard sectored diskettes. Please indicate single or double density.

4.* Microsoft Basic version 5.21 under CP/M single density, soft sectored 8 inch disk.

5. Apple II Applesoft Basic on either 13 or 16 sector diskettes or cassette. This version includes a menu driven utility for entry of Keplerian Data. Options are available for output to screen and/or printer.

6. WØSL IBM program-menu driven. Includes updated version of W3IWI program. This tabulated version shows the coordinates for up to eight satellites in real time, selectable graphics display of world map on which these satellites are moved real time. For IBM-PC, PC Jr., PCXT, or PCAT with small modification. Requires: 128K RAM, IBM color graphics adapter, IBM color monitor, 1 floppy disk drive DOS 2.0 or later with BASICA. Will run as is on DOS 2.X. Requires modification for use on DOS 3.0 as covered in documentation.

7. N4HY Deluxe IBM compatible computer program--Menu driven program for tracking and scheduling of amateur satellites. Incorporates very fast algorithm for finding usable passes of the satellite of your choice. All data is in data files. Satellite pointing angles relative to your location and a new "Window Track" mode are included. The "Window Track" feature identifies mutual "windows" between your location and other specific locations around the world. This program is very user friendly and includes menu driven utilities for data entry. Also available for the 8087.

8. WØSL satellite tracking with graphics-- This program is functionally identical to the version for the IBM PC; however, it has been modified to run on the PC Jr due to differences in the graphics capabilities of the two machines. Special requirements: IBM-PC Jr., 128K of RAM minimum, color monitor, floppy disk drive (5-1/4"), DOS 2.0 or later, Cartridge BASIC.

9. Texas Instruments TI 99/4A cassette only.

10. KØRZ antenna positioning software. Note that this will only work on the Apple II computer with hardware accessories described in ORBIT No. 11. The antenna positioning systems comprise two diskettes of programs.

11. Radio Shack TRS-80 Model 4 for TRSDOS Version 6.0.

12.* Radio Shack TRS-80 Color Computer. Requires 32K machine and extended BASIC--cassette only.

13. New AMSAT VR85 for C64. Datapoint map of 2000. Plots 20 satellites including space shuttle. Includes up to nine transponders with applicable modes for all satellites. Real time or selected increments for all satellites.

14. QUIKTRAK-2064. Much enhanced version of AMS-2064. It uses an advanced algorithm providing exceptional speed in acquiring AOS time. On distribution media, a machine language version is included for C64 owners providing another significant speed gain. Indicate cassette or disk.

15. Atari--disk only.

16. Times-2068--cassette only. Please note that this is the W3IWI tracking program.

17. HP-41 Programmable Calculator-ORBIT I, W3IWI Prediction Program. Original version converted to run on the HP-41. This program will output AZ and EL in real time or all W3IWI parameters in the prediction mode. Special requirements: HP-41C + QUAD MEMORY, HP-41CV, HP-41CX, card reader (desirable but not essential).

18. HP-41 Programmable Calculators - ORBIT II--Same as ORBIT I with the addition of functions from the time module for output of time and the ability of the calculator to read the real time clock and date when in the real time mode. Special requirements: HP-41C + QUAD MEMORY + TIME, HP-41C + TIME, HP-41CX, Card Reader (desirable but not essential).

19. Heathkit H-89 - W3IWI Prediction Program. Original version converted to run on the H-89. Special requirements: H-89, CP/M, M BASIC, floppy disk drive (5-1/4"), (H-17, hard

sectored).

20. "Using Microcomputer Programs for Radio Amateur Satellite Orbital Prediction" by N5AHD. Primarily for Radio Shack, CP/M Based S-100 Bus Computers.

*Note: * indicates no instructions included --software manual recommended.*

AMSAT headquarters notes that you can charge your programs via Mastercharge or Visa by calling (301)589-6062. Prices range from $10 to $30 depending on the program and if you are a member of AMSAT or not. It is best to write for the latest list and be sure to include an SASE.

Getting the Data

Hopefully by now you will have or want to get a satellite tracking program for your computer. But, where do you get the all-important Keplerian data to make that software work? One source of data is from NASA at the Goddard Space Flight Center.

Requests for prediction bulletins should include International Designator and name of the satellite. Send to:

Project Operations Branch (Code 513)
NASA/Goddard Space Flight Center
Greenbelt, MD 20771

The user can request either two line orbital elements or three part prediction bulletins for the satellite of interest. I have found that you only need the two lines for your computer work and the bulletins get to you faster than the three part bulletins.

The following are format explanations for the two line orbital element sets.

PART I, LINE ONE FORMAT:

Col	Description
1	Line number (1)
2	(blank)
3-7	Satellite number
8	U - Not applicable
9	(blank)
10-11	International Designator (Last two digits of launch year)
12-14	International Designator (Launch number of the year)
15-17	International Designator (Piece of launch)
18	(blank)
19-20	Epoch Year (Last two digits of year)
21-32	Epoch (Julian day and fractional portion of the day)
33	(blank)
34-43	First time derivative of the mean motion or ballistic coefficient (depending on the ephemeris type)
44	(blank)
45-52	Second time derivative of mean motion (Field will be blank if doesn't apply)
53	(blank)
54-61	BSTAR drag term (Not needed by the common user)
62	(blank)
63	Ephemeris Type (Specifies the ephemeris theory used to produce the elements)
64	(blank)
65-68	Element Number
69	Check Sum

PART I, LINE TWO FORMAT:

Col	Description
1	Line number (2)
2	(blank)
3-7	Satellite Number
8	(blank)
9-16	Inclination
17	(blank)
18-25	Right Ascension of the Ascending Node
26	(blank)
27-33	Eccentricity (decimal point assumed)
34	(blank)
35-42	Argument of Perigee
43	(blank)
44-61	Mean Anomaly
52	(blank)
53-63	Mean Motion
64-68	Revolution Number at Epoch
69	Check Sum

Figuring Out the Angle

One of the most common questions asked by satellite monitors is how to figure "look angles" (azimuth/elevation) for geostationary satellites.

The following equations and a sample problem will help you find these look angles to accurately point your antenna for best reception.

For the same problem I will use my location in Orange Park, FL, and the Fleetsatcom F1 satellite at 100°W longitude.

My latitude is 30.19° N

Example:

TWO LINE ORBITAL ELEMENTS

```
1 01328U 65032   A 81361.75324961 0.00000070            20661-3 0   7532
2 01328  41.1873 263.1704 0245364 221.6817 136.4995 13.36210358813588
```

My longitude is 81.73° W

The equation for azimuth is:

$$\text{Azimuth} = \tan^{-1} \frac{\tan x}{\sin y}$$

The more complicated elevation angle formula is:

$$\text{Elevation} = \tan^{-1} \frac{\cos x \cos y - .151}{\sqrt{1-(\cos x \cos y)^2}}$$

First, let's figure the values of x and y in our formula. The value X is equal to the difference between the satellite's longitude and your antenna site longitude. Now for our sample problem:

The value of $X=100°W - 81.73°W$
$X=18.27°$

The value Y is equal to the site latitude in degrees, therefore:

$$Y = 30.19°$$

Now substituting x and y into our azimuth formulas, the result is as follows:

$$\text{Azimuth} = \tan^{-1} \frac{\tan 18.27°}{\sin 30.19°} = \frac{.33013761}{.50286909}$$

$$= \tan^{-1} (.65650806) = 33.28°$$

Since Orange Park is east of the spacecraft, I added 180 degrees to the resultant. Therefore my azimuth to the Fleetsatcom F1 satellite is 213.28°.

If you noticed, trig was used for the calculations. If you do not have a scientific calculator that does trig functions, then any well-stocked library will carry books that have trig tables for sine, cosine, and tangent functions. The calculator is the easiest way to go and will help in dividing those decimal numbers.

As for the elevation formula, substituting x and y, I have the following (E=Elevation):

1. Substituting our values of x and y

$$E = \tan^{-1} \frac{\cos 18.27 \cos 30.19-.151}{\sqrt{1-(\cos 18.27 \cos 30.19)^2}}$$

2. Performing our trig cos function

$$E = \tan^{-1} \frac{.94958975 \times .86436258-.151}{\sqrt{1-(.94958975 \times .86436258)^2}}$$

3. Performing the indicated multiplication and squaring our bottom expression

$$E = \tan^{-1} \frac{.82078985 -.151}{\sqrt{1- (.82078985)^2}}$$

4. Performing the indicated subtraction

$$E = \tan^{-1} \frac{.66978985}{\sqrt{1-.67369598}} = \frac{.66978985}{\sqrt{.32630402}}$$

5. Clearing the square root

$$E = \tan^{-1} \frac{.66978985}{.57123027}$$

6. Dividing our decimal numbers and finding the inverse tangent

Elevation = $\tan^{-1}$ 1.1725391
Elevation = 49.5 degrees above the horizon

Now I know some enterprising individual is out there saying, "Boy, that would run great as a computer program!" That individual is right and I hope someone will write one for the public domain and send Satellite Tips and Hints a copy. Wouldn't it be nice if it was done for the Commodore 64?

More Information Sources

There are other sources of Keplerian data needed to feed your satellite tracking program for your computer.

W1AW, the amateur radio station of the ARRL in Newington, CT, has scheduled broad casts that include Keplerian element sets for orbiting amateur radio satellites as well as NOAA weather satellites.

These bulletins are transmitted in voice, Morse code and radioteletype. For more information on these broadcasts contact:

American Radio Relay League
225 Main Street
Newington, CT 06111

Another source of amateur satellite/weather satellite Keplerian information can be found in the pages of the *Amateur Satellite Report*. Just about every issue of that newsletter is packed not only with the latest amateur satellite news but the latest Keplerian data on operational satellites. A one year subscription will cost you $2.00 in the United States, Canada and Mexico. The issues are bi-weekly and mailed First Class. ASR's address is:

Satellite Report
221 Long Swamp Road
Wolcutt, CT 06716

Keplerian elements for amateur radio satellites and special space shuttle-amateur related missions can be found on the AMSAT BBS. If you have a modem you can get the latest information by dialing 1- 512-852-8194. The system uses 300 baud, no parity, 8 bit protocols.

Satellite News, now published in a monthly journal format, has as its major features: Space Objects Digest, Military Space Digest, Space Operations Review and Space Systems Digest.

The features include complete orbital data for all new space objects, a complete list of re-entered space objects including orbital lifetimes and space object identifications, astronaut/cosmonaut flight logs, daily orbits of reconnaissance and manned spacecraft, orbital data for satellites of special interest, and space object catalog numbers.

Also included are details of space craft launches with initial orbital data and transmissions, spacecraft operations, and a unique updated monthly listing of the orbits and locations of geosynchronous spacecraft.

Each issue has technical reviews of spacecraft and their experiments, construction and development including detailed descriptions of major spacecraft and systems, plus specialized sections on Soviet Space activity, military spacecraft launches and space operations of the major powers.

The price is $4.00 US per issue and you can subscribe to as many issues as you wish. If you are going to pay by U.S. check, you must add $2.50 to cover UK bank charges. Overseas readers should pay by International Money Order or by check drawn on a London bank or by direct transfer to UK Girobank Account No. 61 040 1009. Payments in cash in U.S. and UK currency accepted at senders risk. Price includes surface mail postage to all countries.

You can get more information on this unique monthly journal from its publisher,

Geoffrey Falworth
14 Wesley Drive
Banbury OX16 9LW, England

Keeping up with the Soviets

There is a way to keep up with the latest satellite launches from the Soviet Union: monitor the daily TASS bulletin in English on 14.901 from the Cuban Prensa Latina station, CLN451. The station transmits 50 baud/425 Hz normal shift. The TASS bulletin usually runs from 1200-1700 UTC. If any new spacecraft are launched by the Soviet Union, you'll know about it.

The current Salyut 7 space station uses 142.417 MHz FM for voice downlink and listeners on the east coast and eastern portions of the midwest should have no problem hearing the Russians in orbit.

But what do you do when the Russians start scrambling? The Russians aren't using an exotic scrambling technique, just simple voice inversion.

Russian Meteosat Bulletins

On January 15, 1978, the Hydrometeorological Service of the U.S.S.R. began transmitting RTTY over the Global Trunk System of the W.M.O. (World Meteorological Organization) orbital data and predictions for Meteor 2 satellites in the national code FANAS (Forecast for Ascending Node for Automatic Satellites).

These messages are transmitted at 0945 GMT on Tuesday morning on radio frequencies 3330, 5140, 7685, 9190, and 13530 kHz.

The message is in five sections, the first of which gives information on the satellite's orbital period, longitudinal increment and daily data on the ascending node. This takes the form of time as hours, minutes and seconds in GMT and longitude in degrees and decimals of a degree in the specified octant of the globe. First and last ascending nodes are given for odd numbered days.

The second section in plain language (English) gives information on the operating mode of the onboard TV system, the radio frequency of APT transmissions and any change in frequency. The third section gives the time in minutes and seconds measured from the ascending node epoch at which the satellite crosses circles of latitude at 10 degree intervals. It has become customary to provide information only for the part of the orbit over which the TV transmitter will be operating.

The fourth section, which is rarely used, carries, in plain language, an explanation of the content of the FANAS message and definitions of the relevant digital groups of the transmitted message. The final section provides information on latitude in degrees and decimals of a degree (but ignores the hemisphere), longitude within the specified octant, and height above the earth's surface (over the Krasovsky ellipsoid) in tenths of a kilometer, for each even minute measured from the ascending node epoch.

This fifth section may also contain nonregular information on the operation of the onboard set of instruments, changes of operating modes, recommendations, requests, and the time of transmissions from geosynchronous satellites.

Code numbers are provided for the countries of the European community, Japan, U.S.A. and the U.S.S.R. as launching agencies with others held in reserve. Code numbers are also provided for the American ATS and NOAA series of satellites in addition to the Meteor-2 series of the U.S.S.R.

Monitoring the Soviet Satellites

by Mike Smithwick

With help from the powerful Molniya television satellites, even the most modest home satellite TV terminals can eavesdrop on the latest gossip from Moscow, catch the current fads in the Ukraine, or pick up details of the latest in state-of-the-art tractor repair.

The Molniya series of satellites form a communications bridge between Moscow and the western (Siberia) reaches of the Soviet Union. First and foremost it is a television satellite, not at all unlike our own DOMSATs (DOMestic SATellites) such as COMSTAR or SATCOM. But, unlike the DOMSATs, Molniya carries just one television channel mixed with a host of other communication services.

But what really makes Molniya unique is its unusual orbit (dubbed a "Molniya" orbit). With an inclination of 66 degrees, perigee of 600 miles, and an apogee of 40,000 miles the satellite presents itself as a moving target to whoever wishes to watch it, unlike the stationary U.S. satellites.

Orbiting the earth every 12 hours, the satellites reach apogee twice each day, first over central Russia, then over northern Canada. It is this Canadian peak which we are interested in, as this is where the satellites are activated for broadcasting.

Because of the nature of the orbit, 4 satellites are required for a full day's broadcasting. Each is turned on for 6 hours, beginning at 1 hour before apogee and lasting to 5 hours past apogee. At this time the next satellite is rising and is activated for the next 6 hours.

The Canadian apogee permits stateside watchers to "see" the Molniyas which also provide a link for the famous "hot-line" between Washington and Moscow.

Such wide-ranging coverage necessitates both a wide beamwidth on part of Molniya's antennas, and a high transmitter power in order to make up for the lower gain of the antennas. What this all means is that the transmitted signal is comparable to stateside DOMSATs, with a signal strength of about +32 dBw over the entire country. The higher-than-usual output power is also required to compensate for the deficiencies of Soviet micro-

NOVOSTI news (shows map of Lebanon) viewed in U.S. from Molniya satellite (Tom Kuiken)

wave technology. So, while we copy the Molniya signals with our 3 and 4 meter dishes, the Soviet ground stations use 12 meter systems. The television signal may be found slightly below channel 9 on the home satellite receiver at a frequency of 3775 MHz.

Located on a 7.5 MHz subcarrier is the Russian National Radio Service. This can be copied in the same way as the U.S. DOMSAT subcarrier services.

While the video is broadcast in the international SECAM format, standard U.S. NTSC sets can display a surprisingly usable picture. Older sets work better than the newer models since many newer sets lack the fine picture adjustment controls required to steady the picture (leaving the duty to digital circuitry which can't seem to lock up on the Soviet video).

Unfortunately, both the sound and color are incompatible with U.S. standards. The audio is "pulse width modulated" within the video horizontal blanking signal. Simple circuit designs are available to decode the audio, but color would still need a SECAM monitor.

There are 2 audio channels. They can be used for bilingual broadcasts or, more commonly, one channel is used for video cueing instructions from the Moscow studios.

Tracking the Birds

As seen from the ground, the satellites trace out a lopsided "U" in the sky and, as with all satellites, the exact location of the "U" is dependent upon the station's location, time of day, and time of year. To further complicate matters, each Molniya resides in a slightly different orbit from the rest.

The first step is to orient the antenna north towards the sub-satellite point at apogee. For Molniya, this point is Mansel Island, located in Hudson Bay, Canada. Remember to use Great Circle directions and to mark the antenna platform so the dish may be repositioned at a later date.

As if all of this weren't complicated enough, the Molniya orbit is not exactly 12 hours, but instead falls 2 minutes, 10 seconds short. Thus, the satellite will repeat its track 4 minutes 20 seconds earlier each day. As you can see, good records are very important for finding the satellites from one night to the next!

The best time for observation is during the satellite's apogee when it moves the slowest; as much as 45 minutes may lapse before the antenna needs to be readjusted. However, as the bird comes gliding down towards perigee, it will be in view for no more than 10 minutes at a time.

"TV Guide"

The programming on the satellite ranges from humdrum to dull, demonstrating a startling poverty of imagination, much like American television. Of course, for us westerners, no matter what is shown, the glimpse of Soviet society as they see it is quite extraordinary.

My receiving system consists of a 4 meter dish and a 100 degree LNA. This gives me pictures at apogee with no more than 10% snow, often snow-free at perigee. Since the signal is right-hand circularly polarized, a circular feed would improve the picture.

The programming day begins at 21:00 UTC with the Vremyi news broadcast, sponsored by TASS. The schedule is varied, but typically the news will be followed by chess updates, exercise programs, documentaries on Soviet life, and a movie. And, true to form, there is the occasional propaganda piece.

Soviet TV programming rarely ends on the hour or half-hour as ours does; instead, a program might end 17 minutes past the hour, with the next one immediately following.

When a Molniya reaches the end of its operational pass, you should be able to observe the switch-over. Without warning, the programming will suddenly be interrupted with the "ORBITA" logo and a set of color-bars. The pattern will last for about 2 minutes before the spacecraft is turned off.

Watching the Molniya can be both fascinating and frustrating. But remember, the purpose of Soviet television is not to entertain, but to present to the citizens the world as the Soviet government wants them to see it.

Chapter Two

AMATEUR RADIO SATELLITES

One of the most exciting moments of my association with satellites over the years was the first time I communicated through a satellite. Instead of just listening to communications through a satellite, I was participating in the communication process! I rank that experience right up there with my first novice radio contact. Butch Weber (WA4GIF) helped make that thrill possible, for which I will be forever grateful.

Amateur radio operators can have the thrill of communicating through satellites that the amateur community builds themselves. OSCAR, Radio Sputnik, and Iskra amateur radio satellites all make up the exciting world of amateur radio satellites. If you're a ham and you haven't jumped on the amateur satellite bandwagon, you don't know what you are missing. I refuse to get on HF anymore. I don't like the crowds, the expense to put out a decent signal, or to put up with variable propagation conditions.

Amateur satellites offer the ham and the newcomer SWL a lot to be enjoyed. The amateur radio operator will find that operating conditions can be predicted with much more predictability than HF band can. Due to the satellite having a predictable orbit, amateurs can schedule the times they want to communicate with another station. You do not have to worry about propagation; if both you and the other station are within range of the satellite, you can communicate. Amateur space applications are just now starting to be realized by a few amateurs and the years ahead promise to be bright.

For the SWL, amateur radio satellites are the easiest to monitor with existing equipment. You can cut your teeth on these satellites and learn the basics in the process. QSL's often can be obtained from the satellite operators.

Who puts up these satellites, how many are there, and what does the future hold? This first chapter of *Communications Satellites* will tell you.

Amateur Satellite History

In April 1959, a visionary by the name of Don Stoner (W6TNS) suggested in an article in *CQ* magazine that hams ought to build a satellite capable of relaying amateur communications. Don's article fired the imaginations of several amateur radio operators and in 1960 a group was formed in Sunnyvale, California, known as the OSCAR association (*O*rbiting *S*atellite *C*arrying *A*mateur *R*adio).

The easy part of the OSCAR group's project would be to build the satellite; the hard part would be to get it into orbit! An important factor that led to the launch of OSCAR I was the fact that most rockets have more capability than the weight of the satellites they carry; ballast must be carried into orbit to compensate for this extra lift.

Instead of extra weight, why not put an extra satellite into orbit? The OSCARs could go up as ballast and piggyback on the main satellite's rocket launch. While this does have its drawbacks, it has been a very cost effective way to send amateur radio into orbit.

OSCAR I

On December 12, 1961, the Discoverer 36 satellite aboard a Thor-Agena rocket roared off a pad at Vandenberg AFB at 2040 GMT. Also on board was amateur radio's pride and joy, the first amateur radio satellite, OSCAR I. Thousands of hams around the world followed the launch of OSCAR I with interest as the satellite circled the globe transmitting 0.1 watts of telemetry on 144.983 MHz.

The satellite was placed into a 372 X 211 km orbit, inclined 81 degrees. OSCAR I used a monopole antenna and weighed only 10 pounds. The satellite's radio life was only 21 days and it quit transmitting January 1, 1962. The satellite finally decayed on January 31, 1962.

More than 570 amateurs in 28 countries forwarded observations to Project OSCAR on the telemetry information received during the 21 days of operation. The space age had come to amateur radio.

OSCAR II

Project OSCAR's second satellite came on the heels of the first launch barely six months after the launch of OSCAR I. On June 2, 1962, OSCAR II was placed into a 384 X 206 km orbit inclined 73 degrees with a classified USAF payload. The satellite was launched from Vandenberg AFB aboard a

Thor-Agena B rocket at 0032 GMT.

This 10 pound satellite carried basically the same equipment as OSCAR I, although a few improvements had been incorporated. The OSCAR II satellite also transmitted its 144.983 MHz signal through a monopole antenna but only for 18 days. OSCAR II decayed into the atmosphere on June 21, 1962.

OSCAR III

Launched on March 9, 1965, from Vandenberg AFB, OSCAR III was placed into a 924 X 891 km orbit inclined 103.2 degrees. This satellite brought Don Stoner's dream to reality as OSCAR III carried the first amateur relay transponder for communications.

OSCAR III carried a 1 watt, 50 kHz wide transponder with an uplink near 146 MHz and a downlink near 144 MHz. OSCAR III also carried two beacons, one for use as a tracking and propagation experiment and the other relayed spacecraft telemetry. About 1000 amateurs in 22 countries used the limited relay capability on OSCAR III to communicate with each other.

The satellite used partial solar cells for power and transmitted for 15 days. But it did prove that Don Stoner's article in CQ magazine was no dream, but a reality and amateurs could place orbiting relay stations in the sky for amateur use.

OSCAR IV

While the Project OSCAR group was working on OSCAR III, another launch opportunity presented itself to the amateur community aboard a Titan III-C rocket that would carry the little 40 pound satellite into a 21,000 mile circular orbit.

Since Project OSCAR members were involved with OSCAR III, several members of the TRW radio club volunteered to tackle building OSCAR IV. Time was crucial and OSCAR IV was a no frills satellite; it carried a transponder and identification beacon, but no on-board telemetry downlinks.

Unfortunately, the host vehicle never achieved its mission of placing the payloads into 21,000 mile orbit, and amateurs did not achieve their dream of a geostationary satellite for amateur communications purposes. No geostationary orbiting satellite opportunity has been made possible since.

OSCAR IV carried a 144 MHz uplink and a 432 MHz downlink with a bandwidth of 10 kHz. The satellite was placed into a 29,120 km X 169 km orbit inclined 26 degrees. It was launched from Cape Canaveral on December 21, 1965, and operated for almost three months. Some communications were possible with OSCAR IV, but due to attitude stabilization problems OSCAR IV never really performed up to expectations.

OSCAR IV did accomplish one milestone though--the first United States-to-Soviet- Union satellite QSO. Amateur satellites were now truly international in nature.

AUSTRALIA-OSCAR 5

Designed and built primarily by the Astronomical Society and Radio Club in Melbourne, Australia, this satellite was finally launched January 23, 1970, after 15 months of constant frustration for the Aussies and U.S. amateurs. Launched by a Delta rocket from Vandenberg AFB the satellite was placed into a 1476 X 14431 km orbit. The satellite carried no communication transponder, but did carry two telemetry beacons on 29.450 MHz (10 meter band) and 144.050 MHz (2 meter band).

Due to delays in getting OSCAR 5 into orbit, George Jacobs spoke to the COMSAT amateur radio club about forming an east coast version of Project OSCAR. As a result, AMSAT (the Radio Amateur Satellite Corporation) was officially formed March 3, 1969, in Washington, D.C.

At AMSAT, a young engineer was designated project manager of A-0-5 to assist the Aussies in getting the satellite into orbit; Jan King (W3GEY) tackled not only this satellite, but went on to design and construct AMSAT-OSCAR 6, 7, 8, AMSAT Phase III-A, B (OSCAR 10), and Phase III-C. Jan's contributions to the satellite program have been tremendous over the years; he is one of the true giants in amateur radio.

OSCAR 5 was the last of the Phase I satellites which were basically experimental in nature; it was time to move on to Phase II.

OSCAR 6

On October 15, 1972, OSCAR 6 roared into space with the NOAA-2 satellite atop a Delta rocket from Vandenberg AFB. OSCAR 6 captured the imagination of amateurs and school children alike; during its almost five year lifetime, amateur radio operators brought the OSCAR satellite to the nation's classrooms so that children conduct experiments with the satellite.

Launched into a 1453 X 1447 km orbit inclined 101.4 degrees, OSCAR 6 carried the following communication/telemetry capability:

Uplink 145.900-146.000 MHz (100 kHz bandwidth)
Downlink 29.450-29.550 MHz
Telemetry beacon 29.450 MHz

OSCAR 6 had an extremely sensitive transponder; stations running as little as 10 watts on 146 MHz could access the satellite with good signals!

OSCAR 7

Following the success of OSCAR 6, AMSAT built OSCAR 7, the second phase II satellite. The 40 pound satellite was launched on November 15, 1974, aboard a Delta rocket as ballast for NOAA-4. The satellite was placed into a 1457 X 1438 km orbit inclined 101.4 degrees.

With the launch of OSCAR 7, the amateur

satellite service entered a new era: Two transponders for communications were carried into orbit, as was a telemetry encoder to convert the spacecraft status information into teletype, allowing direct printout at 60 wpm at all monitoring stations suitably equipped.

OSCAR 7 was an international space craft in the truest sense; it was built and designed by amateurs in the U.S., Canada, Australia, and Germany. The final assembly of OSCAR 7 was accomplished in Jan King's basement after four years and $60,000.

OSCAR 7 carried the following communication transponders/telemetry beacons:

Telemetry beacons:
29.502 MHz CW, AFSK-RTTY
145.975 MHz CW, AFSK-RTTY
435.100 MHz CW, AFSK-RTTY
2304.000 MHz CW, (not operational due to FCC non-approval)
Uplink: 145.850-145.950 (Mode A)
Downlink:29.400-29.500 (Mode A)
Uplink: 432.150-432.175 (Mode B-High power Mode C-low power)
Downlink:145.975-145.925 (Mode B/C-inverted passband)

OSCAR 7 operated until mid-1981, provided dual communications with OSCAR 6 for ground stations.

OSCAR 8

OSCAR 8 was launched on March 5, 1978, on a Delta rocket from Vandenberg AFB along with the primary payload, Landsat-C. Inserted into a 909 X 898 km orbit, the satellite had an inclination of 98.8 degrees and was built as a stop-gap measure until the first Phase III satellite achieved orbit.

The ARRL donated $50,000 to AMSAT for the construction of OSCAR 8 which carried minimal telemetry and command systems. The original plans called for a Mode A trans ponder only, but the JAMSAT (Japanese AMSAT) folks did provide a Mode J transponder in time for the launch integration phase.

OSCAR 8 contained the following communications transponders/telemetry beacons:

Telemetry:Mode A beacon 29.402 MHz(110 mW)
Uplink: Mode J beacon 435.095 MHz (100 mW)
Downlink: Mode A 145.850-145.950 MHz (1-2 Watts PEP)
Uplink: Mode J 145.900-146.000 MHz
Downlink: Mode J 435.100-435-200 MHz (1-2 Watts PEP)

In mid-1983, OSCAR 8 ceased operating just about the time that AMSAT-OSCAR Phase III-B (OSCAR 10) was launched. OSCAR 8 truly filled its mission as a gapfiller.

RADIO SPUTNIK 1 AND 2

In the mid-1970s when relations with the Soviet Union were good, the U.S. and the USSR were cooperating on several fronts, one of which was space. Plans were set in motion for a joint U.S.-USSR space project, ASTP (Apollo-Soyuz Test Project).

Several Soviet engineers came to this country on tours of NASA's facilities including the Goddard Space Flight Center. Several of the Soviet engineers who also were amateurs contacted some of the AMSAT people who were at Goddard about amateur satellites.

Nothing more was heard of the Soviet amateur programs until an October, 1975, issue of *Radio*, a Soviet electronics magazine, featured an article written by Budin and Fekhel, describing a terrestrial Mode A repeater. In 1977, the Soviets filed with the ITU a notice on the Soviet amateur satellite service. This was followed by a series of articles in *Radio* on the new Soviet RS (Radio Sputnik) satellites the Soviets intended to launch.

On October 26, 1978, a Soviet C-1 booster carrying Cosmos 1045 and RS-1 and 2 into orbit, lifted off from Plesetsk. Each satellite carried a Mode A (146 to 29 MHz) transponder. Since Soviet amateurs are limited to 5 watts transmit power, the RS birds carried very sensitive receivers. These satellites also carried protection devices to shutdown a satellite if a ground station used too high a power to access to satellite since these devices could only be turned back on from the Soviet Union, some passes over the U.S. were inaccessible.

AMSAT PHASE III-A

Black Friday occurred on May 23, 1980, for the amateur satellite community. Everything was set for the first high-orbit amateur satellite to be launched. Phase III-A was carried aboard the second ESA Ariane launch from Kourou, French Guiana, with the Firewheel 1 experiment. Everything was going according to plan--the rocket was flying a nominal course and mission.

Then, at Kourou, disaster struck. The words "non-nominal flying" ..."problem in an engine"... "the rocket is going down" ..."splashdown" rang through the loud speakers at the launch site. Amateurs around the world had lost their first Phase III bird, an event which marked the low point of the amateur satellite service. AMSAT members, however, stuck with the program and now Black Friday is just another memory in the long history of amateur radio.

UOSAT-OSCAR 9

On October 6, 1981, the first amateur scientific satellite was boosted into a 541 x 538 km orbit inclined 97.5 degrees. This non-communication satellite was built by the University of Surrey in England. UO-9 is a very sophisticated satellite that is currently being used for scientific and education purposes.

The primary telemetry frequency is 145.825 MHz NBFM with a power of 450 mW. This beacon can be heard on the simplest of equipment; I have used an ICOM 2AT handheld handy talkie with a rubber duck antenna with satisfactory results.

The data sources available from UO-9 on the 2 meter beacon are:

1200 baud ASCII telemetry
300 baud ASCII telemetry
110 baud ASCII telemetry
45.5 baud RTTY telemetry
12/20 wpm Morse code telemetry
Synthesized speech (digitalker) telemetry
Spacecraft asynchronous interface
Earth imaging system data through a CCD camera

UO-9 also has a 400 mW beacon operating on 435.025 MHz, which is the spacecraft's primary high speed experimental data channel. This beacon is intended for advanced amateur ground stations, scientific experimenters, and the UO-9 ground station network.

There is also a phased HF beacon system for propagation studies experiment on board. This experiment has not been activated because of problems with a stuck gravity gradient boom experiment. If they can succeed in getting the boom unstuck, look for HF activity from UO-9 on 7.001, 14.001, 21.001, and 28.001.

This is a great satellite for the amateur to practice satellite tracking techniques and the digitalker is a fantastic tool for telemetry decoding. This voice synthesizer uses both male and female voices to give telemetry information.

Reception reports of UO-9 trans missions can be sent to the following address for a QSL card:

Dr. M.N. Sweeting
UOSAT Control Centre,
Dept. of Engineering
Univ. of Surrey,
Guilford, Surrey
GU2 5XH, ENGLAND

More information on UO-9 is available from the above address, from AMSAT headquarters, or from the book, *The Satellite Experimenters Handbook* by Martin R. Davidoff, K2UBC, available at most amateur radio stores or through the American Radio Relay League, 225 Main Street, Newington, CT 06111. If you don't have a copy of this one on the shelf for reference, you're missing a lot of valuable information on satellite monitoring.

Radio Sputnik 3-8

On December 18, 1981, the Soviet Union surprised the amateur radio community with the launch from Plesetsk of 6 amateur radio satellites on one launch vehicle, a C-1 booster.

RS3-RS8 utilizes 2 meter to 10 meter mode A transponder making these satellites easy to monitor on a shortwave receiver. These satellites, like their predecessors, use 5 watts of power.

While activity on these satellites has decreased since the launch of OSCAR 10, and some of the birds have gone silent, a number of amateurs continue to use the Radio Sputniks for communications when OSCAR 10 is not in view.

These satellites carry several interesting features that are summarized as follows:

In general, each spacecraft contains two general purpose beacons plus some contain additional communication equipment.

*RS 3 & 4 are experimental in nature; they contain only beacons and no communications packages.

*RS 5 & 7 contain one transponder each (Mode A) and one autotransponder or Robot transponder.

*RS 6 & 8 contain one communication transponder each, (Mode A) only.

Spacecraft telemetry is easy to monitor. These satellites use Morse code to give the satellites current health to ground stations. Beacon frequencies for the RS satellites are as follows:

Satellite	**Beacon #1** (MHz)	**Beacon #2** (MHz)
Radio 3	29.321	29.401
Radio 4	29.360	29.403
Radio 5	29.331	29.452
Radio 6	29.411	29.453
Radio 7	29.341	29.501
Radio 8	29.461	29.502

The general transponders are all Mode A with non-inverting outputs. I have communicated through and listened to these orbiting repeaters and signals are quite audible on my portable receiver with a simple preamp.

The communications transponders for the RS birds are as follows:

Satellite	**Uplink**	**Downlink**
Radio 5	145.910-145.950	29.410-29.450
Radio 6	145.910-145.950	29.410-29.450
Radio 7	145.960-146.000	29.460-29.500
Radio 8	145.960-146.000	29.460-29.500

Radio Sputnik's 5 & 7 offer the licensed amateur operator a little extra fun through the autotransponders or Robot transponders they carry aboard; these devices enable the amateur to contact the satellite directly. If you contact the satellite using the correct transmit sequence in Morse code, an on-board computer will acknowledge your call, assign you serial contact number, and store your call letters and contact number for later downlinking when queried by Soviet ground stations.

Several amateurs have reported receiving RS5 and RS7 QSL cards via these Robot transponders for contacts they have made. The following are the uplink and downlink frequencies used by these robots:

Satellite	Uplink	Downlink
Radio 5	145.826	29.331
Radio 6	145.835	29.341

The following procedure should be used by amateurs for contacting the robot transponders. When the robot is active it will be calling CQ on the downlink frequency. Send a few dits on the uplink frequency; if you hear your dits regenerated on the downlink, you have captured the window which is 2 to 3 kHz wide call the satellite (10 to 30 wpm) using the following format:

RS7 DE (your callsign) AR

If the satellite received your call Radio 7 will answer:

(your callsign) DE RS7 QSO NR 123 OP ROBOT TU FR QSO 73 SK

The digit 123 number represents the robots QSO number that is incremented after each contact that the robot has. If the Robot only gets a partial message you may hear the following: QRZ, QRM, or RPT. If the robot wants you to slow down or send faster, it will send either QRS or QRQ. On these birds clean high speed CW does work the best.

I feel the RS birds represent the best way for SWLs and amateur operators to get experience in listening to HF satellite comms or working an amateur satellite. If you would like to try for a QSL card, send reception reports of the CW telemetry you monitored to:

Radio Amateur Satellite Committee
Radio Sport Federation
Box 88
Moscow, USSR

As mentioned before, some of the Soviet birds are now silent; as of this writing RS 5 is still available for communications use.

Astronaut Dr. Owen Garriott, W5LFL, is holding the clipboard he used to log the stations he talked with during off-duty chats during the Amateur Radio operation on STS-9. The radio is nested in the window immediately to his left, while the antenna is enclosed in the box mounted to the window above and forward of Dr. Garriott (NASA photo).

Iskra Satellites

One of the more obscure Russian satellites lite series is the ISKRA (Russian for "Spark"). Iskra 1 was an aeronautical research satellite launched July 10, 1981, that transmitted on 5,100 MHz.

Iskra 2 was unique in the annals of amateur radio space history. Manually expelled from the airlock of the Russian man space station, Salyut 7, on May 17, 1982, the 62 pound satellite apparently never fulfilled its primary mission of providing a 21 to 29 MHz linear transponder communications package in orbit. The 29.578 MHz beacon was quite audible during the satellite's short orbital life. Iskra 2 reentered the earth's atmosphere on July 9, 1982.

Iskra 3 was also launched from the Salyut; pushed through the airlock by hand by the resident cosmonauts, Iskra 3 suffered the same fate as Iskra 2.

Iskra 3's beacon transmitted on 29.538 MHz and was audible until its reentry into the earth's atmosphere. It is believed that both Iskra satellites contained amateur communications transponders utilizing 15 meters (21.230-21.270 MHz) as an uplink and 10 meters (29.580-29.620 MHz) as a downlink.

Interested monitors should watch the Soviet manned programs for new possible Iskra activities.

Activities in Space

STS-9 was launched on November 28, 1983. On board STS-Spacelab 1 was the first amateur radio space station, carried by Owen Garriott

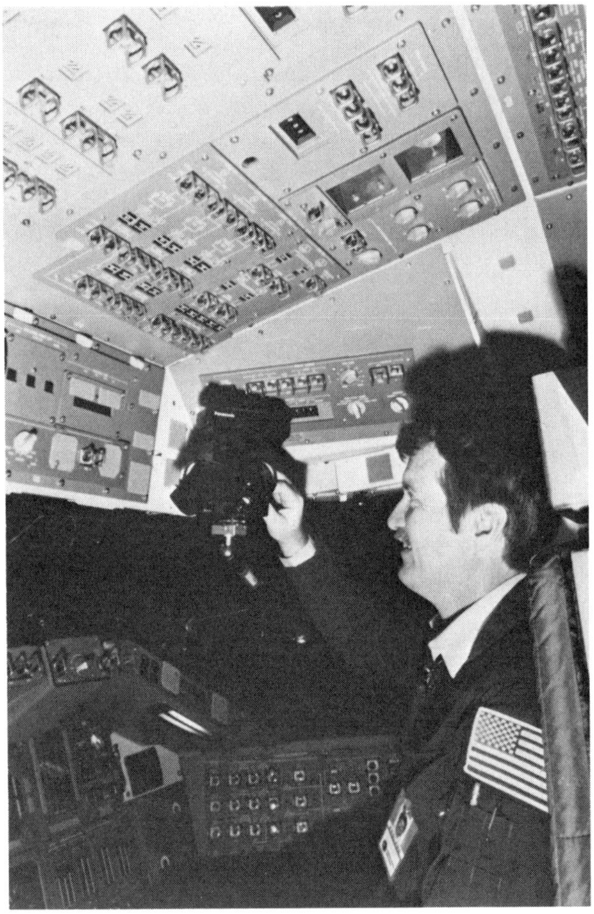

Mission Specialist Anthony W. England, W0!H/ORE, conducts ham operations from STS-51F (NASA photo).

W5LFL. For the first time "average citizens" from localities all over the world had a chance of talking directly to an astronaut in orbit.

Garriott operated his two meter HT from from the aft flight deck of the Space Shuttle orbiter, Columbia. His two meter operation provided amateurs around the world with one of the most historic pileups on the two meter band ever!

Amateurs never really got to talk to Owen; he used a tape recorder to tape transmissions on the 20 different frequencies he was monitoring. Listeners to the W5LFL mission were treated to descriptions of what Owen was seeing out the window of the orbiter and happenings in and around the cockpit by the STS-Spacelab crew members.

Owen used the following downlinks: 145.550 (primary), 145.250, 145.530, and 145.570 MHz. I only heard Owen on his primary frequency.

On July 29, 1985, the second ham-in- space mission roared into space from the Cape aboard Challenger. Tony England, WØORE, aboard with the Spacelab 2 mission, conducted ham operations during the STS-51F mission. Also aboard the shuttle for this mission was John-David Bartoe, W4NYZ.

Tony's ham equipment was tied to the shuttle's power busses, thus enabling WØORE to communicate for longer periods in orbit. Another first on this mission was the transmission of SSTV from the shuttle. The video was converted from the NASA fastscan video onboard Challenger. Hams on the ground received the video on 145.55 MHz. The SSTV signal was interspersed with an MFM CW ID every few minutes.

Tony wanted primarily to use this mission to further education and let young people have the chance to communicate with the Challenger via ham radio. Consequently, there was a series of uplink frequencies that were not published to the general ham populace to facilitate these communications.

WØORE's primary downlink frequency was on 145.550 MHz; uplink channels that were published included 144.91, 144.97, 145.03, and 145.63 MHz.

On October 30, 1985, the shuttle Challenger once again flew into space with ham astronauts aboard. This mission included science astronauts R. Furrer, DD6CF; E. Messerschmid, DG2KM, of Germany; and Dr. Wubbo Ockels, PE1LFO, from Holland.

The ham operations were conducted during the Spacelab D-1 mission of STS-61A. The astronauts carried a VHF/UHF transceiver that was designed and constructed by Bosch of Germany. The transceiver had a power output of 10 watts FM.

Callsign for the mission was DPØSL; operations were conducted on the following frequencies:

	Transmit	Receive
Channel 0	145.450 MHz	437.125 MHz
Channel 1	145.475 MHz	437.175 MHz
Channel 2	145.550 MHz	437.225 MHz
Channel 3	145.575 MHz	437.275 MHz
Channel 4	-----	437.325 MHz
Channel 5	-----	437.375 MHz

Note: Selection of transmit and receive frequencies was separate. There were no dedicated frequency pairs, except for channel 3/3 (normal operation). The modes were: beacon transmission only; beacon operation with automatic recording of incoming calls ("CQ de DPØSL record on tape K" followed by a one minute receiving time with automatic recording of calls on tape); and two-way voice operations.

OSCAR 10

AMSAT-OSCAR 10, launched on June 16, 1983, represents the most sophisticated amateur radio satellite launched to date. After three years and the loss of the Phase III satellite, amateur radio operators around the world celebrated this major accomplishment in amateur radio satellite history.

OSCAR 10 acts as a wideband repeater of amateur radio signals; it has a long expected service life and a higher orbit which allows users longer access time.

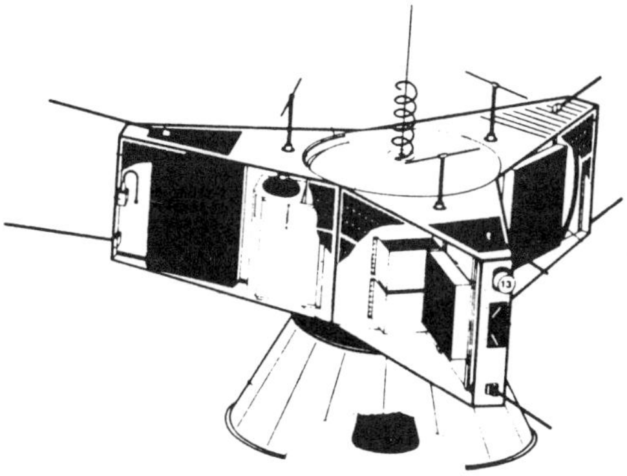

OSCAR 10

AMSAT officials estimate that OSCAR 10 will have a operational lifetime of 7 to 10 years thanks to a spare battery; battery problems are blamed for most satellite failures.

AO-10's frequencies are in the VHF/UHF/Microwave regions:

Mode B

General Beacon	145.810 MHz
Engineering Beacon	145.987 MHz
Uplink	435.025-435.175 MHz
Downlink	145.975-145.825 MHz

Mode L

General Beacon	436.020 MHz
Engineering Beacon	436.040 MHz
Uplink	1269.050-1269.850 MHz
Downlink	436.950-436.150 MHz

The AMSAT calling and net frequency is 145.957 MHz; it is very active and interesting to monitor. Several other interesting frequencies known as Special Service Channels (SSC) are used for experiments, special nets, and other out of the ordinary operations:

SSC L1	145.830 MHz
SSC L2	145.840 MHz
SSC L3	145.972 MHz
SSC L4	145.962 MHz

OSCAR 11

March 1, 1984, saw the launch of the University of Surrey's second scientific satellite, UOSAT OSCAR 11, placed into a 693 X 674 km orbit, inclined 98.2 degrees with the Landsat 5 spacecraft.

UO-11 contains a 145.825 MHz beacon that is nearly identical to UOSAT OSCAR IX's, a newly designed 435.025 MHz beacon; and an improved 2401.5 MHz beacon.

UO-11 contains the following experiments and devices aboard:

Digitalker--
Essentially the same as UO-9's, this 550 word speech synthesizer is used for speaking telemetry.

Navigation Magnetometer--
Used for experimental measurements of earth's magnetic field disturbances.

DSR experiment--
This unit stores data from the CCD imager, particle counter experiment, or computer UART, and outputs it in a checksummed format at 1200, 2400, 4800, and 9600 bps.

DCE--
This experiment will be used to investigate packet radio through UO-11, as well as interface with the navigation magnetometer and the telemetry system for long term storage.

Space dust experiment--
Built by students at the University of Kent to study space dust particles.

CCD camera--
A redesigned version of the one carried aboard UO-9, this camera will allow the user to observe earth as seen from UO-11.

Particle detectors & wave correlator experiment--
Three geiger counters with different electron energy thresholds study the ionosphere D, E, and F layers, similar to those carried on UO-9.

UO-9 and 11 are very interesting satellites to monitor, they can afford the amateur or SWL a chance to conduct scientific experiments using a spacecraft.

Radio Sputnik 10/11

RS-10 and RS-11 were launched from the Soviet launch site Plesetsk at 0724 UTC, June 23, 1987. The launch was letter-perfect, placing the RS's and the primary payload, Cosmos 1861, in a 105 minute orbit inclined 83 degrees.

The desired orbit was attained very precisely: the nodal period is 105.0245 minutes; the orbital increment is 26.3824 degrees west per orbit. Average height is close to 1000 km (621 miles). In comparison to other Oscars, the new RS's are higher than UO-11 (700 km) and AO-8 (900 km), but lower than AO-7 (1400 km). In fact, RS-10 and 11 are much lower than any prior RS's.

RS-1 through RS-8 were very high for low earth orbiters (LEOs) at 1700 km, substantially decreasing their life expectancy since they came close to the lower edges of the Van Allen radiation belts at that altitude. RS-10 and 11 will last much longer at 1000 km; their altitude assures stable orbits for several decades.

NORAD has designated the composite Cosmos 1861/RS-10/RS-11 object 18129; its international designation is 87-54A.

RS-10 and 11 were built at the Tsiolkovskiy Museum for the History of Cosmonautics in Kuluga, an industrial center 180 km southwest of Moscow. The transponders are called BRTK-100 (the Russian equivalent of "equipment for radio amateur satellite communication") and the overall

BEACON STATUS INDICATORS

Channel Number	Status Designators	Meaning/Equations
1A	IS	Telemetry data source sampling period 90 minutes or
	NS	Telemetry data source sampling period 10 minutes
1N		Power supply voltage over sample period where V=N/4 volts
2A	IR	2 meter receiver with -20 dB attenuator in or
	NR	2 meter receiver with -20 dB attenuator out
2N		Output power of 2 meter transmitter where W=N/10 in watts
3A	ID	15 meter receiver with -10 dB attenuator in or
	ND	15 meter receiver with -10 dB attenuator out
3N		Output power of 10 meter transmitter where W = N/10 in watts
4A	IG	15 meter uplink off or
	NG	15 meter uplink on
4N		15 meter receiver AGC voltage where V = N/5 in volts
5A	IU	2 meter receiver off or
	NU	2 meter receiver on
5N		2 meter receiver AGC voltage where V =n N/5 in volts
6A	IW	Special command station channel off or
	NW	Special command station channel on
6N		Special command station AGC voltage where V = N/5 in volts
7A	IK	Output power of 10 meter beacon = 1 watt or
	NK	Output power of 10 meter beacon = 300 milliwatts
7N		Service command parameter, 10 meter mode
8A	IO	Output power of 2 meter beacon = 1 watt or
	NO	Output power of 2 meter beacon = 300 milliwatts
8N		Service command, parameter, 2 meter mode
9A	AS	Status of 1st memory board = off or
	MS	Status of 1st memory board = on
9N		10 meter transmitter temperature where T = N-10 in degrees C
10A	AR	Status of 2nd memory board = off or
	MR	Status of 2nd memory board = on
10N		2 meter transmitter temperature where T = N-10 in degrees C
11A	AD	Special service channel for loading memory is open or
	MD	Special service channel for loading memory is closed
11N		20 volt power supply temperature where T = N-10 in degrees C
12A	AG	Code store memory status is open or
	MG	Code store memory status is closed
12N		9 volt power supply temperature where T = N-10 in degrees C
13A	AU	Output information from memory via 10 meter beacon or
	MU	Output information from memory via 2 meter beacon
13N		Control parameter backup 9 V power supply where V = N/5 volts
14A	AW	Attenuator of 15 meter robot receiver = -10 dB or
	MW	Attenuator of 15 meter robot receiver = 0 dB
14N		IF voltage of 15 meter robot receiver where V = N/5 in volts
15A	AK	Attenuator of 2 meter robot receiver = -10 dB or
	MK	Attenuator of 2 meter robot receiver = 0 dB
15N		IF voltage of 2 meter robot receiver where V =nN/5 in volts
16A	AO	Special command channel 2 meter output power = 1 watt or
	MO	Special command channel 2 meter output power = 300 milliwatts
16N		Robot QSO counter where 00-32 QSOs logged is indicated as 00 and 33; 128 QSOs indicates in the range of 80-99.

project management is in the hands of DOSAAF, a military-related organization whose major mission is the training of pre-draft-age youth in militarily-significant technology.

RS-10 and 11 are apparently identical except with regard to frequency, each using three bands to achieve five distinct modes of operation in addition to its auxiliary robot repeaters. On each unit, 15 meters is used exclusively as an uplink band, 10 meters is used exclusively as a downlink band and 2 meters can be employed as either an uplink or downlink band. The complete bandplan is shown in the next section (Amateur Satellite Frequency Guide).

The transponder power outputs are 5 watts; these birds should be very loud. No operating schedule has thus far been announced. There is some concern that the primary payload, Cosmos 1861, which has a 150 MHz downlink, is interfering with RS-10 and 11's 2 meter receiver. This may drive the operating schedule in favor of modes K and T with their 15 meter uplinks and away from Mode A with its 2 meter uplink.

Reading the Beacons

CW telemetry provides various status indicators and measurements made on the transponders. There are 16 channels sent, each in the format of two alpha characters followed by two numeric characters. The alpha part gives a specific status such as "on" or "off" for a specific feature. The numeric part gives a value for a variable of interest such as temperature of the 10 meter transmitter.

For example, "IG35" means the 21 MHz receiver is off; "35" gives the AGC level on the 15 meter receiver, where the value in volts equals the number sent divided by 5 (7 volts). In the accompanying table, the alpha part of a channel is designated by "A" and the numeric part by "N."

Amateur Satellite Frequency Guide

This unique frequency guide condenses all the current operating (1987) amateur satellite frequency plans in one format. We appreciate the efforts of WA2LQQ and W0RPK BBS.

UOSAT OSCAR 9

Downlinks on 2 meters, 70 cm and 13 cm; no uplinks since this is primarily an educational and research spacecraft

Other downlinks at 7.050, 14.002, 21.002 and 29.502 MHz which are part of the HF propagation experiment.

VHF Downlink: 145.825 MHz
UHF Downlink: 435.025 MHz
UHF Downlink: 2401.000 MHz

UOSAT OSCAR 11

Downlinks on 2 meters, 70 cm and 13 cm; no uplinks since this is primarily an educational and research spacecraft

VHF Downlink: 145.825 MHz
UHF Downlink: 435.025 MHz
UHF Downlink: 2401.500 MHz

AMSAT OSCAR 10

Mode B: 70 cm up/ 2 meters down
Mode L: 24 cm up/ 70 cm down
Beacons: Mode B: General beacon 145.810 MHz; Engineering beacon 145.987 MHz. Mode L: (no longer operational)

Mode B: 435.030 - 435.180 up/ 145.825 - 145.975 down (inverting)

Uplink	Downlink		
	145.987 MHz	-	Engineering beacon
435.030	145.975	-	Passband limit, upper
435.045	145.960	-	General use limit, upper
435.050	145.955		
435.060	145.945		
435.070	145.935		
435.080	145.925		
435.090	145.915		
435.100	145.905		
435.105	145.900	-	Nominal passband center
435.110	145.895		
435.120	145.885		
435.130	145.875		
435.140	145.865		
435.150	145.855		
435.160	145.845	-	General use limit, lower
435.180	145.825	-	Passband limit, lower
	145.810	-	General beacon

Mode L: (No longer operational)

FUJI OSCAR 12

Mode JA: 2 meter uplink/ 70 cm downlink (analog mode for voice and CW)
Mode JD: 2 meter uplink/ 70 cm downlink (digital mode for packet radio)
Beacon: 435.795 MHz

Mode JA: 145.900-146.000 MHz up/ 435.900-435.800 MHz down (inverting)

Uplink	Downlink		
145.900	435.900 MHz	-	Passband limit, upper
145.910	435.890		
145.920	435.880		
145.930	435.870		
145.940	435.860		
145.950	435.850	-	Passband center
145.960	435.840		
145.970	435.830		
145.980	435.820		
145.990	435.810		
146.000	435.800	-	Passband limit, lower
	435.795	-	Beacon

Mode JD: (Digital) Any of four 2 meter channels up yields a single 70 cm channel down

Uplink	Downlink		
145.850	435.910 MHz	-	Channel 1
145.870	435.910	-	Channel 2
145.890	435.910	-	Channel 3
145.910	435.910	-	Channel 4

RADIO SPUTNIK 10

Mode A:	2 meter up/ 10 meters down
Mode K:	15 meters up/ 10 meters down
Mode T:	15 meters up/ 2 meters down
*Mode KT:	15 meters up/ both 10 and 2 meters down
*Mode KA:	15 and 2 meters up/ 10 meters down

** Modes KT and KA are simply combinations of modes K, T and A. Beacons can carry telemetry or robot downlink.*

Mode A: 145.860 - 145.900 MHz up/ 29.360 - 29.400 MHz down

Uplink	Downlink	
	29.357 MHz	- Beacon
145.860	29.360	- Passband limit, lower
145.870	29.370	
145.880	29.380	- Passband center
145.890	29.390	
145.900	29.400	- Passband limit, upper
	29.403	- Beacon

Robot A:

Uplink	Downlink
145.820	29.357 or 29.403 MHz

Mode K: 21.160 - 21.200 MHz up/ 29.360 - 29.400 MHz down

Uplink	Downlink	
	29.357 MHz	- Beacon
21.160	29.360	- Passband limit, lower
21.170	29.370	
21.180	29.380	- Passband center
21.190	29.390	
21.200	29.400	- Passband limit, upper
	29.403	- Beacon

Robot K:

Uplink	Downlink
21.120(?)	29.357 or 29.403 MHz

Mode T: 21.160 - 21.200 up/ 145.860 - 145.900 MHz down

Uplink	Downlink	
	145.857 MHz	- Beacon
21.160	145.860	- Passband limit, lower
21.170	145.860	
21.180	145.880	- Passband center
21.190	145.890	
21.200	145.900	- Passband limit, upper
	145.903	- Beacon

Robot T:

Uplink	Downlink
21.120(?)	145.857 or 145.903 MHz

Mode KT: 21.160 - 21.200 up/ 29.360 - 29.400 and 145.860 - 145.900 MHz down

KT Uplink	K Downlink	T Downlink	
	29.357	145.857 MHz	- Beacon
21.160	29.370	145.860	- Lower Pssbnd limit
21.170	29.370	145.870	
21.180	29.380	145.880	- Passband center
21.190	29.390	145.890	
21.200	29.400	145.900	- Upper Pssbnd limit
	29.403	145.903	- Beacon

Mode KA: 21.160 - 21.200 and 145.860 - 145.900 up/ 29.360 - 29.400 down

K Uplink	A Uplink	KA Downlink	
		29.357 MHz	- Beacon
21.160	145.860	29.360	- Lower Pssbnd limit
21.170	145.870	29.370	
21.180	145.880	29.380	- Passband center
21.190	145.880	29.390	
21.200	145.900	29.400	- Upper Pssbnd limit
		29.403	- Beacon

RADIO SPUTNIK 11

Mode A: 2 meters up/ 10 meters down
Mode K: 15 meters up/ 10 meters down
Mode T: 15 meters up/ 2 meters down
*Mode KT: 15 meters up/ both 10 and 2 meters down
*Mode KA: 15 and 2 meters up/ 10 meters down

* *Modes KT and KA are simply combinations of modes K, T and A. Beacons can carry telemetry or robot downlink*

Uplink	Downlink	
	29.403 MHz	- Beacon
21.210	29.410	- Passband limit, lower
21.220	29.420	
21.230	29.430	- Passband center
21.240	29.440	
21.250	29.450	- Passband limit, upper
	29.453	- Beacon

Robot K:

Uplink	Downlink
21.130	29.403 or 29.453 MHz

Mode T: 21.210 - 21.250 up/ 145.910 - 145.950 MHz down

Uplink	Downlink	
	145.907 MHz	- Beacon
21.210	145.910	- Passband limit, lower
21.220	145.920	
21.230	145.930	- Passband center
21.240	145.940	
21.250	145.950	- Passband limit, upper
	145.953	- Beacon

Robot T:

Uplink	Downlink
21.130	145.907 or 145.953 MHz

Mode KT: 21.210 - 21.250 up/ 29.410 - 29.450 and 145.910 - 145.950 MHz down

KT Uplink	K Downlink	T Downlink	
	29.407	145.907 MHz	- Beacon
21.210	29.410	145.910	- Passband limit, lower
21.220	29.420	145.920	
21.230	29.430	145.930	- Passband center
21.240	29.440	145.940	
21.250	29.450	145.950	- Passband limit, upper
	29.453	145.953	

Mode KA: 21.210 - 21.250 and 145.910 - 145.950 up/ 29.410 - 29.450 MHz down

K Uplink	A Uplink	KA Downlink	
		29.407 MHz	- Beacon
21.210	145.910	29.410	- Passband limit, lower
21.220	145.920	29.420	
21.230	145.930	29.430	- Passband center
21.240	145.940	29.440	
21.250	145.950	29.450	- Passband limit, upper
		29.453	- Beacon

PHASE 3C - Estimated launch date : early 1988

Mode B: 70 cm up/ 2 meters down
Mode JL: 24 cm and 2 meters up/ 70 cm down
Mode S: 70 cm up/ 13 cm down

Rudak: 24 cm up/ 70 cm down
Beacons: Mode B: General beacon 145.8125; Engineering beacon 145.975. Mode JL: General beacon 435.650; Engineering beacon 435.675. Mode S: 2400.640.

Mode B:

Uplink	Downlink	
	145.975 MHz	Engineering beacon
435.425	145.975	- Passband limit, upper
435.435	145.965	
435.445	145.955	
435.455	145.945	
435.465	145.935	
435.475	145.925	
435.485	145.915	
435.495	145.905	
435.505	145.895	- Passband center
435.515	145.885	
435.525	145.875	
435.535	145.865	
435.545	145.855	
435.555	145.845	
435.565	145.835	
435.575	145.825	- Passband limit, lower
	145.8125	- General beacon

Mode JL:

L Uplink	J1 Uplink	Downlink	J2 Uplink	
1269.325		435.975 MHz		L Passband limit, upper
1269.330	145.820	435.970	144.440	J sub-band limit, upper
1269.340	145.830	435.960	144.450	
1269.350	145.840	435.940	144.460	J Sub-band center
1269.360	145.850	435.940	144.470	
1269.370	145.860	435.930	144.480	J Sub-band limit, lower
1269.380		435.920		
1269.390		435.910		
1269.400		435.900		
1269.410		435.890		
1269.420		435.880		
1269.430		435.870		
1269.440		435.860		
1269.450		435.850		L Passband center
1269.460		435.840		
1269.470		435.830		
1269.480		435.820		
1269.490		435.810		
1269.500		435.800		
1269.510		435.790		
1269.520		435.780		
1269.530		435.770		
1269.540		435.760		
1269.550		435.750		
1269.560		435.740		
1269.570		435.730		
1269.575		435.725		L Passband limit, lower
		435.675		Engineering Beacon
		435.650		General Beacon

Mode S:

Uplink	Downlink	
	2400.650 MHz	Beacon
435.610	2400.695	- Passband limit, lower
435.615	2400.700	

435.620	2400.705	
435.625	2400.710	- Passband center
435.630	2400.715	
435.640	2400.725	- Passband limit, lower

Rudak:

Uplink	**Downlink**
1269.675	435.675 MHz - Single channel

AMSAT Radio Nets

The best way to keep track of amateur radio satellite activity is to monitor AMSAT Radio nets. The table included here gives the latest schedules for AMSAT's many worldwide nets.

Net Description	Day	Time	Freq	Notes
International				
AMSAT International	Sunday	1900 UTC	14.282 MHz	1
AMSAT International	Sunday	1800	21.280	1
AMSAT South Pacific	Saturday	2200	14.282	2
National				
AMSAT UK	Sunday	1015 Local	3.780	
AMSAT UK	Mon & Wed	1900 Local	3.780	
SA AMSAT	Sunday	0900 UTC	14.280	
SA AMSAT	Sunday	0900	7.080	
AMSAT UK	Sunday	1000	3.685	3
Regional				
AMSAT NA East Coast, 75m	Wednesday	2000 EDT	3.840	
AMSAT NA Mid-America 75m	Wednesday	2100 CDT	3.840	
Local, USA				
Los Angeles 2m	Wednesday	2000 Local	144.144	
Los Angeles 2m	Daily	0730 Local	144.144	
AMSAT S.Pacific/LA	Sunday	2200 UTC	144.144	4
Cent Cal OSCAR AMSAT net	Tuesday	2000 Local	147.150	5
East Coast 75m/NYC	Tuesday	2000 Local	144.280	6
Colorado AMSAT net	Wednesday	0300 UTC	147.225	
Houston Area AMSAT net	Tuesday	2200 Local	145.450	7
Chicago Area AMSAT net	Wednesday	1930 Local	146.880	8
Miami Area AMSAT net	Tuesday	2000 Local	146.925	9
South Africa				
SA AMSAT	Sunday	0900 UTC	145.650	
UK				
Scotland: Paisley	Daily	0900 Local	144.???	

NOTES:

1. Net inactive. Reactivation when conditions improve.
2. This net may return to 21.280 MHz as conditions dictate.
3. Back-up frequency is 7.084 MHz.
4. Links to 443.525, 146.655, and 223.720 MHz.
5. Two-meter simulcast of HF net by W6SP.
6. Two-meter simulcat of HF net by WA2LQQ.
7. Back-up frequency is 146.700 MHz.
8. PL 1B required for access.
9. Rebroadcast of East Coast 75m net.

Chapter Three

WEATHER SATELLITES

One of the most reliable, non-voice satellites that can be decoded by amateur satellite hobbyists are the weather satellites. Monitoring these satellites can keep the hobbyist instantly informed of world weather conditions without waiting for the evening weather newscast. Satellite imagery is fascinating to watch and can be used as an educational aid to teach students scientific fundamentals of physics and meteorology.

This chapter will include information on geostationary weather satellites as well as solar orbiting satellites launched by the Russians and the United States. Information was provided by several amateur weather satellite buffs and the staff of NOAA/NESS, specifically Robert Popham. I am indebted to Grant Zehr (WA9TFB), Stephen Padar, Tom Van Kuiken, and Paula Murphy for their assistance.

Numerous satellites have been launched since the first weather satellites were placed into orbit in 1960. The United States has always led the way in weather satellite technology. The tables in Appendix D show some U.S. satellites of the first two decades and gives data on their capabilities.

Current Orbiting Weather Satellites

NOAA Satellites

The TIROS N/NOAA A-G satellites series was introduced with the launch of the protoflight satellite TIROS N on October 13, 1978. This NASA funded satellite was followed into order by NOAA-6 (NOAA-A before launch) on June 27, 1979, and NOAA-7 on June 23, 1981. The two satellite system has been considered operational since that time.

NOAA has been receptive to requests from other agencies to use a national resource (NOAA satellites) to further space research as long as the primary environmental mission of the satellite was not impacted.

In 1975, NASA asked NOAA to permit them to modify three satellites in the NOAA A-G series to carry Search and Rescue (SAR) demonstration hardware. NOAA concurred in this request recognizing the potential advantages of using a space platform in a

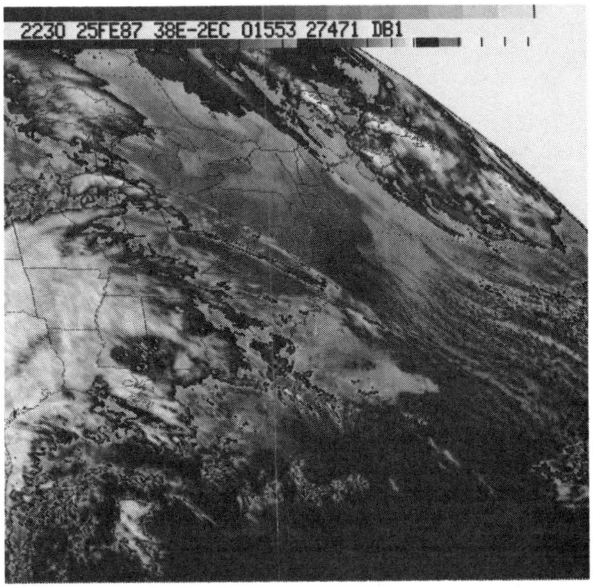

cost effective approach to meeting high seas and terrestrial search and rescue, distress, alerting, and locating activity.

The system flown on the NOAA satellites provides the data base necessary to evaluate this potential. The SAR program is a cooperative effort of the United States, Canada, France, and the USSR.

A NASA experiment aimed at making long-term determinations of the components of earth's incoming and outgoing radiation will fly on the NOAA-F and G satellites. Flight of these Earth Radiation Budget Experiment (ERBE) instruments in sun synchronous orbit will be in conjunction with flight of duplicate instruments on a dedicated NASA research and development satellite in an orbit that is not sun synchronous.

NOAA will add an instrument for operational long-term monitoring of the global distribution of ozone with a first flight likely on NOAA G. This instrument, the Solar Backscatter Ultraviolet Radiometer/2 (SRUV/2), is an adaptation of a similar instrument flown on Nimbus weather satellites.

The primary sensors for the NOAA E-J satellites are:

1. A TIROS Operational Vertical Sounder (TOVS), a three instrument system consisting

of:

a. The High Resolution Infrared Radiation Sounder (HIRS/2), a 20 channel instrument making measurements primarily in the infrared region of the spectrum. The instrument is designed to provide data that will permit calculation of the temperature profile from the surface to 10mb; water vapor content in three layers in the atmosphere; and total ozone content. The design is based on the HIRS instrument flown on the Nimbus satellite. It is also identical to the instruments flown on the TIROS N and NOAA A-D satellites.

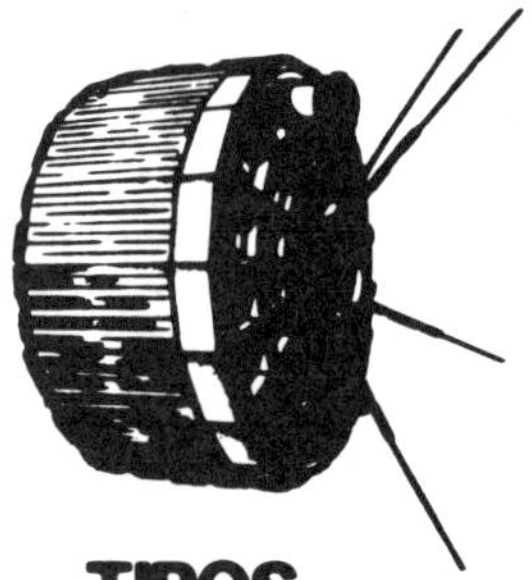

TIROS

b. The Stratospheric Sounding Unit (SSU) employs a selective absorption technique to make measurements in three channels. The spectral characteristic of each channel is determined by the pressure in a carbon dioxide gas cell in the optical path. Eight or nine of these satellites are planned for the TIROS-N, NOAA A-J series.

c. The Microwave Sounding Unit (MSU) is a four channel Dicke radiometer, making passive measurements in the 5.5 mm oxygen band. This instrument, unlike those making measurements in the infrared region, is little affected by clouds.

Data for the TOVS will be available as part of the HRPT transmission and on the spacecraft beacon transmission, frequently referred to as the Direct Sounder Broadcast (DSB).

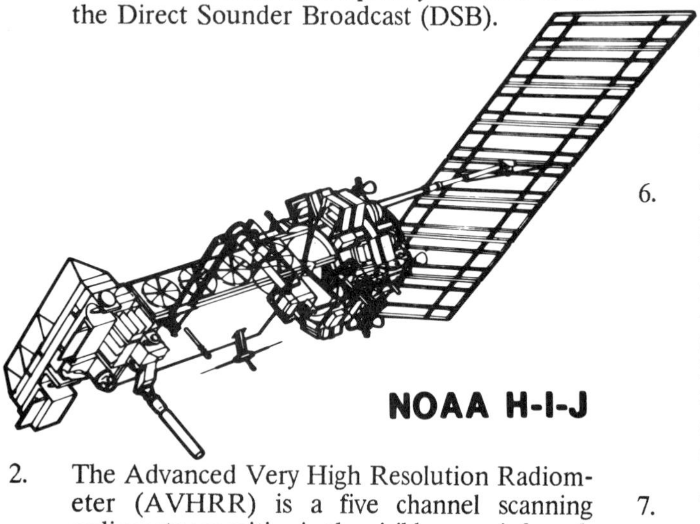

NOAA H-I-J

2. The Advanced Very High Resolution Radiometer (AVHRR) is a five channel scanning radiometer sensitive in the visible, near infrared, and infrared regions. This instrument will provide data for central processing, APT, and HRPT outputs. HRPT data (all spectral regions) will be transmitted at full resolution (1.1 km); while APT (two selectable spectral regions) will have reduced resolution (4 km).

3. The Space Environmental Monitor (SEM) data will also be included in the HRPT and beacon transmissions, consisting of two separate instruments and a data processing unit. The components are:

a. The Total Energy Detector (TED) measures a broad range of energetic particles from 0.3 to 20 keV (thousand electron volts) in 11 bands.

b. The Medium Energy Proton and Electron Detector (MEPED) through NOAA C; senses protons, electrons, and ions with energies from 30 keV to several tens of MeV (million electron volts).

c. The High Energy Proton and Electron Detector (HEPAD) through NOAA C; senses protons and alphas from a few hundred MeV up through relativistic particles above 840 MeV. This instrument is now flying on the NOAA and GOES satellites, first launched on GOES 4 in 1980.

4. The Data Collection System (DCS) is a random access system to acquire data from fixed and free-floating terrestrial and atmospheric platforms. Platform location will be possible by ground processing of the doppler measurements of the carrier frequencies. Data collected from each platform, which includes identification as well as environmental measurements, are included in the HRPT and beacon transmissions.

5. The Solar Backscatter Ultraviolet Radiometer (SMUV/2) will be used starting with NOAA G. The SBUV/2 is a non-spatial-scanning, nadir viewing instrument designed to measure scene radiance in the spectral region around 160 nanometer (nm) to 400 nm. The data gathered will be used to determine the vertical distribution of ozone in the earth's atmosphere, total ozone in the atmosphere, and solar spectral irradiance.

6. The Search and Rescue (SAR) system is a random access system to acquire data from Emergency Locator Transmitters (ELT) (121.5 and 243.0 MHz) and Emergency Position Indicating Radio Beacons (EPIRB) carried on general aviation aircraft and some classes of marine vessels. The received data is retransmitted on a discrete frequency to ground stations where it is processed to locate the origin of the distress signal.

7. The Earth Radiation Budget Experiment (ERBE) is a two part instrument consisting of:

a. A fixed field-of-view instrument with four

viewing radiometers and one shuttered sun viewing radiometer. The earth viewing radiometers consist of pairs of broadband (0.2 to 50+ micrometer wavelength) and shortwave (0.2 to 5 micrometer) monitors.

b. A scanning instrument comprising three earth scanning channels differing only in spectral response characteristics, the total radiation channel sensitive to 0.2 to 50 micrometers; the shortwave channel sensitive to 0.2 to 5 micrometers; and the longwave channel sensitive to 5 to 50 micrometers, with 3 degrees field of view.

The ERBE is planned for flight as an experiment on NOAA F and NOAA G satellites.

Now that you have an account of the instrument complement onboard U.S. NOAA satellites, here is a description of the satellites and the frequencies that they use:

Spacecraft: Total weight--2,200 lbs.
(excludes expendables)
Payload: Weight, including tape recorders--
850 lbs.
Spacecraft size: 165 inches in length
74 inches in diameter
Solar array: 7.8 ft x 16.1 ft: 125 square feet
515 watts, end of life at worst solar angle
Power requirements: Full operation - 475 watts
Reserved - 40 watts
Attitude Control
System:0.2 degrees in all axes
0.14 degrees determination
Orbit: 833; 870 km nominal, sun synchronous
Launch vehicle: Atlas E/F
Lifetime: 2 years planned
Communications:
Command Link 148.560 MHz

TIROS-N Spacecraft

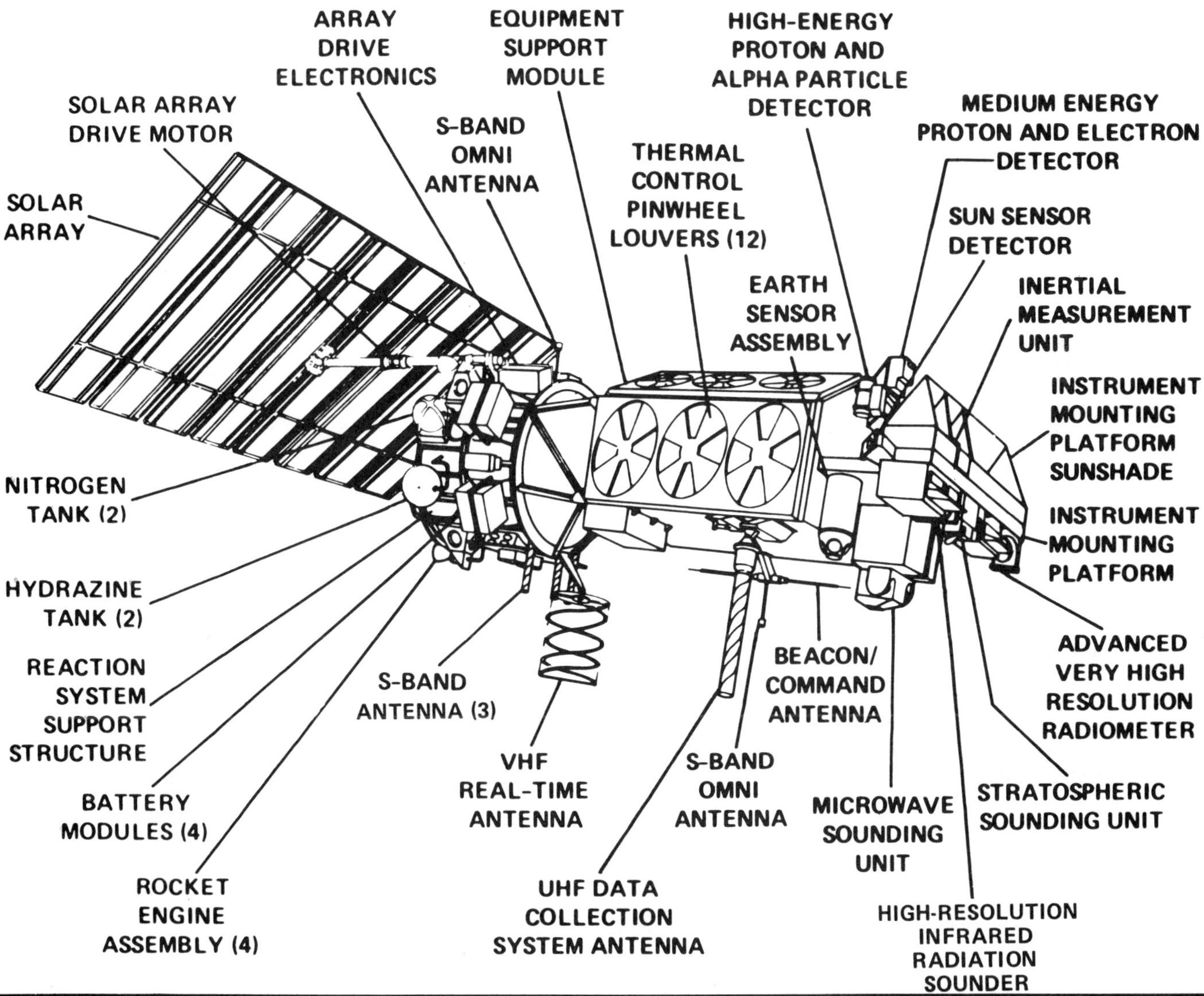

NOAA 7, afternoon picture, summer '82 (Grant Zehr, WA9TFB)

Beacon- 136.770; 137.770 MHz
S-Band - 1698; 1702.5; 1707 MHz
APT - 137.50; 137.620 MHz
DCS - (uplink) 401.650 MHz
SAR - 1544.5 MHz
SAR - (uplink) 121.5; 243.0; 406 MHz

Current NOAA Satellite Status

NOAA-8's attitude systems failed June 11, 1984, and the spacecraft is tumbling. No systems are being used operationally, but there appears to be some hope of recovery. The NOAA-8 beacon is being left on and the NOAA-6 beacon is being turned off during conflicts between the two spacecraft.

NOAA-8 Statistical Data:

Launched March 28, 1983
International Designator: 1983-022A
Period: 101.3 minutes
Equator Crossing: 1536 Local (sun) time
HRPT frequency: 1698.0 MHz
Channels 2 (0.725-1.10) and 4(10.5-11.5)
are transmitted continuously.
Orbital Inclination: 98.73 degrees
Daytime (morning) descending orbit
APT frequency: 137.50 MHz
DSB frequency: 136.77 MHz

NOAA-7 has a serious power regulation problem caused by a loss of power supply shunt loads. NOAA 7 is the primary afternoon spacecraft, crossing the equator northbound in daylight at about 1536 local (sun) time each day.

NOAA-7 Statistical Data:

Launched June 23, 1981
International Designator: 1988-059A
Period: 102.0 minutes
Equator Crossing 1536 Local (sun time)
HRPT frequency: 1707.0 MHz
Channels 2 (0.725-1.10) and 4 (10.3-11.3)
are transmitted continuously
Orbital Inclination: 98.97 degrees
Daytime (afternoon) ascending orbit:
APT frequency: 137.620 MHz
DSB frequency: 137.770 MHz

NOAA-6 has been reactivated to replace NOAA-8 and is the primary morning operational spacecraft.

NOAA-6 Statistical Data:

Launched June 27, 1979
International designator: 1979-57A
Period: 101.21 minutes
Equator Crossing 0739 Local (sun time)
HRPT frequency: 1698.0 MHz
Orbital Inclination: 98.67 degrees
Daytime (morning) descending orbit
APT frequency: 137.50 MHz
DSB frequency: 136.770 MHz

According to NOAA/NESS personnel, the next generation of weather orbiting satellites, called "NOAA Next," has been postponed for budgetary reasons; instead, three more of the Advanced TIROS-N type satellites will be procured to continue the series into the 1990's. These three---NOAA K, L, and M--- will carry an Advanced Microwave Sounder Unit (AMSU). The VHF beacon systems will also be retained. Probable launch date for NOAA-K will be late 1989 or early 1990.

DMSP Air Force Orbiting Weather satellites will be covered in the U.S. Military chapter.

SARSAT

Search and Rescue Satellite-Aided Tracking

In July of 1985, an 11 nation international conference announced an agreement to institute an improved satellite search and rescue system to locate and relay distress calls worldwide.

The international conference has approved the use of 406 MHz for Emergency Locator Beacons. Payloads carried by American NOAA and Soviet Cosmos navigational satellites will now be able to detect emergency beacons anywhere on earth and then relay encoded distress information when the satellite comes within range of ground stations.

While the old system merely sends a beep, the new

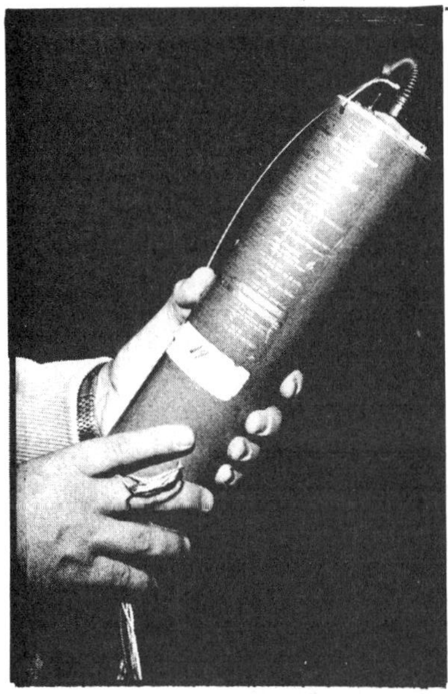

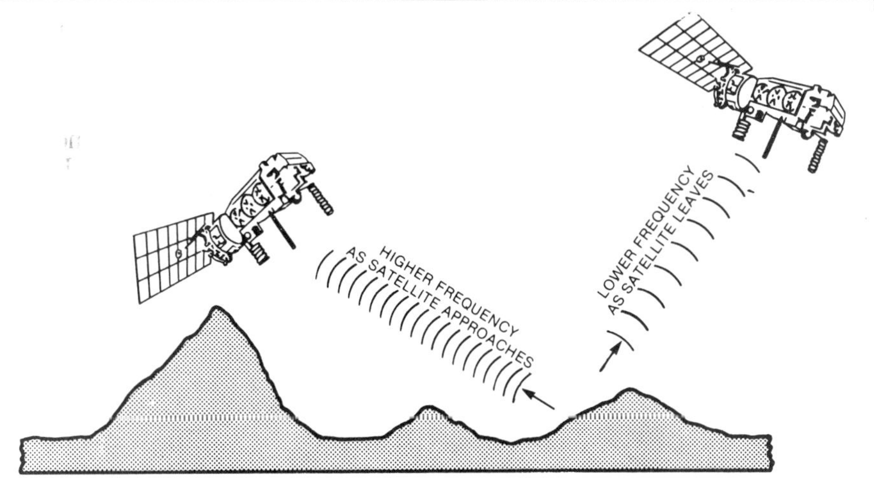

Illustration of Doppler effect

(Above) A typical Emergency Position-Indicating Radio Beacon (EPIRB) for ocean-going ships such as tankers and passenger vessels. (Below) A typical ELT for private planes.

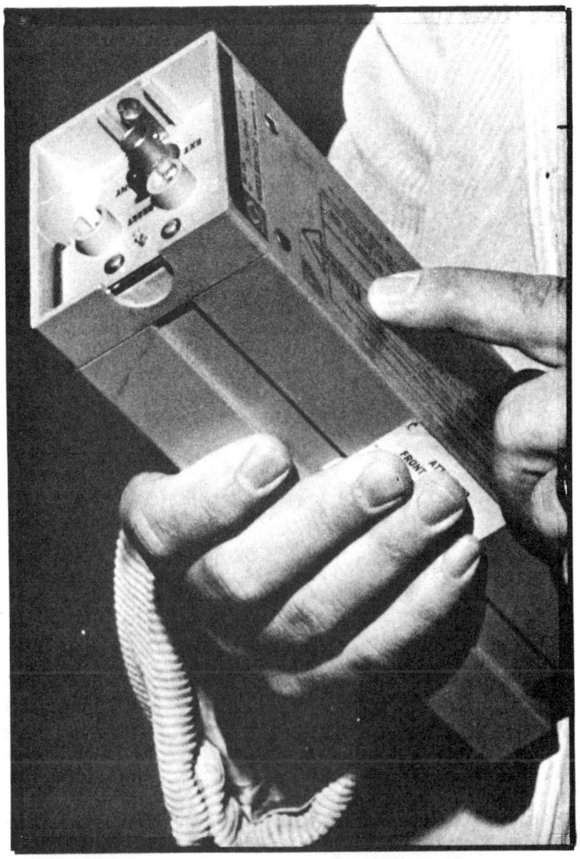

system sends a coded message which can tell ground stations who or what is in trouble and may even provide additional details about location.

The search and rescue system is called SARSAT and consists of space and ground components. The key elements are:

(1)A satellite-borne receiver and frequency translation repeater for Emergency Location Transmitter (ELT)/Emergency Position Indicating Radio Beacons (EPIRB) bands (121.5, 243 and 406 MHz). This system is used for regional coverage/alerting;

(2)Local user terminal (LUT) which receives the relayed ELT/EPIRB signals and processes the Doppler data to earth to aid in locating the transmitting platform;

(3)Operational and experimental ELT and EPIRB systems;

(4)A satellite-borne receiver and processor for the experimental 406 MHz ELT/ EPIRB transmissions (provided by Centre National d'Etudes Spatiales--CNES--France). The unit is in many ways equivalent to the Data Collection System (DCS) provided by CNES and flown on all NOAA weather satellites;

(5)Mission Control Centers for coordinating activities, processing global experimental data and coordinating search activity support.

When an emergency occurs on a vehicle carrying ELT or EPIRB, the unit is activated either automatically or manually. The weak (100 mW) signal, used for homing by search forces, is amplified and amplitude modulated so that it can be recognized by listening to an aircraft band receiver.

With the SARSAT system, the signal is received by the satellite when within range of the transmitter, then translated and rebroadcast to any LUT within view of the satellite. Because of the relative velocity difference between satellite and platform, the signals received at the satellite are Doppler shifted.

At the LUT, special processing of the received signals after time tagging allows the weak transmission of the Doppler shifted signal to be processed to determine the position of the transmitter. This information is then provided to a Rescue Coordination Center for their use in locating the stricken vessel or aircraft.

The new 406 MHz system will work in the same manner as the DCS system. The relatively strong (5 watts) signal is transmitted on a periodic basis. The

Doppler shifted frequency received on the satellite is measured and time tagged. This information is both transmitted in real time at 1544.5 MHz and stored on board the satellite so that regional and global coverage are obtained.

The data received either at the LUT or, through the NOAA ground complex, at the Mission Control Center is processed to determine platform location and, through its coded transmission, platform identification. This information is then sent to the Rescue Coordination Center for their operations.

The SARSAT system recorded the distress beacon signals of the Air India jetliner that crashed off the Irish coast killing all 329 people aboard early in 1985. Coordinates of the disaster were relayed to a ground station in France within minutes of the disaster.

Meteor Satellites

Information provided by Grant Zehr (WA9TFB)

Prior to 1977, the Russian Meteor weather satellites were turned on only within the borders of the USSR. During the summer of 1977, it was discovered by Charles Vermillion's Direct Readout Applications Group at NASA that the Russian Meteor APT System was transmitting video at 137.3 MHz on Meteor 2-3.

Another Russian satellite transmitting high resolution pictures at 137.150 MHz in a 240 line/minute video format was received intermittently while passing over the United States.

Radio Amateurs who regularly track the Meteor weather satellites have made valuable discoveries. The Soviets have launched a dual weather satellite program. The Meteor class 1 satellites have been monitored by amateurs in North America on an irregular basis during the last five years; these satellites appear to be experimental in nature and developments are probably flown on these spacecraft. Orbital parameters for the Meteor class 1 satellites normally are: Period: 97.5 minutes and Inclination: 97.8 degrees Scan Rate: 240 lines per minute. Meteor 1-30 launched June 18, 1980, used several frequencies to transmit weather date; these included 137.120 and 137.150 MHz. Meteor 1-31, launched July 10, 1981, used the frequency of 137.130 MHz to transmit weather satellite imagery.

Meteor class 2 satellites provide regular APT service over the United States. The following class 2 satellites have transmitted weather satellite APT imagery:

Meteor 2-3	77-117A	137.300 MHz
Meteor 2-4	79-21A	137.300 MHz
Meteor 2-5	79-95A	137.300 MHz
Meteor 2-6	80-73A	137.400 MHz
Meteor 2-7	81-43A	137.300 MHz
	(13 black bars)	

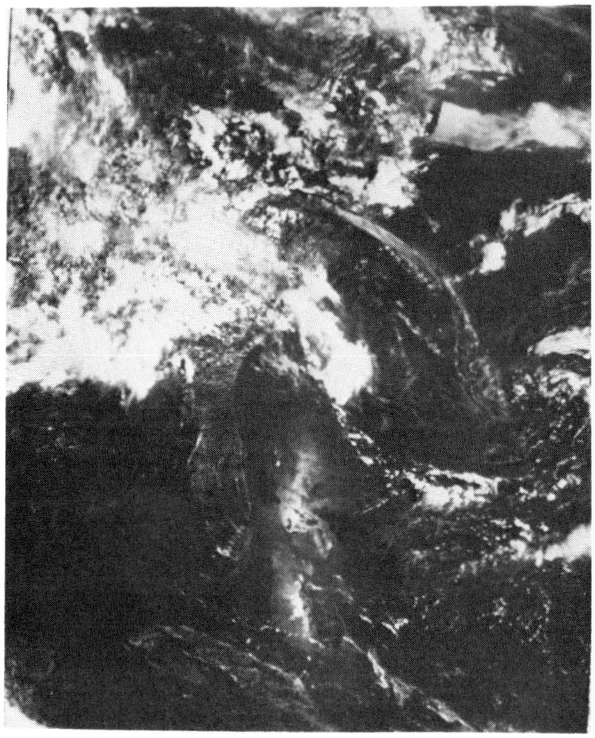

METEOR 2-8, Florida, Cuba, and Bahamas (Grant Zehr, WA9TFB)

Meteor 2-8	82-25A	137.850 MHz
	(14 Black Bars)	
Meteor 2-9	82-116A	137.300 MHz
	(17 Black Bars)	

Marciano Righini (I4MY) has pointed out that the black and white bars that are visible along the edge of the Soviet APT imagery (created by sync pulses) can be used to ID a particular satellite sending the pictures. Greg Roberts (ZS1BI) suggests that these black and white bars may also contain data on satellite aperture setting and grey scale calibration.

There are differences between U.S. and Russian weather satellites. First, the Meteor 2 sends a single picture versus the two (Visible/IR) that the NOAA birds send. Users will note that the class 2 satellites have a FM deviation of ∓10 kHz, so a 30 kHz wide IF will probably give optimum results. Due to the loss of video subcarrier during synchronization pulses, monitors will need to have a local oscillator to control FAX machine speed. A stable variable frequency source that is tunable from 2280-2520 Hz should give the monitor satisfactory results.

To assist monitors now equipped to monitor the NOAA satellites, the following are characteristics of the Meteor class 2 weather satellite series:

Carrier Frequency: 137.300 (Primary), 137.400, and 137.850 MHz
Modulation: Frequency modulation (FM)
Deviation: ∓10 kHz
Transmitted Output Power: 5 watts
Subcarrier: 2400 Hz (but this does drift

somewhat)

Subcarrier modulation: Amplitude Modulation (AM)

Scan Speed: 120 lines/minute (Meteor class 1 - 240 lines/minute)

Picture Phasing: 90-100 milliseconds of AM subcarrier at a frequency of 256 Hz

Monitors should be aware of other possible Meteor weather satellites; these frequencies include: 137.200, 137.620, 461.500, 464.000, 466.500, 468.500, and frequencies around 1000 MHz.

Geostationary Weather Satellites

By far, more information and pictures of better quality and usability are available to the hobbyist equipped to monitor the geostationary weather satellites than for the polar orbiting birds. The cost for equipment has gone down while the technology has improved. Geostationary satellite monitoring is the ultimate in weather satellite displaying.

GOES

Provision of timely global weather information, including advance warning of developing storms, is the primary function of the U.S. GOES meteorological program. To provide more complete information, Hughes constructed for NOAA a new generation of geostationary satellites known as GOES D,E,F, now part of the Global Weather Watch system.

NOAA's earlier GOES spacecraft (1,2,3) and their predecessors, the Synchronous Meteorological Satellites (SMS), employ a visible/infrared spin scanner radiometer (VISSR) to produce two-dimensional imagery from the visible and infrared spectral regions. The VISSR senses radiation from the earth and its atmosphere and from 36,000 km above the equator, the VISSR transmits black and white, television-like images of one third of the earth's surface every 30 minutes.

The GOES D,E,F utilize an improved VISSR incorporating atmospheric sounding. The new sensor, VAS, measures a vertical-temperature-versus-altitude cross section of the atmosphere from which the altitudes and temperatures of the clouds can be determined and a three dimensional picture of their distribution can be drawn for more precise weather prediction.

Using GOES imagery, meteorologists can measure the frame-to-frame movement of selected clouds at different altitudes to obtain their wind direction and speed and to better understand atmospheric circulation patterns.

The GOES satellites also carry space environment monitors (SEM) to investigate solar particle emissions and assist in studying the effects of solar activity on earth's telecommunications systems. The SEM detects solar protons, alpha particles, solar electrons, solar X-rays, and magnetic fields.

Weather photo from the GOES-1 satellite (Ford Aerospace)

Synchronous Meteorological Satellite-1, the predecessor to the later operational Geostationary Operational Environmental Satellite (GOES), was launched in May 1974 (Ford Aerospace).

A data collection system on GOES receives and relays environmental data sensed by widely dispersed surface platforms such as river and rain gauges, seismometers, tide gauges, buoys, ships and automatic weather stations. Platforms transmit sensor data to the satellite at regular intervals upon interrogation by the satellite, or in an emergency alarm mode whenever a sensor receives information exceeding a preset level is triggered by the sensor. These transmissions occur on a variety of frequencies from 401.700 to 401.9471 MHz.

The primary transmission from the GOES satellites that is of interest to the amateur is WEFAX (Weather Facsimile), both a communications subsystem and a service provided by and through the GOES.

WEFAX was first introduced as a communications relay experiment on Applications Technology Satellite ATS-1, later included on ATS-3, and subsequently refined for incorporation on the GOES spacecraft. It involves acquisition and processing of satellite data on the ground and retransmission of these data back through the geostationary spacecraft to relatively-low-cost receiving stations within the receiving range of the spacecraft. Thus, the original data source of WEFAX products may or may not be the same spacecraft from which the WEFAX transmissions are received.

The GOES-1 brought a new dimension to weather forecasting when it was launched into space in October 1975. With it, the United States had a new "quick look" at weather patterns every 30 minutes, a marked improvement over the previous photos taken every 12 hours by Polar orbiting satellites (Ford Aerospace).

GOES WEFAX transmissions are made at S-band (1691.0 MHz). Automatic Picture Transmission

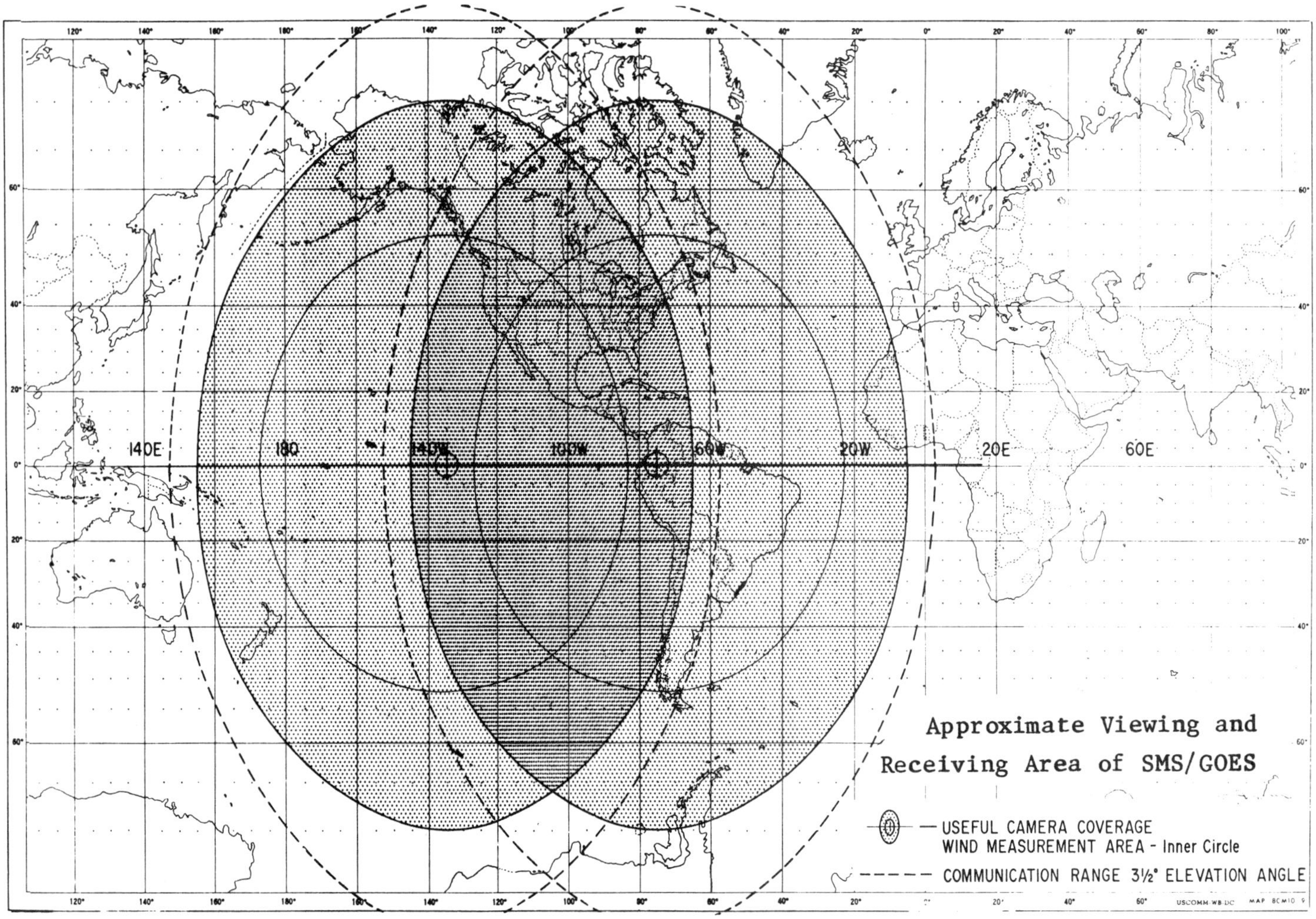

(APT) ground stations, the most common ATS WEFAX data recipients, require modifications in order to receive GOES S-band WEFAX broadcasts. These simple modifications allow 137 MHz APT receiving equipment to receive WEFAX broadcasts at minimum cost.

In 1979, NOAA/NESS established a dedicated WEFAX experimental service from a satellite located at 107 degrees west. This service, referred to as GOES-CENTRAL, uses an earlier GOES spacecraft no longer suit able for imaging.

The following is a brief description of the capabilities and constraints affecting GOES WEFAX products and schedules.

Transmission Schedules

GOES-West and GOES-East

With certain exceptions, WEFAX data are transmitted in ten-minute time slots at half hourly intervals, 24 hours per day. These ordinarily follow the 20 minute VISSR image acquisition periods. Broadcasts are staggered between GOES-EAST and GOES-WEST; that is, both satellites are not broadcasting WEFAX data at the same time.

In theory, there are 48 ten-minute time slots (two per hour) available on each spacecraft for WEFAX broadcast; in practice, however, six ten-minute time slots are reserved for operation of the trilateration ranging system. In addition, a block of time is reserved for the twice-yearly spacecraft eclipse periods centered on spacecraft midnight.

Consequently, only about 35 ten-minute time blocks are available for WEFAX broadcasts. These are not distributed evenly throughout a 24 hour period because of the constraints noted above.

GOES-Central

Since there is no VISSR operation on the GOES-CENTRAL, WEFAX broadcast can be made anytime during the hour. Daily exceptions to this schedule occur during a block of time reserved for ground stations' preventive maintenance and the twice-yearly spacecraft eclipse periods.

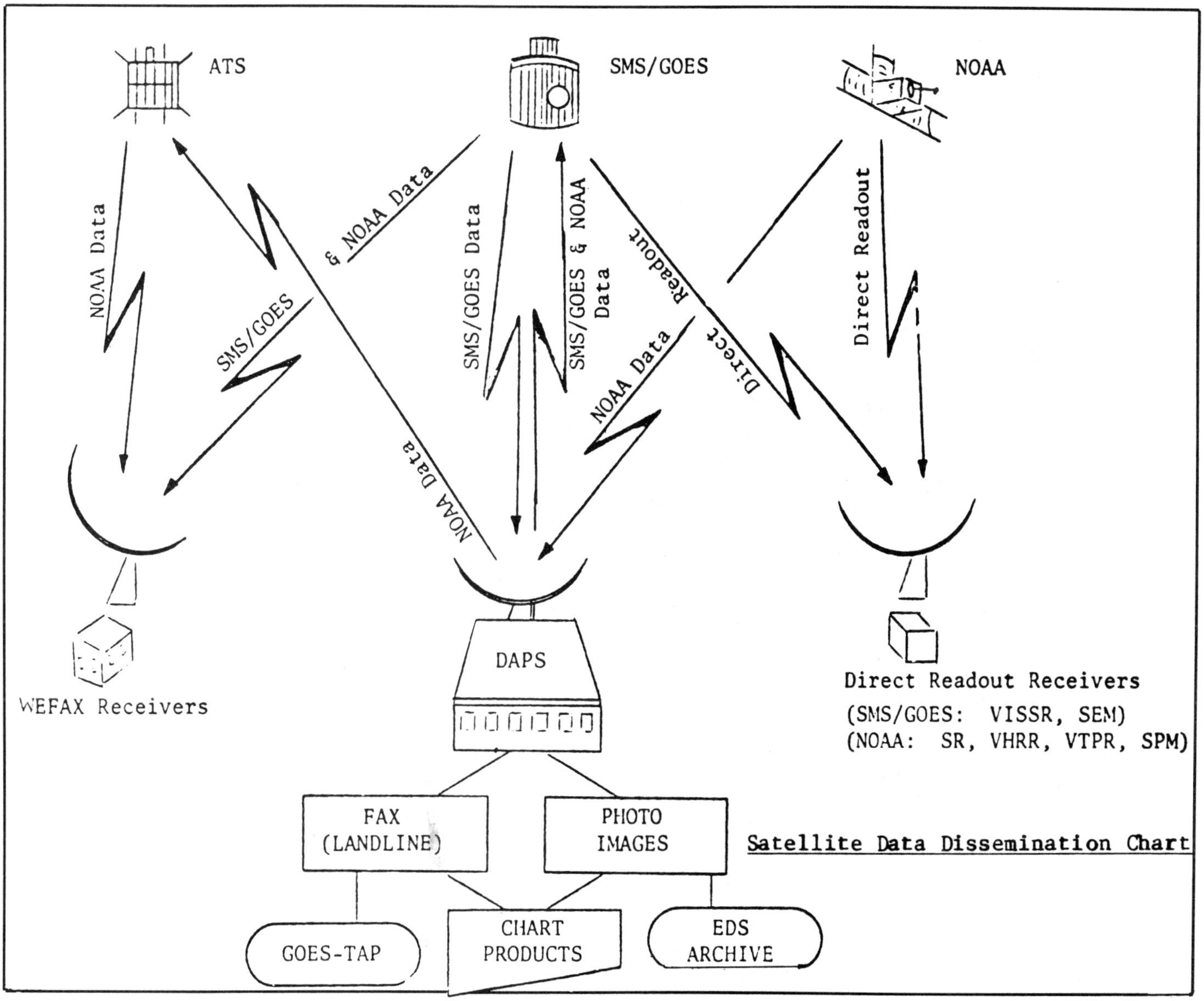

Satellite Data Dissemination Chart

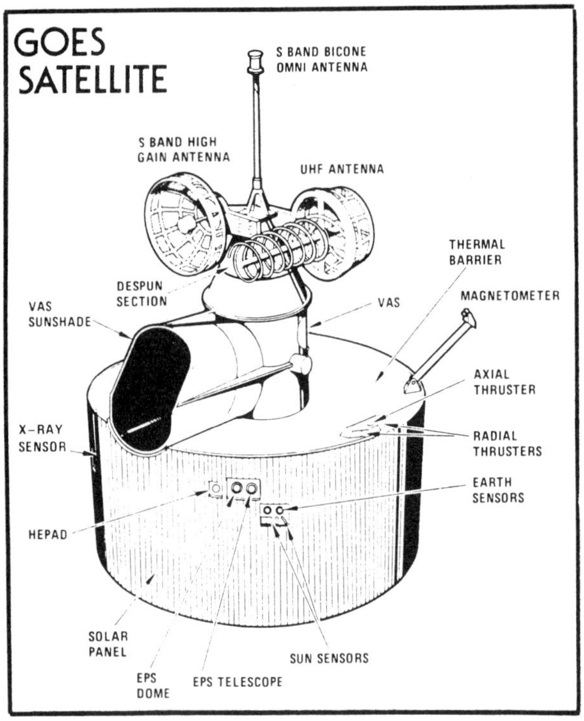

WEFAX broadcast schedules are contained in operational messages transmitted once each day via each of the three spacecraft; in addition, they are often listed in W-series APT information notes distributed by the NOAA coordinator for weather facsimile services, a recent copy of which is reprinted in the appendix so that readers can see the scope of information available on the GOES satellites.

VISSR Imagery

The majority of products transmitted on WEFAX are derived from VISSR instruments on board the GOES-EAST and GOES-WEST satellites. All of these imagery products are transmitted as pairs of related data sectors, that is, each broadcast consists of two sector images related by time or geography.

Polar Imagery

Approximately one-third of the imagery transmitted on GOES WEFAX are derived from low resolution visible and IR observations recorded on-board NOAA's polar orbiting satellites; these are processed, mapped and reformatted for WEFAX transmission. Polar stereographic mapped quadrants, from the northern and southern hemispheres, together with mercator projection mapped data, are used.

Imagery contained in these broadcasts are compiled from observations acquired on several orbits

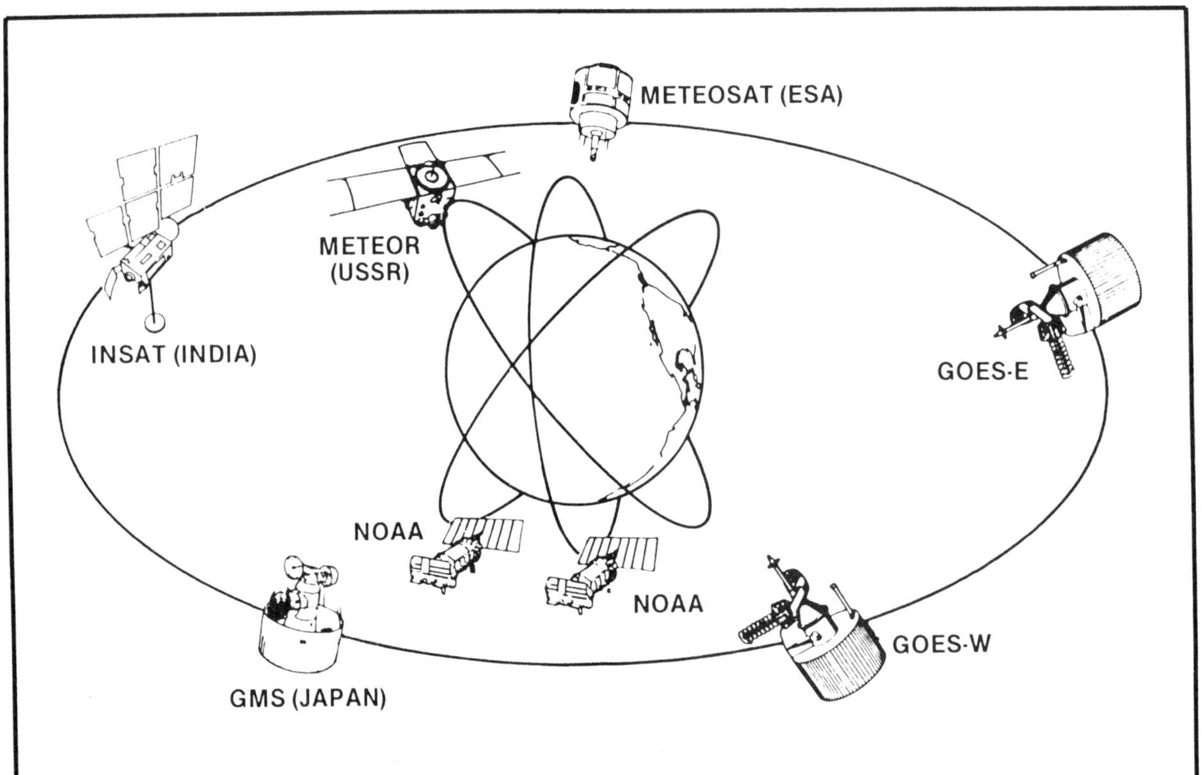

Typical Weather Satellite Orbits

GOES-7, launched in February 1987, replaces the failed GOES-East (NASA photo)

Yet to come, the new generation GOES satellites will significantly advance the capability for measuring the earth's atmosphere, its surface, its cloud cover, and the electro-magnetic environment (Ford Aerospace).

and are not synoptic data; however, such observations afford global views of cloud features, including areas of the Arctic and Antarctic, which cannot be observed from geostationary satellites.

National Meteorological Center (NMC) Charts

NMC charts which are broadcast via WEFAX are computer generated meteorological charts, first introduced on the GOES-CENTRAL as a test program. Now you no longer have to wait for the evening news to take a look at the current weather map.

TBUS Bulletins

The APT predict (TBUS) bulletins are a US national practice code form defined in the World Meteorological Organization (WMO) *Manual on Codes, Volume II*. These bulletins are broadcast daily for each of the TIROS-N (NOAA) series polar-orbiting satellites. Each bulletin contains ephemeris data for orbital predictions.

Obtaining Additional Information

Requests for information on ground equipment requirements, data reception, and WEFAX schedules and products may be addressed to:

Coordinator, Weather Facsimile Services
National Environmental Satellite, Data
and Information Service
National Oceanic and Atmospheric
Administration
E/SP21, WWB, Room 806
Washington, D.C. 20233

GMS/METEOSAT

Currently in geostationary orbit over Europe and the Asian/Pacific regions, these satellites perform basically the same functions as the US GOES satellite program. They send WEFAX data on the same frequency as the US GOES satellites (1691.0 MHz) as well as 1694.5 MHz.

RF and Video Format Data

Operating frequency: 1691.0 and 1694.5 MHz
Modulation: FM
Deviation: ∓9 kHz
Ground signal level: -134 dBm
Video mode: APT
- Video subcarrier: 2400 Hz
- Video modulation: AM
- Black level: minimum 4%
- White level: maximum 90%
- Line rate: 4 Hz (240 lines per min.)

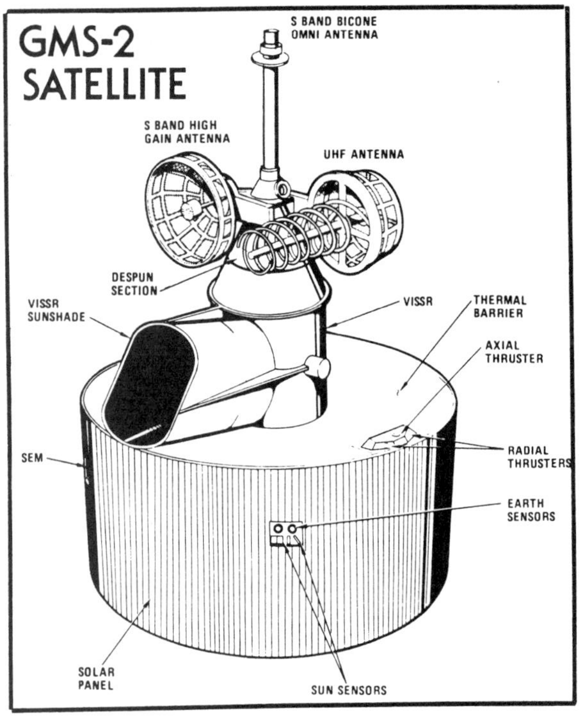

Current (1985) Geostationary Weather Satellite Status

GOES-EAST (GOES-5) failed July 29, 1984. It formerly provided direct readout of VISSR data (from GOES 6) on 1687.1 MHz and WEFAX on 1691.0 MHz. The spacecraft is located at 75° W; its inclination is near zero.

GOES-CENTRAL (GOES-2) - WEFAX services are on 1691.0 MHz. The spacecraft is currently located at 113° W rather than the normal 107° W; its inclination angle is 2.9 degrees.

GOES-WEST (GOES-4) - direct readout of "stretched" VISSR data is on 1687.1 MHz (from the GOES-1 satellite) and WEFAX is on 1691.0 MHz. The spacecraft is located at 138° W; its inclination angle is 1.1 degrees. Users are advised of a weakening S-band transponder on GOES-4 with less than favorable signal to noise ratio.

GOES-6 is the current imaging satellite. There is no WEFAX or "stretched" VISSR data services from this spacecraft. Data from the VAS instrument (i.e.- images) are downlinked to Wallops CDA and then relayed to users through GOES-5. The spacecraft is now located at 108° W. Its inclination angle is near zero.

GOES-1 failed in January 1985. Data was relayed through GOES-4 for direct readout "stretched" func-

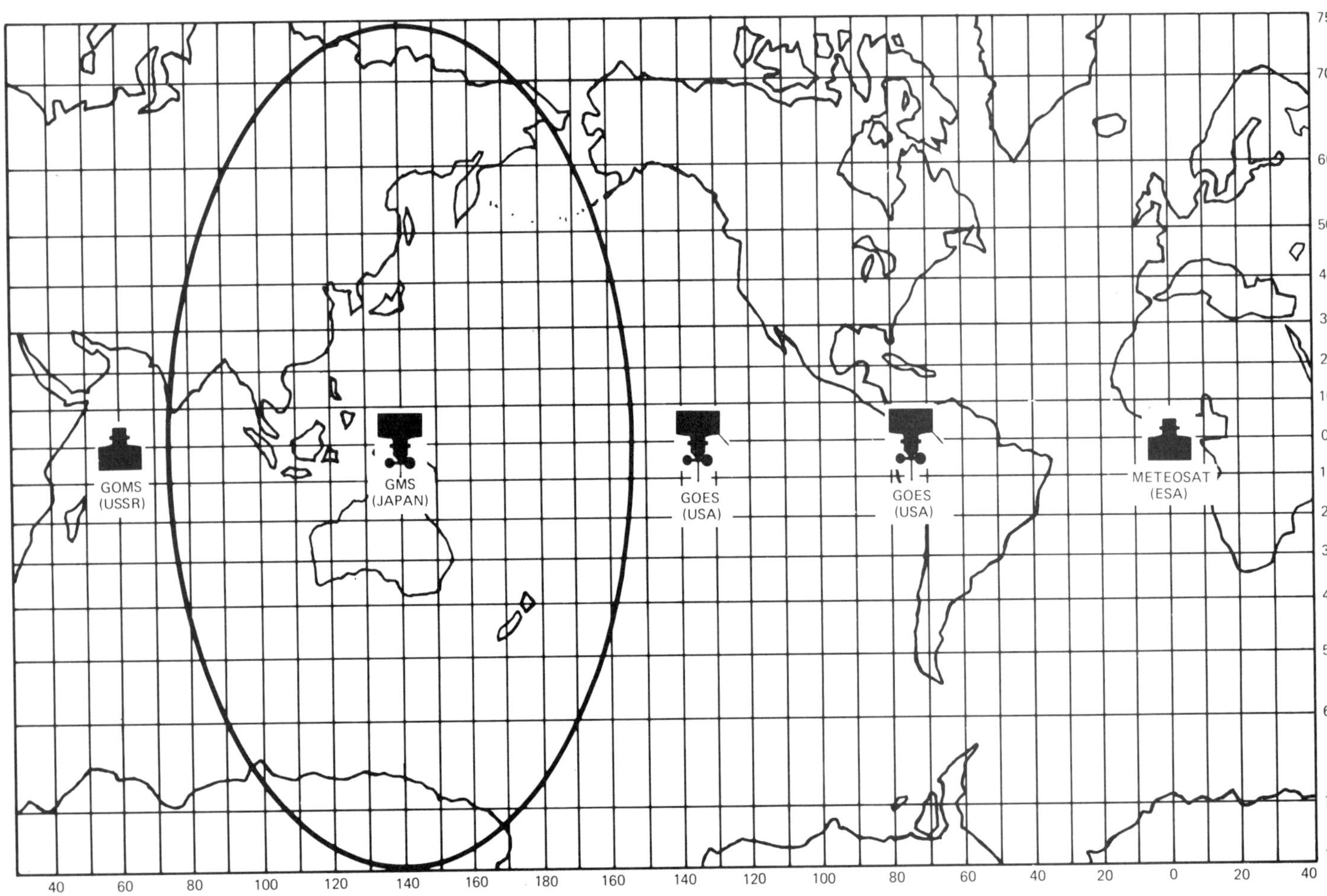

Proposed Positions Of World Weather Watch Geostationary Satellites

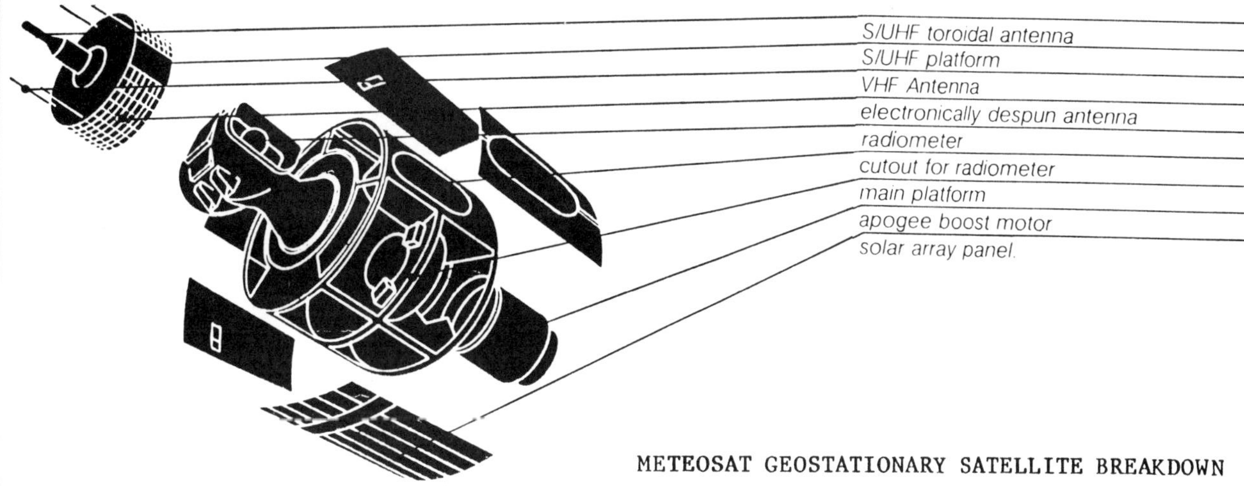

METEOSAT GEOSTATIONARY SATELLITE BREAKDOWN

tion users (VISSR). The spacecraft is located at 130° W; its inclination angle is 4.7 degrees.

GOES-3 is presently drifting westward at a rate of one-half degree per day; it passed the position of GOES-CENTRAL (GOES-2) November 24th, 1984. Because of the weakening GOES-4 S-band transponder, GOES-3 will eventually replace the GOES-4 as the GOES-West spacecraft.

GOES-6 is the current imaging satellite. There is no WEFAX or "stretched" VISSR data services from this spacecraft. Data from the VAS instrument (i.e.- images) are downlinked to Wallops CDA and then relayed to users through GOES-5. Until recently, the spacecraft has been located at 108° W with an inclination angle near zero.

In Febuary of 1987, GOES-H was successfully launched and is now designated GOES-7. It is placed in the GOES-East position, and GOES West (GOES-6) is being returned to the western position at 135° W. GOES-7 is in a circular orbit 22,237 miles above the earth and has been transmitting images since March 24, 1987. It also carries a search and rescue experimental package.

GMS-1 AND 2: GMS-2 suffered a partial failure of the imaging system in early 1984. GMS-1, in stand-by, was moved to 135° E to provide WEFAX, HRFAX and imaging and GMS-2 was moved to 145° E. Japan's GMS-1 also suffered a progressive failure of its imaging capability and by mid-June 1984 was unable to image south of the equator.

After an apparent complete imaging failure early in July, 1984, imaging and dissemination was transferred back to GMS-2, still located at 145° E. Due to its partial failure, imaging is taking place only every six hours at synoptic times. GMS-1 has not been moved from 135° E to make way for GMS-3 launched August 2, 1984 from Japan. WEFAX data is available on 1691.0 MHz.

METEOSAT 1 AND 2: Both METEOSAT 1 and 2 continue to be operated simultaneously to provide image acquisition and dissemination (METEOSAT 2) and data collection (METEOSAT 1). Both satellites can be operated a full 24 hours per day and are presently technically capable of fulfilling all mission requirements except during eclipse periods. METEOSAT 1 is maintained on station between 9° and 11° E; METEOSAT 2 is located between 1° E and 1° W.

Chapter Four

DOMESTIC COMMUNICATIONS SATELLITES

That phone call you made long distance to Grandma at Christmas, the *USA Today* newspaper at your doorstep, the $500.00 your dad wired you for college--all probably connected via our domestic communications satellite network.

Much as you would turn on your shortwave or scanner and monitor the action in these bands, some monitors are now tuning in the satellite bands and enjoying the action that can be heard and seen.

A Little Background

Radio signals least subject to inter ference and best suited to carry large volumes of communications traffic are those in the microwave frequency range. They travel in straight lines and are not deflected appreciably by the earth's atmosphere. If someone in London tried to beam a microwave signal to New York, it would never get there; it would disappear into space or dissipate into the ocean.

Over short distances this problem has been solved by erecting interlinking repeater stations every 30 miles or so employing dish-type antennas which "catch" the beamed signal and retransmit it towards the next station in the chain.

The placing of such a chain across one of the world's major oceans presents some obvious, insurmountable engineering difficulties. Even then, you would have only point-to-point communications; not everyone would be able to hear it.

What is needed is one omnidirectional relay point in the middle of the ocean; With the advent of the space age, it has become possible to have such a "tower in the sky"--without building the tower!

Satellites launched into circular orbit, moving at the same speed the earth rotates, appear to be stationary in the sky. From their positions in space these satellites can "see" about one third of the earth's surface and any stations in that area with their antennas turned toward one of these satellites communicate among each other.

Early satellites received signals beamed in their direction by earth stations, amplified these signals and then retransmitted them on different frequencies.

Present-day satellites have much more sophisticated antenna systems enabling them to re-use frequency ranges several times, concentrating "beams" on different parts of the earth that are high traffic areas, and transmitting in opposites senses of polarization (vertical and horizontal).

It is not our intent to go into great detail about TVRO equipment; There are many outstanding books already written on this subject (see Appendix A). Our intent is to inform you of what is up there, how to locate it, and where to listen for it. For satellite locations see the Appendix section. Keep in mind that domestic satellite broadcasting is a business; if you think shortwave international broadcasters change their frequencies often, just sit back and watch different services come and go on the comsat transponders!

COMSTAR Satellites

When two domestic communications satellites were launched into synchronous orbit on 1976 a new era in long distance calling began for the nation's telephone customers. Relaying calls back and forth across the nation, the COMSTARs are in joint use by AT&T and GTE Satellite Corp. These spin-stabilized satellites, plus two others built by Hughes Aircraft, are owned by COMSAT General Corporation, Washington, D.C., and leased to AT&T who, in turn, operate seven earth stations located near San Francisco, Chicago, Atlanta, Scranton, Los Angeles, Tampa, and Honolulu.

Each COMSAT can relay more than 18,000 two-way high quality telephone calls simultaneously and is designed to provide reliable communications for continental US, Alaska, Hawaii, and Puerto Rico.

The first COMSTAR was launched May 13, 1976 and started commercial service in early July; it was stationed in orbit at 128° W. The second COMSTAR was launched July 22, 1976 and began service the following September; it was stationed at 94° W. The third satellite was launched June 29, 1978 and was stationed at 87° W. The fourth satellite was launched February 21, 1981 and was positioned at 127° W for service.

Each COMSTAR spacecraft has 24 transponders, and each transponder is capable of receiving 1200 one way calls and relaying them back to earth on another frequency. To conserve frequencies, twelve of the transponders receive and transmit horizontally polarized signals and twelve handle vertically polarized signals. This signal polarization technique allows each frequency to be used twice without mutual interference, thus doubling the capacity of the satellite system.

In addition, each satellite carries a millimeter wave experimental package to conduct communications tests on 19040 and 28560 MHz. TT & C signals are downlinked on 3947.5 and 3952.5 MHz.

The following channelization plan is currently used by the COMSTAR satellites:

COMSTAR SATELLITES

Transponder Number	Polarization V=Vertical H=Horizontal	Frequency Range (MHz)	Center Freq. (MHz)
1 (1)	V	3702-3738	3720
1 (2)	H	3722-3758	3740
2 (3)	V	3742-3778	3760
2 (4)	H	3762-3798	3780
3 (5)	V	3782-3818	3800
3 (6)	H	3802-3838	3820
4 (7)	V	3822-3858	3840
4 (8)	H	3842-3878	3860
5 (9)	V	3862-3898	3880
5 (10)	H	3882-3918	3900
6 (11)	V	3902-3938	3920
6 (12)	H	3922-3958	3940
7 (13)	V	3942-3978	3960
7 (14)	H	3962-3998	3980
8 (15)	V	3982-4018	4000
8 (16)	H	4002-4038	4020
9 (17)	V	4022-4058	4040
9 (18)	H	4042-4078	4060
10(19)	V	4062-4098	4080
10(20)	H	4082-4118	4100
11(21)	V	4102-4138	4120
11(22)	H	4122-4158	4140
12(23)	V	4142-4178	4160
12(24)	H	4162-4198	4180

COMSTAR D1/D2 are no longer in use. COMSTAR D3/D4 have been colocated at 76° W. They no longer carry video and are being used for telephone and data relay exclusively.

Galaxy Satellites

With the manufacture of the Galaxy system for its wholly owned subsidiary, Hughes Communication, Inc., Hughes Aircraft Company has entered the commercial satellite communications marketplace.

Three satellites have been launched for use by Hughes:

Galaxy 1 Launched June 28, 1983 (134° W)
Galaxy 2 Launched Sept. 22, 1983 (74° W)
Galaxy 3 Launched Sept. 21, 1984 (93.5°W)

The proposed Galaxy 4 will occupy the orbital position at 137° W.

Each satellite has 24 transponders. Instead of conventional leasing arrangements, Hughes Communications offers cable programmers the unique opportunity to purchase transponders on the Galaxy 1 satellite for the life of the satellite. Eighteen transponders have been sold and Hughes plans to reserve the other six for preemptible programming.

The Galaxy HS-376 satellite has a diameter of 85 inches and is only 109 inches high stowed in the launch vehicle. In orbit, the aft solar panel deploys, doubling the output power, and the antenna reflector erects for a combined height of 269 inches--the equivalent of a two story building. The solar array of K7 cells pro duces 990 watts of dc power; two nickel cadmium batteries furnish power during solar eclipse. With a full load of 300 pounds of station-keeping fuel, Galaxy weighs 1141 pounds.

The transmit and receive beams are created by a 72 inch grid antenna with two reflecting surfaces, one sensitive to vertical polarization, the other to horizontal. Separate microwave feed networks are used for the different polarizations.

Galaxy's signal strength is 34 dBW with each transponder using a multicollector traveling wave tube amplifier (TWTA) with a nine watt output. There are six spare TWTAs. The satellites have a design lifetime of six years.

The following is the bandplan used by the Galaxy C-band satellites.

GALAXY C-BAND SATELLITES

Transponder Number	Polarization V=Vertical H=Horizontal	Frequency Range (MHz)	Center Freq. (MHz)
1	H	3702-3738	3720
2	V	3722-3758	3740
3	H	3742-3778	3760
4	V	3762-3798	3780
5	H	3782-3818	3800
6	V	3802-3838	3820
7	H	3822-3858	3840
8	V	384203878	3860
9	H	3862-3898	3880
10	V	3882-3918	3900
11	H	3902-3938	3920
12	V	3922-3958	3940
13	H	3942-3978	3960
14	V	3962-3998	3980
15	H	3982-4018	4000
16	V	4002-4038	4020
17	H	4022-4058	4040
18	V	4042-4078	4060
19	H	4062-4098	4080
20	V	4082-4118	4100
21	H	4102-4138	4120
22	V	4122-4158	4140
23	H	4142-4178	4160
24	V	4162-4198	4180

COMSTAR antenna

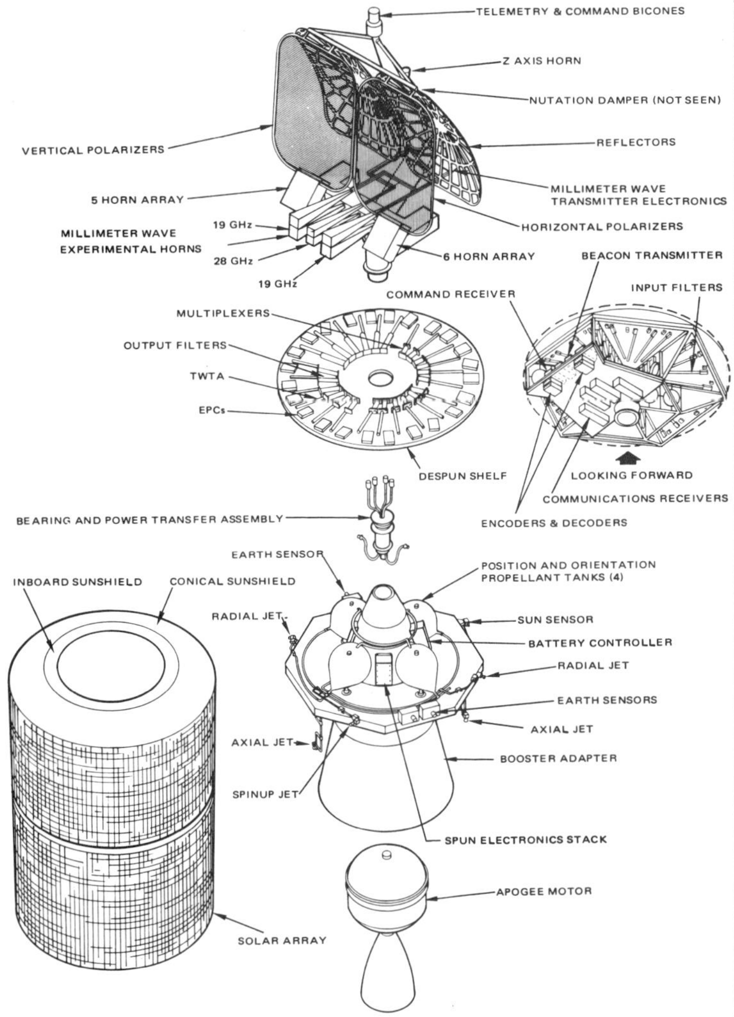

COMSTAR components

Twelve transponders on Galaxy 2 satellite are leased to MCI for their long distance telephone service.

The Future

At present Galaxy has proposed a Ku-band satellite system, designated Galaxy H5-H7, operating with an uplink of 14.0-14.5 GHz and downlink of 11.7-12.2 GHz. Orbital locations will be 73°W, 75°W, 93°W respectively. Each satellite will carry 16 transponders and they are expected to be launched in 1986, 1987 and 1988.

GTE Satellites

Although GTE Corporation got its start in the business of communication satellites with AT&T operating the COMSTAR series of satellites, they plan to branch out with two new series of satellite, Spacenet and Gstar.

Spacenet

Two of GTE's Spacenet satellites have already been launched. Spacenet 1 was launched by the Ariane 1 rocket on May 23, 1984 and placed into its orbital slot at 120° W.

Spacenet 2 was launched November 11, 1984 and placed into geostationary orbit at 69° W. Spacenet 3 is expected to be launched in March, 1985 and will occupy the 91° W position in the Clarke belt. These satellites are expected to have an orbital lifetime of 7.5 years. The following is the Spacenet bandplan for each of the satellites 24-transponders.

P.R.
Dianne Whitaker — CONTEL

SPACENET SATELLITES			
Transponder Number	Polarization V=Vertical H=Horizontal	Frequency Range (MHz)	Center Freq. (MHz)
1	H	3702-3738	3720
2	H	3742-3778	3760
3	H	3782-3818	3800
4	H	3822-3858	3840
5	H	3862-3898	3880
6	H	3902-3938	3920
7	H	3942-3978	3960
8	H	3982-4018	4000
9	H	4022-4058	4040
10	H	4062-4098	4080
11	H	4102-4138	4120
12	H	4142-4178	4160
13	V	3722-3798	3760
14	V	3802-3878	3840
15	V	3882-3958	3920
16	V	3962-4038	4000
17	V	4042-4118	4080
18	V	4122-4198	4160
19	H	11706-11774	11740
20	H	11786-11854	11820
21	H	11866-11904	11900
22	H	11946-12014	11980
23	H	12026-12094	12060
24	H	12106-12174	12140

Gstar A1/A2/A3

Three Ku-band satellites have been proposed for launch starting sometime in 1985. Geostationary orbital positions for the A1/A2/A3 satellites are 103°/105°/101° W respectively.

The following is a breakdown of uplink and downlink frequencies that will be used by the Gstar satellites.

GSTAR SATELLITES						
Tr No	Uplink Center Freq.	*	Uplink Range (MHz)	Downlink Center Freq.	*	Downlink Range (MHz)
1	14030	V	14003-14057	11730	H	11703-11757
2	14091	V	14064-14118	11791	H	11764-11818
3	14152	V	14125-14179	11852	H	11825-11879
4	14213	V	14186-14240	11913	H	11886-11940
5	14274	V	14247-14301	11974	H	11947-12001
6	14335	V	14308-14362	12035	H	12008-12062
7	14396	V	14369-14423	12096	H	12069-12123
8	14457	V	14430-14484	12157	H	12130-12184
9	14044	H	14017-14071	11744	V	11717-11771
10	14105	H	14078-14132	11805	V	11778-11832
11	14166	H	14139-14193	11866	V	11839-11893
12	14227	H	14200-14254	11927	V	11900-11954
13	14288	H	14261-14315	11988	V	11961-12015
14	14349	H	14322-14376	12049	V	12022-12076
15	14410	H	14383-14437	12110	V	12083-12137
16	14471	H	14444-14498	12171	V	12144-12198

***Polarization:** V=Vertical/H=Horizontal

The following is a breakdown of telemetry, tracking and control frequencies that will be used by the Gstar satellites:

Vertical command uplink:	14484-14500 MHz
Horizontal command uplink:	14000-14017 MHz
Command frequency: 14015 MHz	horizontal polarity
Horizont. telem. downlink:	12184-12200MHz
Tracking beacon-horizontal:	12198 MHz
Vertical telemetry downlink:	11700-11717MHz
Tracking beacon-vertical:	11702 MHz

Satcom Satellites

The RCA domestic communications satellite system is the first COMSAT system to provide a broad range of commercial services in the United States. RCA American Communications, Inc. (Americom) started serving customers in December, 1983, using leased channels on Telesat Canada's ANIK A-2 satellite.

On December 12, 1985, RCA launched the first of its own satellites, RCA Satcom 1; Satcom 2 followed on March 26, 1976.

The RCA Satcom spacecraft were unique among domestic communications satellites; their polarization diversity and frequency inter-leaving are now the industry standard. Each of 24 transponders is capable of carrying the equivalent of 1500 two-way telephone calls, or 64 million bits per second of computer data, or one high quality video channel and several audio sub-channels.

The launches of Satcom 2-R and Satcom 4 brought several refinements: the power of six of the transponders has been increased to 8.5 watts (from 5.5 watts), the solar array system has been increased to 90 square feet (from 75) to supply the higher power; four square transponders are on board; larger fuel supply should extend station-keeping life to ten years from the expected eight years of Satcom 1 and 2.

SATCOM SATELLITES				
Satellite	Intl Desig.	Launch Date	Orbital Position	Current Status
Satcom 1	75-117A	12/12/75	136W	No longer in service
Satcom 2	76-29A	3/26/76	119W	No longer in service
Satcom 3	79/101A	12/7/79		Never reached geo
Satcom 3R	81-114A	11/20/81	131W	Cable Net I satellite
Satcom 4	82-04A	1/16/82	83.5W	Network/ occasional video
Satcom 5	82-105A	10/27/82	143W	Turned over to Alascom, Inc; now known as Aurora 1
Satcom 1R	83-30A	4/11/83	139W	Network/ occasional video
Satcom 2R	83-94A	9/9/83	72W	No video

SATCOM BAND PLAN			
Transponder Number	Polarization V=Vertical H=Horizontal	Frequency Range (MHz)	Center Freq. (MHz)
1	V	3702-3738	3720
2	H	3722-3758	3740
3	V	3742-3778	3760
4	H	3762-3798	3780
5	V	3782-3818	3800
6	H	3802-3838	3820
7	V	3822-3858	3840
8	H	3842-3878	3860
9	V	3862-3898	3880
10	H	3882-3918	3900
11	V	3902-3938	3920
12	H	3922-3958	3940
13	V	3942-3978	3960
14	H	3962-3998	3980
15	V	3982-4018	4000
16	H	4002-4038	4020
17	V	4022-4058	4040
18	H	4042-4078	4060
19	V	4062-4098	4080
20	H	4082-4118	4100
21	V	4102-4138	4120
22	H	4122-4158	4140
23	V	4142-4178	4160
24	H	4162-4198	4180

RCA Satcoms 1-4 use 3701.0 and 4199.0 MHz for TT & C; the remaining use 3700.5 and 4199.4 MHz.

The Future

In November, 1983, RCA applied to the FCC for authorization to construct, launch, and operate three hybrid satellites in the 1989-1992 time frame to serve new telecommunications markets. These new spacecraft, incorporating the latest advances in technology, will contain 24 ten-watt C-band transponders and 16 50-watt Ku-band transponders.

The orbital slots requested for the new spacecraft, 61°, 63° and 65° W, would permit RCA Americom to serve the contiguous 48 states and the Caribbean basin, notably Puerto Rico and the U.S. Virgin Islands.

As part of this application, the company has indicated it would, at its option, launch and co-locate three C-band and three Ku-band satellites instead of hybrid models.

Satcom 6 will use the same band plan as previous Satcom satellites and will be located at 67°; expected launch date is May, 1986. Satcom 7-9 are expected to be launched in the 1989-1991 time frame and will be located at 65° W, 63° W, 61° W. Lifetime of these satellites is ten years.

Satcom K-2, with 16 transponders of 45 watts power, is the first of three spacecraft designed and built by RCA Americom to provide a variety of communications services for television program distribution and business networking.

With their 45-watt Ku-band transponder power, these new Ku band satellites can deliver high quality television signals into antennas as small as one meter in diameter. And because Ku-band satellite signals are not affected by terrestrial microwave signals, antennas can be located anywhere in line-of-sight to the satellite, even in major metropolitan areas.

Eight of the Satcom K-2 transponders will be used by the NBC Television Network for program distribution to affiliates and for backhaul services into the network. In addition, Hubbard Broadcasting and its United States Satellite Broadcasting, Inc. subsidiary, have leased four transponders for distribution of the Conus News Service and the Conus Washington Direct Service.

RCA Americom is offering two transponders to earth stations to commercial broadcast stations in the lower 48 states. To date, more than 500 stations have accepted this earth station offer.

The remaining two transponders on Satcom K-2 have been reserved by AT&T Communications, Inc. allowing them to further expand their satellite services.

Satcom K-1 is being marketed for satellite master antenna television, direct-to-home television distribution and private business networking.

Satcom K-2 has been assigned an orbital slot of 81° west, Satcom K-1, 85° and Satcom K-3, scheduled for launch in 1987, 67° west longitude.

The next set of Ku-band satellites (K4-K6) will be launched in the 1989-1991 time frame; these satellites will occupy the same orbital positions as the K1-K3 and will also use the same frequencies.

As mentioned before, RCA has the option of using hybrid satellites instead of the separate C- and Ku-band satellites; these would be a mix of SATCOMS 7-9 and K4- K6.

Telesat of Canada

Telesat of Canada, a consortium of private (largely telephone companies) industry and government, funds an extensive network of geostationary satellites. Anik (Eskimo for "Brother") satellites have progressed through four generations since the first launch in 1972. Once in orbit these satellites carry a Telestar label and serial number.

Telesat of Canada was created by the Canadian parliament in 1969. The following is a breakdown on each of the Canadian COMSATS.

Anik A1/A2/A3

Three Anik A satellites were built for Telesat Canada by Hughes Aircraft. Anik 1 was launched from Cape Canaveral November 9, 1972; orbital station of that satellite was 114° W. A second Anik A was launched April 20, 1973, and was stationed at 109° W. Anik A3 was launched May 7, 1975, and is currently stationed at 114° W. Anik A1 and A2 are no longer in operation.

Currently, Anik A3 serves as a telephone relay satellite but will be replaced by the Anik D series in the near future.

ANIK A

Diameter:
190 cm (75 in.)

Overall Height:
341 cm (134 in.)

Weight:
297 kg in orbit
(654 lb)

First flight spacecraft

Anik launch, Nov. 9, 1972

Anik A Satellites

Tr No	Frequency Range (MHz)	
1	3702-3738	
2	3742-3778	
3	3782-3848	TT&C frequencies:
4	3822-3858	4198.0 MHz
5	3862-3898	4199.0 MHz
6	3902-3928	
7	3942-3978	All transponders are
8	3982-4018	horizontally polar-
9	4022-4058	
10	4062-4098	
11	4102-4138	
12	4142-4178	

Anik B

Telesat's second generation COMSAT is known as Anik B. This satellite pioneered 12 GHz service and was the first operational satellite of its kind. Hybrid in nature, this satellite also uses the C-band plan of Anik A satellites and the same TT & C frequencies.

Anik B was built by RCA and launched on December 16, 1978. It currently occupies the 109° W slot in geostationary orbit. Design lifetime for Anik B, also known as Telesat 4, is seven years.

The 12 GHz frequency plan is as follows:

		Anik B
TR.	13	11704-11776 MHz
	14	11784-11856 MHz
	15	11864-11936 MHz
	16	11944-12016 MHz
	17	12024-12096 MHz
	18	12104-12176 MHz

Anik C/C2/C3

ANIK C

Transponder Number	Frequency Range (MHz)
1	11703-11757
2	11716-11770
3	11764-11818
4	11777-11831
5	11825-11879
6	11838-11892
7	11886-11940
8	11899-11953
9	11947-12001
10	11960-12014
11	12008-12062
12	12021-12075
13	12069-12123
14	12082-12136
15	12130-12184
16	12143-12197

Anik C spacecraft are Canada's first totally dedicated commercial telecommunications satellites to use 12/14 GHz Ku-band frequencies. These satellites will realize a 100% increase in telecommunications capa city over the first Anik A spacecraft utilizing the Hughes spacecraft HS-376 design.

The Anik C satellite is serving Canada with audio, video, and data telecommunications services; in addition, Anik C2 is carrying five channels of DBS programming for USCI and C3 is carrying ten channels of video programming.

Two Anik C spacecraft have been launched and

TELSTAR 3

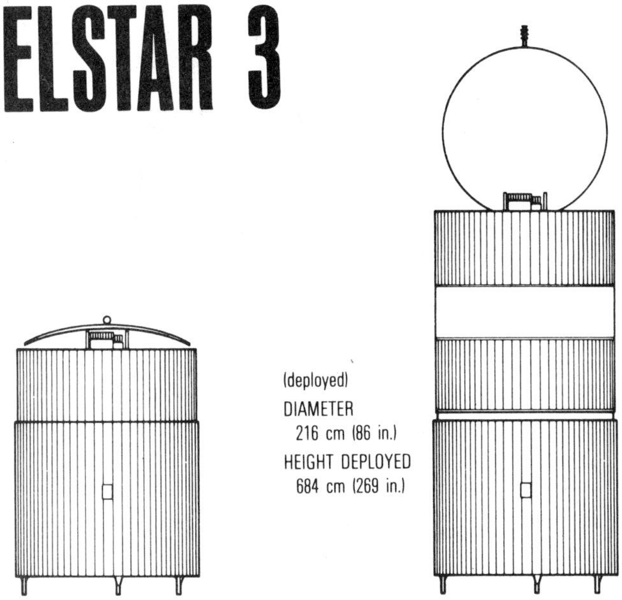

are currently operational:

Anik C-2	112.5° W
Anik C-3	117.5° W

The third satellite in the series (Anik C1) will occupy the 107.5° slot. The following is a frequency breakdown for the Anik C satellites (transponder polarities unknown at this writing, but possibly the same as GTE satellites):

Anik D

Two new Anik satellites, designated Anik D1/D2, will replace the three Anik A satellites. On August 26, 192, Anik D1 was launched into its orbital operating position at 104.5° W and Anik D2 will be positioned at 109° W. These satellites use the typical 24 channel C-band assignments and polarization of each transponder is the same as the WESTAR 4/5 and Galaxy series of satellites.

Using the Hughes Aircraft HS-376, the Anik D satellites will have a design lifetime of nine years.

Telstar 3 Satellites

AT&T's new generation of satellites that are replacing the Comstar series are the Telstar 3 spacecraft, built around Hughes Aircraft's successful HS 376 design.

The Telstar satellites will have a ten year lifetime as opposed to seven years for the earlier Comstars. This life extension is due chiefly to the introduction of solid state power amplifiers in place of traveling wave tubes for 18 of the 30 high power transmitters, and improved nickel cadmium batteries.

Telstar 3-D rises from the cargo bay of earth-orbiting Discovery (Rockwell Int'l)

The Telstar spacecraft are spin stabilized, a concept developed by Hughes. The HS 376 spacecraft utilizes two solar cell covered concentric cylinders, the outer of which is extended downward is space for nearly six feet to greatly increase the solar generating power capacity over the model it replaces. These solar cells develop 917 watts at beginning of life. A folding dual reflector which deploys in space, is located at the forward end of the spacecraft.

These spacecraft are 86 inches in diameter with a

Westar II launch October 10, 1974

stowed configuration of 108 inches in height which increases to 269 inches after deployment. Weight in orbit is 1438 pounds.

Like their predecessors, the Telstar 3 satellites use the 6/4 GHz band, utilizing the same channelization plan as the Comstar satellites.

Telstar 301, also known as 3A, was launched on July 29, 1983 from the ETR in Florida; its orbital position is 96° W.

The other orbital locations for the Telstar satellites are:

Telstar 304 (3D)	76° W
Telstar 302 (3B)	86° W
Telstar 303 (3C)	125° W

Westar Satellites

With the launching of Westar 1 and 2 in 1974, Western Union became the first company to place domestic communications satellites into orbit. Westar 3, a 12-transponder satellite identical to the first two Westars, was launched in 1979. Westar IV, Western Union's first 24 channel satellite was launched February 26, 1982; Westar 5 was launched June 9, 1982 also carried 24 transponders into orbit; Westar 6, one of two satellites lost by the Space Shuttle in early 1984, and was recovered by the shuttle.

The Westar satellites' current status are as follows:

Westar 1	99° W	No longer active
Westar 2	123.5° W	No longer active
Westar 3	91° W	Active
Westar 4	98.5° W	Active; replaced Westar 1
Westar 5	123° W	Active; will move to 119.5
Westar 6	On ground	Awaiting reassignment

The systems control earth station, located in Glenwood, New Jersey, performs tracking and telemetry functions, making periodic "station-keeping" adjustments so that the satellites maintain constant position relative to Earth.

The Westar system, which includes seven major earth stations, is integrated with Western Union's 10,000 mile terrestrial microwave network. This landline satellite wideband transmission system provides a variety of communications services for business, government and the general public.

Western Union has many customers that utilize their satellites including: Public Broadcasting, National Public Radio, Robert Wold Company, Hughes TV, ABC, CBS, SIN, AP Radio, Group W, RKO, and Cable Network News. All of these services transmit audio and/or video over Westar.

The print media also uses the Westar satellites; the *Wall Street Journal* sends complete editions via facsimile techniques for regional printing of its eastern, western, and midwestern editions.

Text and pictures of *U.S. News & World Report* are transmitted via Westar, as are color photos and text of the midwestern and western editions of *Time*, *People* and *Sports Illustrated*.

Westar Satellite Data

Westar 1,2,3

Overall Height:11.3 feet, including antenna
Diameter: 6.3 feet
Weight of satellite in orbit: 655 pounds
Transmission channels: 12 transponder, each with 36-MHz usable bandwidth
Solar array: Will yield more than 262 watts after seven years in orbit.
Coverage "footprint": 48 contiguous state- 33 dBW
Puerto Rico- 26 dBW EIRP
Alaska, Hawaii- 27 dBW EIRP
Relay capacity: 1,500 one-way voice circuits, or one full-color TV signal with audio, or data at up to 60 megabits per second (each transponder). Satellite manufacturer: Hughes Aircraft
Launch vehicle: Thor-Delta 2914 (NASA)

BANDPLAN OF WESTAR 1,2,3

Tr No	Frequency Range (MHz)	Center Freq (MHz)
1	3702-3738	3720
2	3742-3778	3760
3	3782-3818	3800
4	3822-3858	3840
5	3862-3898	3880
6	3902-3938	3920
7	3942-3978	3960
8	3982-4018	4000
9	4022-4058	4040
10	4062-4098	4080
11	4102-4138	4120
12	4142-4178	4160

All signals carried on these satellites are horizontally polarized.

Tracking, telemetry and control signals are received from the satellites on 4198.250, 4198.750, 4199.250 MHz.

Westar 4 and 5

Overall Height: 22.1 feet
Diameter: 7.2 feet
Weight of satellite in orbit: 1,290 pounds
Transmission channels: 24 transponders, each with 36 MHz of usable bandwidth.
Solar array: 686 watts of prime power after ten years operation.
Coverage "footprint": 48 contiguous states- 34.0 dBW EIRP
Alaska- 31.0 dBW EIRP
Hawaii- 28.2 dBW EIRP
Puerto Rico- 27.2 dBW EIRP
Relay capacity: 2,400 one-way voice circuits, or one full-color TV signal with audio, or data up to 60 megabits per second (each transponder).
Satellite Manufacturer: Hughes Aircraft
Manufacturer's designation HS-376G
Launch vehicle: Thor-Delta 3910 plus PAM-D (NASA)

Assembly of first flight spacecraft

WESTAR Antenna

Telemetry, tracking, and command equipment in Glenwood N.J. station

WESTAR antenna coverage

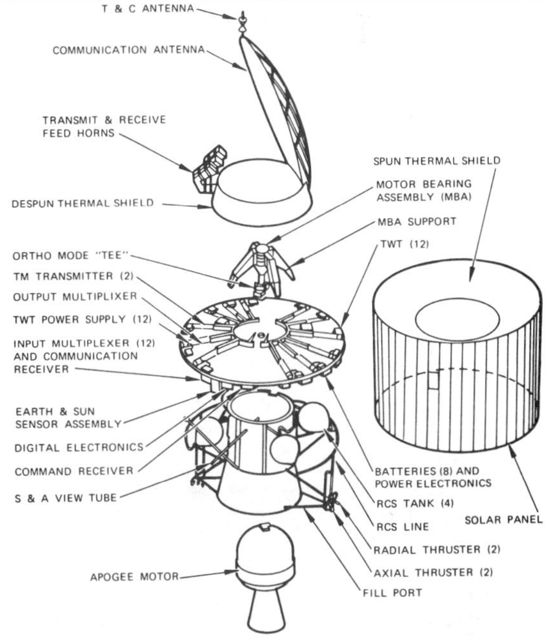

WESTAR elements

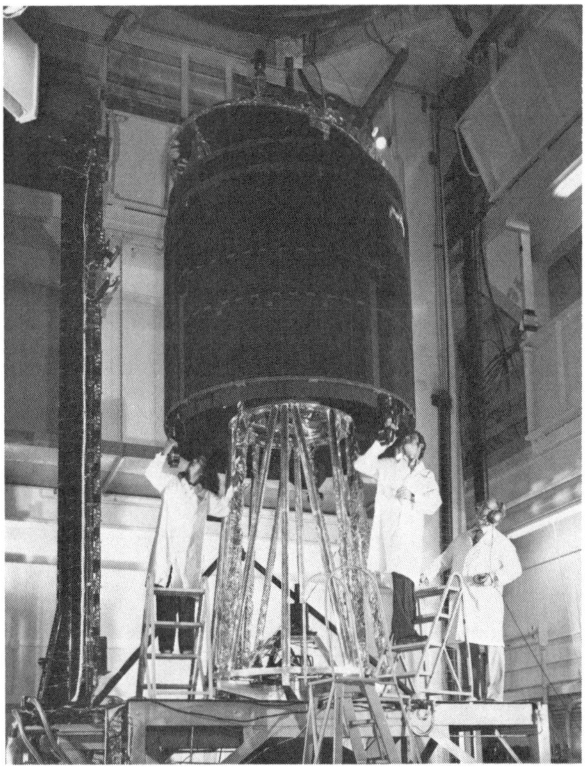

Western Union's Westar IV (Hughes Aircraft)

WESTAR IV IS DOUBLE THE SIZE OF EARLIER WESTARS

WESTAR 4 AND 5 BANDPLAN

Tr No	*	Frequency Range (MHz)	Center Freq (MHz)
1	H	3702-3738	3720
2	V	3722-3758	3740
3	H	3742-3778	3760
4	V	3762-3798	3780
5	H	3782-3818	3800
6	V	3802-3838	3820
7	H	3822-3858	3840
8	V	3842-3878	3860
9	H	3862-3898	3880
10	V	3882-3918	3900
11	H	3902-3938	3920
12	V	3922-3958	3940
13	H	3942-3978	3960
14	V	3962-3998	3980
15	H	3982-4018	4000
16	V	4002-4038	4020
17	H	4022-4058	4040
18	V	4042-4078	4060
19	H	4062-4098	4080
20	V	4082-4118	4100
21	H	4102-4138	4120
22	V	4122-4158	4140
23	H	4142-4178	4160
24	V	4162-4198	4180

*V=Vertical/H=Horizontal

TT & C for the Westar 4 and 5 satellites are on 3701 and 4199 MHz.

Westar 6-18

All these Westar satellites will be C-band satellites like Westar 4 and 5. It is unknown at this time if shuttle-recovered Westar 6 will be relaunched for use at its orbital slot or not. The following information for the Westar satellites is for 1985-2000.

Westar 6	17.8° W
Westar 7	96° W
Westar 8	98.5° W
Westar 9	81° W
Westar 10	93° W
Westar 11	91° W
Westar 12	79° W
Westar 13	99° W
Westar 14	123° W
Westar 15	130° W
Westar 16	93° W
Westar 17	91° W
Westar 18	87° W

Westar A-F

Western Union will soon join the rest of the Ku-band satellites with uplinks between 14.0-14.5 GHz and downlinks between 11.7-12.2 GHz. Each satellite will carry 16 transponders and have a design lifetime of 10 years. Orbital locations for these satellites are: 93° W, 91° W, and 87° W; launch dates are in the 1988-1999 time frame.

Western Union Earth Stations

1. *Glenwood, New Jersey (near New York)*
Serves the metro areas of Boston, Buffalo, Philadelphia, Baltimore, Wilmington, New York, Pittsburgh and Washington, D.C. Performs tracking, telemetry and command functions and directs station-keeping adjustments for Westar satellites.

2. *Estill Fork, Alabama (near Atlanta)*
Serves the Atlanta metro area, and beams antenna adjustment signals to the Westar satellites.

3. *Lake Geneva, Wisconsin (near Chicago)*
Serves the metro areas of Milwaukee, Detroit, Cleveland, St. Louis, Chicago, Indianapolis, Dayton, Columbus and Minneapolis/St. Paul.

4. *Steele Valley, California*
Serves the Los Angeles area.

5. *Cedar Hill, Texas (near Dallas)*
Serves the metro areas of Kansas City, Dallas and Houston. Performs as on-shore terminal for Westar satellite service to offshore drilling rigs in the Gulf of Mexico and beams antenna adjustment signals to the Westar satellites.

6. *Sky Valley, California (near San Francisco)*
Serves the San Francisco area.

7. *Issaquah, Washington (near Seattle)*
Serves the Seattle area.

SBS Satellites

On November 15, 1980, SBS-1 was launched and began service in March, 1981 as the first Satellite Business Systems all-digital 12/14 GHZ satellite. SBS-1 began delivering voice, teleconferencing, computer-to-computer communications and electronic mail and document distribution services to corporate customers via private intracompany networks.

The second satellite was launched on September 24, 1981 and located at 97° W; SBS-3 was launched November 11, 1982 and located at 95° W; SBS-4 was launched by the STS-41D Space Shuttle crew on September 5, 1984 and placed into geostationary orbit at 89° W. SBS-1 was placed into geostationary at 100° but will be shifted to 99° W in the near future.

Satellite Business Systems, a private company jointly owned by IBM, COMSAT General, and Aetna Life and Casualty, was formed to own and operate the first 12/14/ GHZ (Ku-band) satellite business communications system in the United States. Among the service's benefits are long-distance telephone cost savings of up to 30 percent over conventional long line costs.

Astronaut Dale A. Gardner, wearing the manned maneuvering unit (MMU), approaches the spinning Westar VI satellite over Bahama Banks. The end effector of the remote manipulator system (RMS), controlled by Dr. Anna L. Fisher inside Discovery's cabin, awaits its duty at right.

The system includes small RF ground terminals of which Hughes Aircraft is building 100. Each terminal is equipped with a 5.5 or 7.7 meter dish designed to be mounted on the rooftop or adjacent grounds of the customer's facilities.

The SBS satellites are the first of Hughes Aircraft's newest series of communications satellites (HS-376). The spacecraft has two concentric solar panels to improve output power of the satellite. Design mission of each satellite is seven years.

Each satellite has ten 43-MHz-wide transponders using 20 watt multicollector traveling wave tubes. Signal strength exceeds 43.7 dBW in the primary eastern coverage zone. With low noise, solid-state FET microwave integrated circuit techniques used in the redundant receivers, weighted beam receive sensitivities range from G/T +2.0 to -2.5 dB/k over the most densely populated regions of the United States. The satellites operate in a time division multiple access (TDMA) mode.

Each satellite has capacity for 12,500 two-way phone calls or 10 color TV transmissions simultaneously. Data bit rate is 480 megabits per second.

The following is a breakdown of the channelization plan used on the first four SBS satellites:

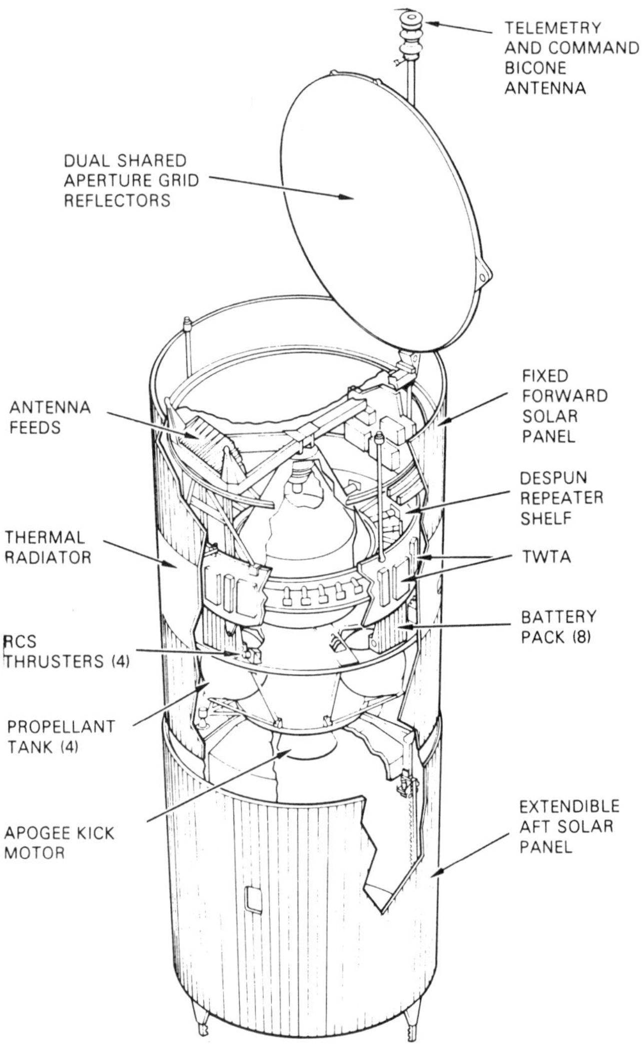

LOCATION OF WESTAR IV UNITS

SBS 1-4		
Tr No	Frequency Range (MHZ)	Center Freq (MHZ)
1	11703.5-11746.5	11725
2	11752.5-11795.5	11774
3	11801.5-11844.5	11823
4	11850.5-11893.5	11872
5	11899.5-11942.5	11921
6	11948.5-11991.5	11970
7	11997.5-12040.5	12019
8	12046.5-12089.5	12068
9	12095.5-12138.5	12117
10	12144.5-12187.5	12166

The TT & C channels for the SBS satellites are: 4197.25, 4197.75, 11701, and 12199 MHZ.

Satellite Business now has plans for five more SBS satellites with improved capabilities. SBS-5 and 6 will be 14-transponder models and will increase the usable bandwidth from the former 430 MHZ to 870 MHZ. Due for launch in 1986 (SBS 5) and 1987 (SBS 6), orbital positions will be 124° W and 99° W, respectively. Uplinks will be from 14.0-14.5 GHZ and downlinks will be from 11.7-12.2 GHZ.

SBS 7/8/9 will be improved even more and will carry 19 transponders for a bandwidth of 785 MHZ. Uplink and downlink frequency ranges for these satellites will be the same as SBS 5/6. Orbital locations for SBS 7/8/9 will be 93° W, 97° W, and 95° W, respectively.

In 1984, NBC concluded a deal with SBS to carry network programming until they can transfer their network operations over to RCA-Ku satellites to be launched in 1986/87.

Future Domestic Geostationary Satellites

Based on information filed with the FCC and ITU, the following satellite systems have been proposed or approved for launch. A quick glance at future systems shows that the decade of the 80s should aptly be called the decade of geostationary satellites.

ABC 1 & 2

Managed by Advanced Business Communications, two satellites in orbit and one ground spare will make up this business satellite communications system. ABC 1 (83° W) will be launched in December 1986 and ABC 2 (130° W) will be launched in February 1987.

Uplink/downlink frequencies are 14.0-14.5 GHZ/11.7-12.2 GHZ, respectively. Each satellite will have 20 transponders for a total usable bandwidth of 864 MHZ.

ASC 1/2/3/4/5

American Satellite Company's entry proposes an initial orbital constellation of ASC 1/2/3.

ASC 1 (81° W) was launched on August 27,1985; ASC 2 (128° W) will be launched in March 1986; ASC 3 launch will occur sometime in 1987 if the FCC approves their application and will be located at 101° W.

Uplink/downlink for this hybrid satellite are as follows:

5.925-6.425 GHz/3.7-4.2 GHz (C-band)
14.0-14.5 GHz/11.7-12.2 GHz (Ku-band)

The C-band communication package will carry 18 transponders and the Ku-band package will have 6 transponders. These satellites will have 1296 MHZ of usable bandwidth and a 7.5 year lifetime.

American Satellite's follow-on satellites for 1989-1990 reflect advanced technology that manufacturers hope to have at that time. ASC 4/5's orbital location is still pending FCC approval for a 93° W orbital slot.

These satellites are tri-band hybrids with uplinks/downlinks as shown:

5.925-6.425/3.7-4.2 GHZ
14.0-14.5/11.7-12.2 GHZ
29.5-30.0/19.7-20.2 GHZ

Aurora 1/2/3

Alascom, Inc., part of the RCA group, now has control of this RCA SATCOM series; Aurora 1, a standard 24-transponder C-band communication package, was covered in the RCA SATCOM section under the designation SATCOM 5. Aurora 2 and 3 will be identical in design and frequency coverage and, if approved by the FCC, these satellites will occupy 137° and 141° W orbital slots, respectively.

CCC 1 and 2

Columbia Communications Corp. has filed for two hybrid satellites to operate on their behalf in geostationary orbit using the following uplinks/downlinks:

5.925-6.425 GHz/3.7-4.2 GHz
14.0-14.5 GHz/11.7-12.2 GHz

Each satellite will have 24 transponders for a total usable bandwidth of 1728 MHZ. CCC 1 will be located at 62.5° W and CCC 2 will be located at 147.5° W. No launch dates have been set at this writing.

Comstar K1/K2/K3

COMSAT General will enter the domestic Ku-band market with three satellites that will have a 10 year lifetime. Uplinks for these satellites will be from 14.0-14.5 GHZ and downlinks will be from 11.7-12.2 GHZ. Each satellite will have 16 transponders and a total usable bandwidth of 864 MHZ. Orbital positions for these satellites will be 93° and 101° W. No launch dates have been confirmed at this writing.

DIGISAT 1A/1B/2

Digital Telesat's three satellite system will see launches in the 1988 time frame. It is uncertain whether the Digisat 1A/1B will comprise two separate satellites operating from the same orbital slot (57° W) or whether they were registered as a hybrid package. The uplinks and downlinks for 1A/1B are:

1A 14.0-14.5 GHz/11.7-12.2 GHz
1B 5.925-6.425 GHz/3.7-4.2 GHz

Digital Telesat also has a second satellite to be launched in 1988 and will be located at 134° W using the same uplinks and downlinks as Digisat 1A.

Equastar 1 and 2

Equatorial Corp. is now planning on launching

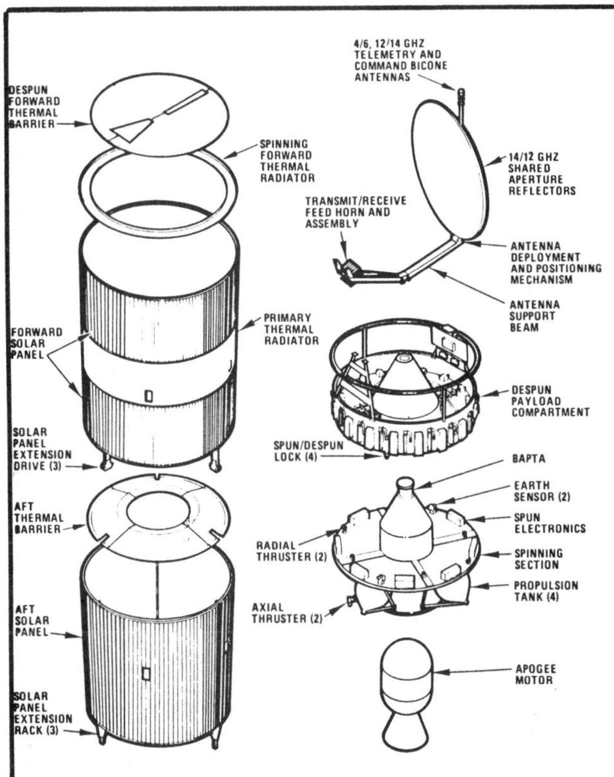

SBS exploded view.

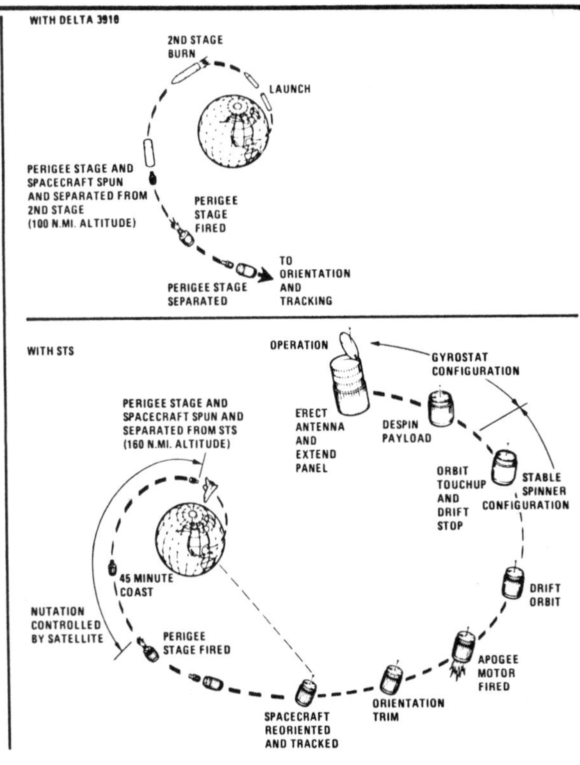

SBS launch sequences.

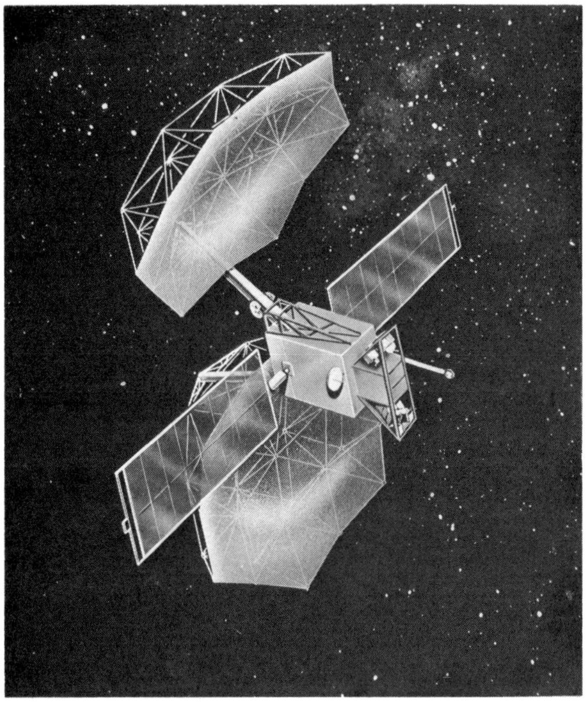

Canada's Mobile Satellite (MSAT) spacecraft (Canada Dept of Communications)

two 24-transponder C-band satellites in the 1987/1988 time frame. Orbital positions for Equastar 1 and 2 are 93° and 122° W, respectively, and they will use regular C-band uplink/downlink frequencies (5.925-6.425/3.7-4.2 GHZ).

Ford 1/2/3

Ford Aerospace will enter the geostationary COMSAT market in 1988 with three hybrid COMSATS, each containing 48 transponders and total usable bandwidth of 1728 MHZ. Uplink/downlinks are:

5.925-6.425 GHz/3.7-4.2 GHz
14.0-14.5 GHz/11.7-12.2 GHz

Orbital positions for Ford 1/2/3 are 101°, 93°, 73° W, respectively.

Geostar 1/2/3

Geostar Corp. will provide personal navigation satellites starting in 1987. They will launch three satellites into geostationary orbit that will be located at 70°, 100° and 130° W. Each satellite will have two transponders and two uplink/downlink communications packages as follows:

Uplinks(Center frequency) 1618 MHZ/5175 MHZ
Downlinks (Center freq) 2492 MHZ/5125 MHZ

Each satellite will have a usable bandwidth of 32 MHZ.

Mobilesat 1 and 2

Mobile Satellite Corp. has filed a proposal with the FCC to launch two satellites in 1987 to provide communications for aircraft traffic control, ground mobiles and cellular 'phone traffic.

Mobilesats will provide a variety of services including one and two-way radio communications among airborne, land and water-based mobile units. These satellites will be located at 85° and 125° W, respectively, and will serve all 50 states, Puerto Rico and Canada.

Voice and data transmission services will be offered as well as innovative services such as telephone interexchange trunking, position location and international air traffic control surveillance.

Mobilesat satellites are proposed to operate on the following allocations:

821-825	land mobile-satellite uplink
845-851	land mobile-satellite uplink
866-870	satellite to land mobile downlink
890-896	satellite to land mobile downlink
1543.5-1558.5	satellite-air and land mobile downlink
1645-1660	air and land mobile-satellite uplink
11.7-12.2 GHZ	satellite-ground station downlink
14.0-14.5 GHZ	ground station-satellite uplink
14/12 GHZ	TT & C frequencies

These satellites will be most interesting if approved by the FCC and only reflects the fact that scanner listening will continue to change and probably involve satellite listening in the years to come.

Msat

Canada's MSAT (*M*obile *Sat*ellite) is proposed to provide a more effective and reliable two-way radio and radiotelephone service to supplement today's short range terrestrial mobile communications systems. It would reach virtually all Canada to serve those traveling on foot, by land vehicles, airplane, or ship and be able to link them with anyone served by radio or telephone system.

The system, operating in the 806-890 MHZ band, will include the capacity for data collection from platforms, much like the DCS used by U.S. weather satellites. An experimental extremely-high-frequency communications package for use by National Defence is also being studied.

The elements of the MSAT system are: a satellite in geostationary orbit; a partially integrated spare spacecraft on the ground; ground control facilities; and a family of terminals for mobile and fixed installations.

A study of feasibility began in 1980, with a launch projected for late 1987 or early 1988. If the MSAT system is implemented, experiments and pre-operational services would be carried out for seven years.

Cost to the user would be comparable to mobile radio or mobile telephone rates in major cities. Its benefits would be measured not only in dollars, but in health and safety as well as in the knitting together of sparsely populated areas.

The satellite will carry two 9-meter dishes and the maximum in-orbit dimension from sail tip to sail tip is 30 meters. The spacecraft's solar power system is capable of producing 3 Kw of power minimum and the design lifetime is seven years.

RSI 1/2/3/4

Rainbow Satellite plans to launch four Ku-band satellites starting in 1985. These satellites will use 14.0-14.5 GHZ uplinks and 11.7-12.2 GHZ downlinks. RSI-1 (132° W) and RSI-2 (79° W) will use sixteen 54-MHZ transponders for a total bandwidth of 864 MHZ. These satellites will have a design lifetime of eight to ten years.

RSI-3 (75° W) and RSI-4 (93° W) will use twenty 43-Mhz transponders for a total bandwidth of 860 MHZ and will have a design lifetime of eight to ten years.

RSI-1 and 2 will be launched in 1985 and 1986, respectively, and RSI-3 and 4 will be launched in 1987.

Skylink 1 and 2

These satellites will be launched by Skylink, Inc. in the 1987 time frame. They will use 800 MHZ uplinks and 900 MHZ downlinks with a total usable bandwidth of 8 MHZ divided between two transponders.

Typical C-band earth terminal for the cable TV industry

Spotnet C1/C2 & K1/K2/K3/K4

National Exchange, Inc. will launch their satellites into the geostationary Clarke belt sometime in 1987.

The Spotnet C1/C2 satellites will utilize C-band uplinks and downlinks as follows: 5.925-6.425 GHZ/3.7-4.2 GHZ with earth satellite carrying 24 transponders for a total usable bandwidth of 864 MHZ.

Spotnet K1-K4 satellites will utilize the Ku-band uplink and downlink frequencies of 14.0-14.5 GHZ and 11.7-12.2 GHZ; each satellite will contain 24 transponders with a total usable bandwidth of 864 MHZ.

These satellites will be co-located at 101° W and 75° W with each location containing two of the Ku-band satellites and one of the C-band satellites.

Usat 1/2/3/4

Starting in 1985, U.S. Satellite Systems will launch the first of four Ku-band satellites into geostationary orbit; they plan to launch one satellite each year after 1985. These satellites will contain 20 transponders with a total bandwidth of 860 MHZ.

Uplink/downlink frequencies are: 14.0-14.5 GHZ/11.7-12.2 GHZ. Usat 1/2/3/4 will be located at 85°/120°/75°/83° W with a design lifetime of eight years.

Chapter Five

INTERNATIONAL SATELLITES

Recently, a good friend of mine asked me, "Larry, where have all the point-to-point stations that relay international telephone calls on shortwave gone?" I said, "Satellites, my dear man, satellites." The next logical question was, "How can I still monitor them?" I answered, "You can't unless you get a TVRO setup." Now, I've seen folks wince in my day, but the mere mention of satellites to this fellow almost made him cry.

The message is clear: If you want to continue to hear interesting utility DX that you used to hear on shortwave, you are going to have to move up to the satellites to hear them. All kinds of services are finding that satellites are much more reliable than HF, from the U.S. domestic satellites to the international communications satellites which we will cover in this chapter.

Monitors of video transmissions on these satellites should be prepared to encounter different TV systems used by other countries. Additionally, INTELSAT satellites use lower power levels and half transponder channels so some modification of the TVRO setup will be necessary and a multiple-standard TV monitor will be required.

More details can be found in sources listed in Appendix A at the end of this book.

The current status of the geostationary ring is outlined in the Satellite Location Guide in the Appendix.

Arabsat 1 & 2

The Arabsat satellites are medium sized, multi-mission satellites designed by Aerospatiale and built by Ford Aerospace. Arabsat was built for the Arab Satellite Communications Organization of the Arab League of Arab States. The League is headquartered in Riyadh, Saudi Arabia.

The 22 members of the Arab Satellite communications Organization that will use the two Arabsat satellites now in orbit are: Algeria, Bahrain, Egypt, Iraq, Jordan, Kuwait, Lebanon, Libya, Morocco, Mauritania, Oman, Palestine, Qatar, Saudi Arabia, Sudan, Somali, Syria, Tunisia, United Arab Emirates, Yemen (Arab Republic), and Yemen (Democratic Republic).

The satellite weighs 2,580 lbs (Ariane launch) and 2,800 lbs (shuttle launch). In-orbit weight of the satellite is 1,230 lbs. The main body is 7.4 feet (2.26 meters) by 5.4 feet (1.64 meters) by 4.9 feet (1.49 meters). The solar panels span 67.9 feet (20.7 meters) tip-to-tip with arrays deployed.

Arabsat will transmit telephone, telex, facsimile, television, and data within the 22 Arabsat member states.

The payload consists of the following: 25 4/6 GHz (C-band) transponders, providing 8,000 simultaneous telephone circuits or equivalent traffic; and one 4/2.5 GHz (C/S band) transponder, providing semi-direct television transmission to community receivers with three meter antennas. The satellites use three-axis stabilization and are designed for on-orbit lifetimes of seven years.

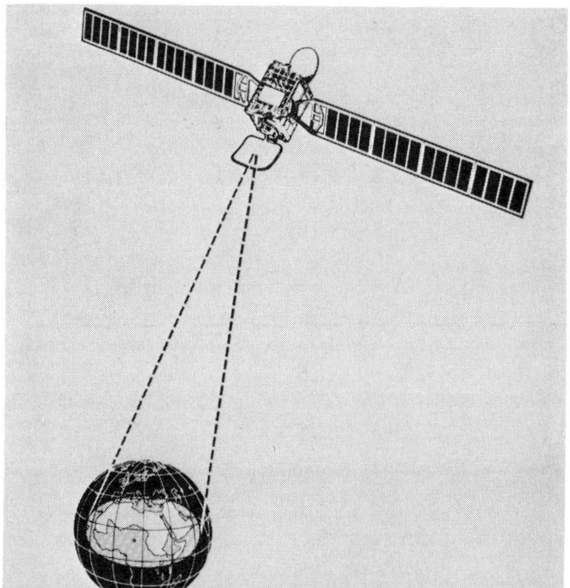

ARABSAT Communications Coverage -- 22,300 miles above the equator (Ford Aerospace)

ARABSAT comsat (Aerospatiale)

ARABSAT TRANSPONDERS

Tr No	Freq Range (MHz)	Tr No	Freq Range (MHz)
	2544-2651; and	13	3940.75-3973.75
1	3718.75-3751.75	14	3959.25-3992.25
2	3737.25-3770.25	15	3977.75-4010.75
3	3755.75-3788.75	16	3996.25-4029.25
4	3774.25-3807.25	17	4014.75-4047.75
5	3792.75-3825.75	18	4033.25-4066.25
6	3811.25-3844.25	19	4051.75-4084.75
7	3829.75-3862.75	20	4070.25-4103.25
8	3848.25-3881.25	21	4088.75-4121.75
9	3866.75-3899.75	22	4107.25-4140.25
10	3885.25-3918.25	23	4125.75-4158.75
11	3903.75-3936.75	24	4144.25-4177.25
12	3922.25-3955.25	25	4162.75-4195.75

Arabsat 1 (85-15A) was launched on Friday evening February 8, 1985, at 6:22 pm EST from French Guiana on an Ariane rocket. The satellite was placed into geostationary orbit at 19° E. Arabsat 2 was launched on shuttle mission 51G aboard the space shuttle Discovery on June 17, 1985. The payload was released by the shuttle crew on June 18 and placed into geostationary orbit at 26° E. A third satellite is being used as a ground spare.

Monitors equipped to monitor C/S band transmissions in the Middle East are encouraged to send reports to channel occupancy on the Arabsats to the author in care of the publisher.

AUSSAT 1/2/3

After a much publicized debate over C- band versus Ku-band, the Aussies have decided on the Ku-band.

Aussat-1 was launched on the STS 51I (85-76B) mission on August 27, 1985, and was positioned at 156° E. Aussat-2 (85-109C) was launched on the STS-61B mission on November 27, 1985, and was positioned at 160° E. The third Aussat will be launched in 1990 and will be positioned at 164° E.

These Hughes HS-376 satellites will provide communications to remote Australian sites and other services such as air traffic control relay to aircraft flying over the Australian interior.

Each satellite will contain four DBS channels and use the 14.0-14.5 GHz band for uplinks and 12 GHz for downlinks. Each DBS channel will use a 30-watt transponder and there will be eleven additional 12 watt channels.

The Aussat-1 satellite started regular transmissions on Australia Day, January 26, 1986. The satellite uses the world's first "B-MAC" (Multiplexed Analogue Component, Type B) on the Homestead and Community Broadcast Satellite Service (HACBSS). Advantages of this system include high quality television reception; six digital sound channels; a data channel; and more reliable reception, even under extreme climatic conditions. The system will also be able to, when implemented, transmit high definition television signals.

The system is capable of carrying stereo and mono sound broadcast plus television with stereo sound, teletext and a data channel which could provide emergency warnings, special educational programs and other services.

Information has not been received as of this writing on frequency transponder usage or polarity information.

Brazilsat 1

On February 8, 1985, an Arianespace Ariane 3 rocket was launched from Kourou, French Guiana, carrying the Brazilsat 1 and the Arabsat 1 satellites.

The Brazilsat was built by Canada's Apar Aerospace, Ltd. and is being used by Brazil's state run company Embratel for national telephone, television, telex, and data transmission. Embratel authorities expect to lease excess capacity of this satellite to neighboring countries.

Brazilsat weighed 2,513 (1,140 kg) at launch and has an in-orbit weight of 1,479 lbs (671 kg). Brazilsat 1 was located at 65° W and Brazilsat 2 will be located at 70° W. No frequency information or transponder plans for these C-band satellites has been received thus far.

CS (Sakura)

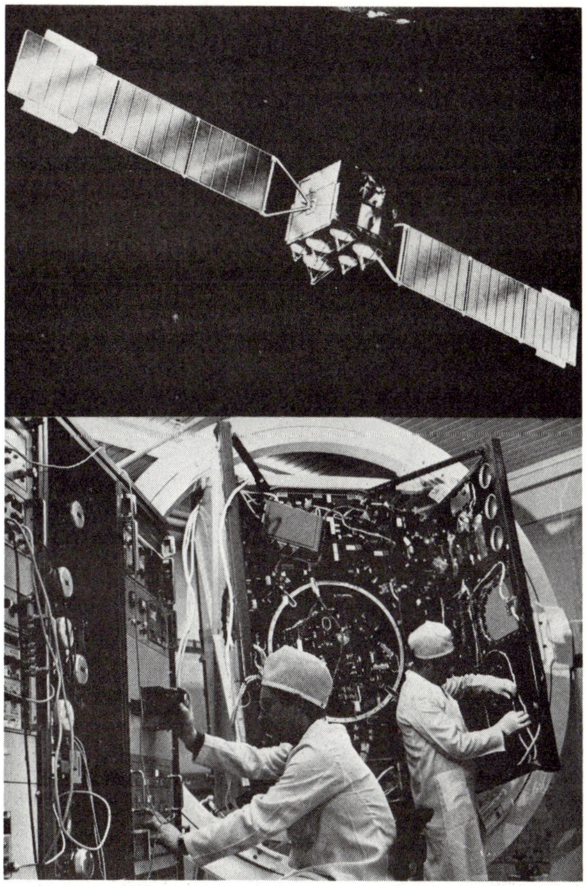

ECS 1-5 (AEG-Telefunken)

CS

Sakura

The Japanese Sakura COMSAT was launched December 15, 1977 by a Delta 137 rocket from the Kennedy Space Center to perform the first experiments using quasi-millimeter waves on satellite communications.

Presently maintained at 150° E, Sakura is used by the Radio Research Laboratories of the Ministry of Post and Telecommunications and the Nippon Telegraph and Telephone Public Corporation to conduct a variety of communications experiments.

The Sakura communications packages use the following bandplan:

C-Band

Uplink 5.925-6.425 GHz/Downlink 3.72-4.18 GHz. 460 MHz bandwidth(2 transponders)

Quasi-millimeter Band

Uplink 27.55-30.05 GHz/Downlink 17.75-20.25 GHz.1200 MHz bandwidth(6 transponders)

CS 2A & 2B

The CS-2A & 2B satellites are comparable to Sakura in capacity. These satellites were launched into geostationary orbit using the Japanese N-II vehicle and placed into geostationary orbit at 130° and 135° E. CS-2B serves the system as an in-orbit spare with CS-2A as the primary space-craft.

The principle objectives are to offer domestic communication service for public and official purposes using quasi-millimeter waves and microwaves as well as to develop technology for communications satellites. Specific objectives include:

- Reservation of a communication link in an emergency;
- Establishment of communication circuits for remote islands;
- Establishment of communications circuits for temporary use; and
- Development of technology for domestic production concerning communication satellites.

The satellites are now managed by the Telecommunications Satellite Corporation of Japan. CS-2A was launched on February 4, 1983 and placed into orbit at 130° E; CS-2B was launched on August 5, 1983 and placed into into orbit at 135° E. The following is a breakdown of the communications packages on board each satellite:

Communications:

Uplinks: 5.925-6.425 GHz/27.55-30.05 GHz
Downlinks:

1) 3735-3915 MHz
2) 3985-4165 MHz
3) 17810-17940 MHz
4) 18090-18220 MHz
5) 18370-18500 MHz
6) 18650-18780 MHz
7) 18930-19060 MHz
8) 19210-19340 MHz

TT&C:

2286.5 MHz @ 1 watt
3820 MHz
4080 MHz
3950 MHz
19450 MHz

Eutelsat (ECS 1-5)

The successful operation of OTS (Orbital Test Satellite) has vindicated the major design innovations proposed for the European Communications Satellite (ECS) system, which include:

- Use of the newly allocation 14 and 11 GHz satellite frequency bands to avoid the congested 6 and 4 GHz bands used by other communication satellites;
- Orthogonal polarization to enable the system capacity to be doubled by frequency reuse. In addition, time division multiple access (TDMA) and speech interpolation (DSI) will also contribute to increasing system capacity.

With the decision to implement the ECS Program, Europe is on the threshold of a new era in telecommunications. The ECS program is currently being carried out under the direction of ESA (European Space Agency) with EUTELSAT (the organ of the European Telecommunications Administrations) as the final customer.

Meteorological pictures returned from INSAT 1A (Ford Aerospace)

The ECS mission is to relay a major portion of inner-European telephone traffic throughout the next decade. These satellites will also provide channels for TV transmission within the European Broadcasting Union (EBU) countries as well as business traffic over special data channels. The project is being carried out by an international industrial group and British Aerospace is the prime contractor.

The ECS satellites are using six frequency ranges with two reuses of these ranges via different polarizations. In addition, ECS satellites carry an antenna configuration that permits four different antenna beams: Eurobeam (EB) - wide coverage; Spot beam east (SE); Spot beam west (SW); and Spot beam Atlantic (SA). Each of these beams are assigned to transponders as follows:

Eurobeam(EB)--Tr. 1X, 3X 1Y, 3Y
Spot beam east(SE)--Tr. 2X, 5X, 2Y
Spot beam west(SW)--Tr. 1X,3X,4X,6X,
1Y,3Y,4Y,6Y
Spot beam Atlantic(SA)--Tr. 5X, 5Y

The following is a breakdown of the transponders carried on the ECS satellites:

EUTELSAT (ECS)

Tr No	Frequency Range (MHz)	Center Freq (MHz)
1X	10950-11033.334	10991.667
1Y	same as 1X; different polarization	
2X	11033.333-11116.667	11075.0
2Y	same as 2X; different polarization	
3X	11116.666-11200	11.158.333
3Y	same as 3X; different polarization	
4X	11450-11533.334	11491.667
4Y	same as 4X; different polarization	
5X	11533.333-11616.667	11575.0
5Y	same as 5X; different polarization	
6X	11616.666-11700.000	11658.333
ESS(S1)	12500.0-12583.334	12541.667

ECS-1 was launched June 16, 1983 and parked at 13° E; ECS-2 was launched August 4, 1984 and parked at 10° E; ECS-3 will be launched in 1985 and parked at 7° E. ECS 4 and 5 are spares scheduled for later launches.

Insat-1B

The first generation Indian National Satellite System (INSAT-1) represents India's first step towards implementing operational space systems for identified socio-economic objectives and national requirements. It is intended for:

- Long distance multimode communications;
- Direct TV broadcasts to rural areas;
- Networking of terrestrial TV transmitters and

INSAT I (Ford Aerospace)

All India Radio.

INSAT-1A was launched on April 10, 1982 but failed in orbit. INSAT-1B was launched on August 31, 1983 and has replaced 1A at 74° E. This satellite uses the following frequencies to transmit a variety of services:

INSAT-1B

Tr No	Freq Range	Center Freq
	2555-2591 2599-2635 MHz	(Indian DBS serv & All India Radio & TV networking)
1)	3172-3748 MHz	3730 MHz
2)	3752-3788 MHz	3770 MHz
3)	3792-3838 MHz	3770 MHz
4)	3832-3868 MHz	3850 MHz
5)	3872-3908 MHz	3890 MHz
6)	3912-3948 MHz	3930 MHz
7)	3962-3998 MHz	3970 MHz
8)	4002-4038 MHz	4010 MHz
9)	4042-4078 MHz	4060 MHz
10)	4082-4118 MHz	4100 MHz
11)	4122-4158 MHz	4140 MHz
12)	4162-4198 MHz	4180 MHz
		All India TV relay
	4034.55 MHz	VHRR Radiometer
	4038.1 MHz	DCP downlink
	402.75 MHz	DCP uplink

INSAT-1B carries a VHSSR weather satellite type device similar to our GEOS weather satellites with a facsimile downlink for these weather satellite pictures in the 4 GHz region.

Morelos 1 & 2

Shuttle mission 51-G saw the deployment of the first Mexican satellite, Morelos 1, on June 17, 1985.

OTS Orbital Test Satellite

MORELOS 1 & 2 TRANSPONDERS

TR No	Freq Range (MHz)	Tr No	Freq Range (MHz)
1)	3705-3741	12)	3985-4021
2)	3729-3800	13)	4025-4061
3)	3745-3781	14)	4049-4120
4)	3785-3821	15)	4065-4101
5)	3809-3880	16)	4105-4141
6)	3825-3861	17)	4129-4200
7)	3865-3901	18)	4145-4181
8)	3889-3960	19)	11703.5-11795.5
9)	3905-3941	20)	11801.5-11893.5
10)	3945-3981	21)	11899.5-11991.5
11)	3969-4040	22)	11997.5-12089.5

Morelos 1 (85-48B) has 18 transmit and receive C-band transponders and transmit beams for four Ku-band channels through the spacecraft's primary antenna. A small planar array is being used for Ku-band uplink channels.

The Hughes HS376 satellite was acquired by Mexico's Secretariat of Communications and Transportation in 1982.

Morelos 2 (85-109B) was launched aboard shuttle mission 61B on November 11, 1985, and placed into orbit where it will be allowed to drift for the next two years into its orbital position. Mexican officials indicated Morelos-2 would be better utilized by placing it in an extended drift to save fuel, rather than bringing the spacecraft on line operationally with Morelos 1.

The Morelos satellite system is intended to provide voice, video, data, and facsimile services for remote areas of the country as well as between cities. There are about 14,000 small villages in Mexico without telephone services.

Mexico's contract with Hughes aircraft includes development and installation of a tracking, telemetry and command station near Mexico City to operate the system.

OTS-2

The Orbital Test Satellite was designed around needs defined by the European Conference of Posts and Telecommunications (CEPT) and the EBU. Launched May 11, 1978, OTS-2 is capable of carrying 3,000 telephone calls and two television channels. It is located at 10° E. The satellite uses the following uplinks and downlinks:

Communications:

Uplinks	14.455-14.460 GHz
	14.153-14.193 GHz
	14.243-14.363 GHz
Downlinks	10.950-11.200 GHz
	11.450-11.470 GHz
	11.490-11.530 GHz
	11.580-11.700 GHz
	11.792.5-11.797.5 GHz

TT&C:

	137.050 MHz
	11.510 GHz
	11.575 GHz
	11.640 GHz

Palapa Satellites

Palapa, a name signifying national unity, was chosen by Indonesian President Suharto in July 1975. The name symbolizes the fulfillment of a vow for unity first expressed by Gajah Mada, a revered national hero of the 14th century who served as a prime minister of the Kingdom of Majapahit. He had vowed not to partake of palapa, a national delicacy, until the goal of national unification was achieved.

On August 17, 1976, the nation became electronically unified when a Palapa satellite, launched the previous month, started service over the world's largest archipelago amid national ceremonies celebrating the republic's 31st year of independence.

Indonesia's second satellite was launched in March 1977. Palapa A1 is stationed at 83° E and Palapa A2 is stationed at 77° E. These satellites built by Hughes Aircraft are identical in design to the Anik A and WESTAR 1/2/3 series satellites. They are operated by Perumtel of Indonesia.

Palapa is a 12 transponder satellite with an average capability of 6,000 voice circuits or 12 simultaneous color TV channels or any combination of the two. Design lifetime of these satellites is seven years.

The Palapa A1/A2 satellites use the following bandplan:

PALAPA A1/A2		
Tr No	Freq Range	Center Freq
1)	3702-3738 MHz	3720 MHz
2)	3742-3778 MHz	3760 MHz
3)	3782-3818 MHz	3800 MHz
4)	3822-3858 MHz	3840 MHz
5)	3862-3898 MHz	3880 MHz
6)	3902-3938 MHz	3920 MHz
7)	3942-3978 MHz	3960 MHz
8)	2982-4018 MHz	4000 MHz
9)	4022-4058 MHz	4040 MHz
10)	4062-4098 MHz	4080 MHz
11)	4102-4138 MHz	4120 MHz
12)	4142-4178 MHz	4160 MHz

All signals on these satellites are horizontally polarized. TT&C channels are 4198.6 and 4199.1 MHz.

Palapa B1/B2/B3

Palapa B is the second series of Indonesian communications satellites designed and built by Hughes Aircraft. Based on the Hughes HS-376 design and featuring an eight year lifetime, they are similar to the Telesat Canada Anik D satellites.

The 24 transponders on these satellites double the capacity of the earlier Palapa A models and each transponder will have a 10 watt output, providing Perumtel with 24 color TV channels. Signal strength for Indonesia will be 34 dBW, but these satellites are designed to deliver voice, video, telephone, and high speed data services to the Phillipines, Thailand, Malaysia, and Singapore as well as Indonesia.

Palapa B1 was launched on June 18, 1983 from the space shuttle and assumed its orbital location at 108° E; Palapa B2 was launched from the space shuttle on February 6, 1984 and, due to a failure of the PAM, left in low earth orbit. The spacecraft was subsequently rescued by the STS-51A crew and returned to space.

Palapa B2's orbital slot of 113° E remains a question mark as far as filling it with another Palapa satellite or the one that was lost. Palapa B3 is supposed to be launched in 1985 and is assigned to 118° E once in geostationary orbit.

Telecom 1A & 1B

These French government satellites carry communications packages for broadcasting, telecommunications and the French military. When fully operational they will perform three main missions:

- Wide-band, high-speed data links between various locations of a French company;
- An "Overseas Territories Telecommunications Link" for telephone and television links between France and her overseas territories;
- Routing of classified government communications.

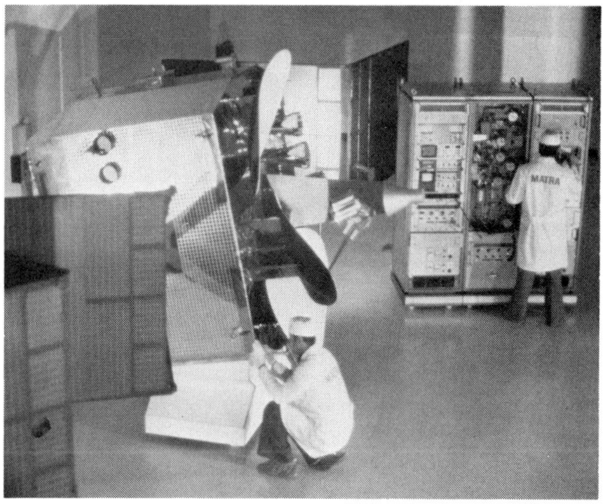

Telecom 1A and 1B

Data transmission is the main mission of Telecom 1 satellites; like the U.S. SBS, this mission is performed in the 12/14 GHz band occupying six repeaters for 36 MHz of useful bandwidth. Each repeater has a capacity of 25 Mbits/sec. (about 350 channels of 64 kbits/sec.). Telecom 1 will permit:

- The routing of the electronic mail and the connection of office management terminals;
- High speed transfer of files between computers;
- Videotransmission and videoconferencing; remote printing of daily newspapers.

The Telecom 1 will use the following 12/14 GHz frequencies:

Uplink: 14.00-14.250 GHz
Downlink: 12.5-12.75 GHz

The French overseas territories telecommunications links will consist of telephone traffic and television programs between France and the French overseas territories Martinique, Guadeloupe, Guiana, St. Pierre and Miquelon and La Reunion in order to replace repeaters now used on Intelsat and Symphonie satellites.

Four repeaters in the 6/4 GHz band can each route one TV program or 1,000 telephone channels.

The Telecom 1 will use the following 4/6 GHz frequencies:

Uplink: 5.925-6.425 GHz
Downlink: 3.7-4.195 GHz

Two repeaters in the 8/7 GHz band will be used for governmental missions on the following frequency bands:

Uplink: 7.980-8.095 GHz
Downlink: 7.255-7.370 GHz

EARLY BIRD

Diameter:
71 cm (28 in.)

Panel Height:
59 cm (23 in.)

Weight:
34 kg in orbit
(76 lb)

Telecom 1A was launched August 4, 1984 and located at 8° W; Telecom 1B was lost due to Ariane rocket malfunction in May 1985. It decayed into the atmosphere October 23, 1985.

Intelsat System

In 1962, the Congress of the United States passed and President John F. Kennedy signed into law the Communications Satel lite Act. In keeping with the mandate of the act, COMSAT was formed in February 1963. A year and a half later, Intelsat came into being.

In June 1965 Intelsat 1, better known as Early Bird, was placed into service, establishing the first satellite pathway between the United States and Europe making possible live transoceanic TV transmission.

The Intelsat satellite system has grown to where it is now the dominant carrier of all international communications, handling about two-thirds of all transoceanic traffic.

The Intelsat system consists of high capacity communications satellites stationed in geostationary orbits 22,240 miles above the equator. From these positions, each satellite serves an area larger than one-third of the earth's surface, thereby providing global coverage.

These satellites and more than 213 earth station antennas operating with them provide more than 976 pathways, or direct communications links, among over 135 countries, territories, and possessions. A growing number of countries are also building earth stations within their boundaries, using Intelsat satellites to improve their domestic communications.

Monitoring and control of the system's satellites is the responsibility of the Intelsat Spacecraft Technical Control Center in Washington, D.C. Stations provided by Intelsat members including two COMSAT-operated TTC&M stations located at the Andover, Maine, and Paumalu, Hawaii perform these functions.

TTC&M stations track satellites after launch, during transfer orbit, and throughout their lives in space; it is through them that commands are issued when it is necessary to alter satellite performance.

Intelsat 1 (Early Bird)

Intelsat 1 was the world's first commercial communications satellite. Launched April 6, 1965, it was placed into service in June 1965; it established the first satellite pathway between the United States and Europe. Early Bird made live television possible for the first time, operating in continuous fulltime service for more than three and a half years.

The satellite used its toroidal beam pattern antenna to deliver point-to-point communications in the heavy corridor between the U.S. and Europe. If television transmission was required, all voice and data traffic had to be relinquished; the satellite could not perform multiple station access. It could transmit only 240 voice circuits or one TV channel. This satellite is no longer used for communications.

Early Bird used the following frequency capability:

INTELSAT II satellites launched in 1967

INTELSAT II launch from Cape Kennedy

Communications:

Uplink 6288-6314 MHz Downlink:4068-4104 MHz
Uplink 6377-6403 MHz Downlink:4148-4174 MHz

TT&C:

Command channels: 6289/6402 MHz
Telemetry channels: 4104/4138 MHz
Tracking Beacons: 136.4104/4104 MHz
137.4138/4138 MHz

Intelsat 2

These satellites were placed over the Atlantic and Pacific Oceans during 1967, introducing multipoint communications capability and extending satellite communications to more than two-thirds of the earth.

These satellites could also carry 240 voice circuits or one TV channel at any one time, but in the earth coverage mode, they could produce signals in both the northern and southern hemispheres.

This series of Intelsats used the same booster and launch facilities as the Early Bird launch.

The following Intelsat 2 satellites were launched:

Intelsat 2 F1	October 26, 1966
Intelsat 2 F2	January 11, 1967
Intelsat 2 F3	March 22, 1967
Intelsat 2 F4	September 27, 1967

Intelsat 2 satellites used the following frequencies:

Communications:

Uplink 6282-6408 MHz;Downlink 4057-4133 MHz

TT&C:

Command channels: No data
Telemetry channel: 136 MHz
Tracking channel: 136 MHz

Intelsat 3

The Intelsat 3 series of satellites established a

INTELSAT IV

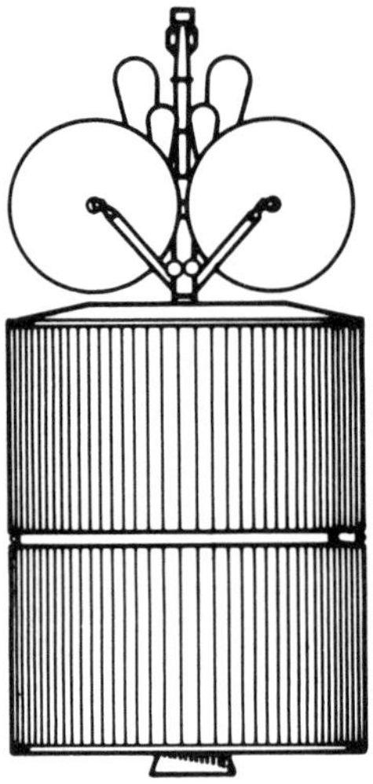

Diameter:
238 cm (93 in.)

Overall Height:
531 cm (209 in.)

Weight:
595 kg in orbit
(1313 lb)

INTELSAT 4A TRANSPONDERS

Tr No	Freq Range (MHz)	Center Freq	Half Tr Centers	Downlink Ant.
1A(1)	3707-3743	3725	3715/3734	NE/SE SB/EH
1B(2)	3707-3743	3725	3715/3734	NW/SW SB/WH
2A(3)	3747-3783	3765	3755/3774	NE/SE SB/EH
2B(4)	3747-3783	3765	3755/3774	NW/SW SB/WH
3A(5)	3787-3823	3805	3795/3824	NE/SE SB/EH
3B(6)	3787-3823	3805	3795/3854	NE/SE SB/WH
4A(7)	3827-3863	3845	3835/3854	NE/SE SB/EH
4B(8)	3827-3863	3845	3835/3854	NW/SW SB/WH
5A(9)	3867-3903	3885	3875/3894	NE/SE SB/EH
6 (11)	3907-3943	3925	3915/3934	Global beam
7A(12)	3947-3993	3975	3955/3974	NE/SE SB/EH
7B(13)	3957-3993	3975	3955/3974	NW/SW SB/WH
8 (14)	3997-4003	4015	4005/4014	Global beam
9A(15)	4037-4073	4055	4045/4054	NE/SE SB/EH
9B(16)	4037-4073	4055	4045/4054	NW/SW SB/WH
10(17)	4077-4113	4095	4085/4104	Global beam
11A(18	4177-4153	4135	4125/4144	NE/SE SB/EH
11B(19)	4117-4153	4135	4125/4144	NW/SW SB/WH
12(20)	4157-4193	4175	4165/4184	Global beam

SB=Spot beam/EH=East Hemispheric/WH=West Hemispheric

global system of communication satellites when they were placed into service over the Atlantic, Pacific and Indian Oceans in 1969. They introduced simultaneous transmission of all forms of communications including telephone, television, telex, data, and facsimile.

These satellites could handle up to 1,200 circuits and four TV channels without interruption to the other services.

These TRW manufactured satellites were launched from the Cape on a McDonnell/ Douglas three-stage, long-tank Delta rocket.

The Intelsat 3 series of satellites were launched as follows:

Intelsat 3 A	September 18, 1968 Launch failure (F1)
Intelsat 3 F2	December 18, 1968
Intelsat 3 F3	February 5, 1968
Intelsat 3 F4	May 21, 1969
Intelsat 3 E	July 25, 1969 Launch failure (F5)
Intelsat 3 F6	January 14, 1970
Intelsat 3 H	July 23, 1970 Launch failure (F8)

This series of satellites used the following frequencies and ranges:

Communications:

Uplink 5930-6155 MHz;Downlink 3705-3930 MHz
Uplink 6195-6420 MHz;Downlink 3970-4195 MHz

TT&C:

Command channel: 6175 MHz
Telemetry: 3933-3967 MHz (Telemetry phase modulated the beacon)
Tracking: 3933-3967 MHz

Intelsat 4

The Intelsat 4 satellites were placed into service between 1971 and 1975 making the system even more flexible and versatile by introducing the spot

INTELSAT IV A antennas undergo inspection

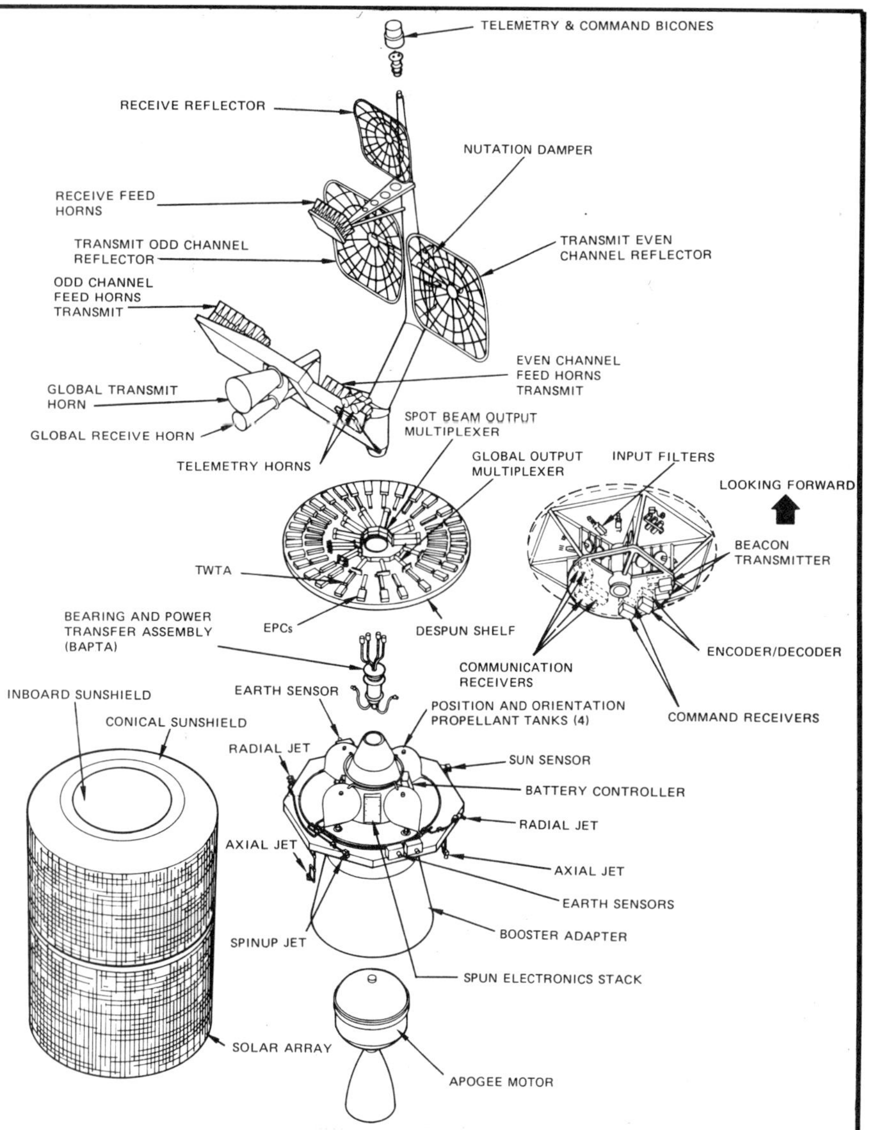

Satellite components

beam concept.

Using 12 transponders with bandwidths of about 36 MHz, the satellite has two earth- coverage receive antennas and two earth-spot-beam (narrow beams) transmit antennas (50- inch diameter parabolic dish mounted on the despun control mast). The aiming of the spot beams is controlled from the ground and have a 4.5 degree beamwidth to provide high-gain coverage in selected areas.

These satellites, using the earth coverage antenna, can relay up to 6,000 half-duplex channels or 12 color TV channels of programming.

The following is a list of Intelsat 4 spacecraft that were launched:

Intelsat 4 F2	January 25, 1971
Intelsat 4 F3	December 19, 1971
Intelsat 4 F4	January 22, 1972
Intelsat 4 F5	June 13, 1972
Intelsat 4 F7	August 23, 1973
Intelsat 4 F8	November 21, 1974 3° E; in use
Intelsat 4 F6	February 20, 1975 Launch failure
Intelsat 4 F1	May 22, 1975 50° W; in use

The Intelsat 4 uses twelve independent repeaters or half-transponders to relay television, etc. The following is the frequency breakdown and channelization for the Intelsat 4 satellites.

Communications:

Uplink 5930-6420 MHz;Downlink 3705-4194 MHz

TT&C:

Command: 6175 MHz
Telemetry: 3947.5/3952.5 MHz
Tracking: 3947.5/3952.5 MHz

INTELSAT 4

Tr No	Freq Range (MHz)	Center Freq	Half Tr Centers	Downlink Ant.
1	3707-3743	3725	3715/3734	WS/Global
2	3747-3783	3765	3755/3774	ES/Global
3	3787-3823	3805	3795/3824	WS/Global
4	3827-3863	3845	3835/3854	ES/Global
5	3867-3903	3885	3875/3894	WS/Global
6	3907-3943	3925	3915/3934	ES/Global
7	3957-3993	3975	3955/3974	WS/Global
8	3997-4033	4015	4005/4014	ES/Global
9	4037-4073	4055	4045/4054	Global only
10	4077-4113	4095	4085/4104	Global only
11	4117-4153	4135	4125/4144	Global only
12	4157-4193	4175	4165/4184	Global only

WS=Western spot beam/ES=Eastern spot beam/ Global=global coverage beam

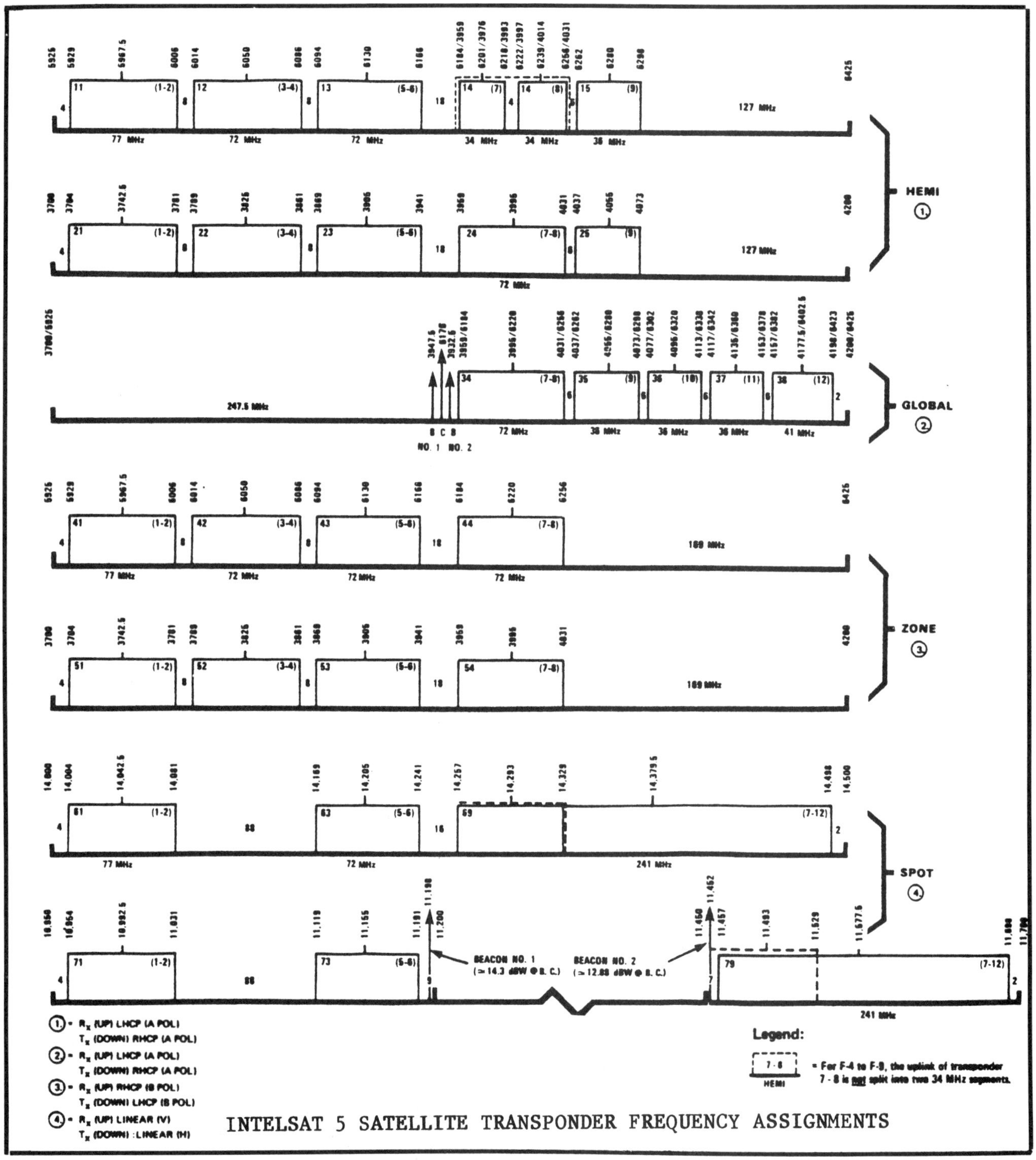

INTELSAT 5 SATELLITE TRANSPONDER FREQUENCY ASSIGNMENTS

Intelsat 4A

The dramatic increase in overseas calls has required the size and power of the international communications satellites to be expanded.

Six new Hughes Intelsat 4As incorporate new antenna technology to provide twice the capacity of Intelsat 4; this twofold increase in capacity is attained by reusing the same frequency through antenna spot beam separation.

The antenna provides coverage of land masses on both sides of the Atlantic basin with four spot beams and has sufficient isolation between the eastern hemispheric and western hemispheric beams to use the same frequencies in the east and west. Beam separation by directional antenna permits this double use of the frequency, significantly increasing satellite communications capability within an allocated frequency band.

While the initial requirement for the Intelsat 4A was for Atlantic service, careful attention was given during the design of the satellite to provide for service over the Indian Ocean as well.

The Intelsat 4A has 20 transponders (individual radio repeaters) compared to 12 aboard each Intelsat 4. Intelsat 4As over the Atlantic Ocean use four transponders for wide beam coverage, focusing on an area slightly larger than one-third of the world's surface which is visible to the satellite. The remaining 16 transponders are used in the spot beam mode to concentrate on small selected areas.

Intelsat 4As over the Indian Ocean have the ability to switch two of their narrow beam channels to wide beam channels to cover the large number of countries they serve.

The Intelsat 4A has an overall height of 275 inches and a diameter of 93 inches; The spacecraft solar panels, covered with nearly 17,000 solar cells, provide primary power of 600 watts. Liftoff weight of the satellite, which has a design lifetime in orbit of seven years, was about 3,335 lbs. The launch vehicle for the Intelsat 4A was the Atlas-Centaur rocket.

The following is a summary of Intelsat 4A satellites that have been launched:

Satellite	Launch Date	Location
Intelsat 4A F1	9/25/1975	31° W
Intelsat 4A F2	1/29/1976	1° W
Intelsat 4A F4	5/26/1977	21.5° W
Intelsat 4A F5	9/29/1977	Launch failure
Intelsat 4A F3	1/6/1978	179° E
Intelsat 4A F6	3/31/1978	175° E

Intelsat 5 & 5A

Flight models five to nine are equipped with an L-band maritime payload for the INMARISAT program making it possible for the satellite to handle ship-to-shore communications.

Satellites 10 to 12 are in reality improved Intelsat 5As awaiting the next generation, Intelsat 6, in 1986.

Intelsat 5 is the most powerful civil telecommunications satellite ever built with twice the capacity of Intelsat 4A. Dual polarization techniques make it possible to reuse frequencies; spatial separation, controlling the size, shape and angle of incidence of signals, allows the same frequency to be used for different geographical locations. The utilization of 11/14 GHz is another new feature. Intelsat 5 is the first Intelsat satellite to be stabilized about three axes instead of by rotation.

The satellites are 21 feet high and 50 feet in length when the solar arrays are deployed. The breakdown of the Intelsat satellites are as follows:

Intelsat 5:

21 transponders 4/6 GHz
(28 for maritime version)
6 transponders 11/14 GHz
12,000 simultaneous telephone circuits
2 color TV channels

Intelsat 5A:

32 transponders 4/6 GHz
6 transponders 11/14 GHz
14,000 simultaneous telephone circuits
2 color TV channels

The following Intelsat 5 series satellites have been launched to date:

Satellite	Date Launched	Location
Intelsat 5 F2	12/6/1980	27.4° W
Intelsat 5 F1	5/23/1981	57° E
Intelsat 5 F3	12/16/1981	24.5° W
Intelsat 5 F4	3/4/1982	34.5° W
Intelsat 5 F5	9/28/1982	63° E
Intelsat 5 F6	5/19/1983	18.5° W
Intelsat 5 F7	10/19/1983	60° E
Intelsat 5 F8	3/5/1984	53° W
Intelsat 5 F9	6/9/1984	Launch failure
Intelsat 5A F10	3/22/1985	26.8° W
Intelsat 5A F11	6/30/1985	27.5° W
Intelsat 5A F12	9/28/1985	?

The Intelsat 5 and 5A satellites have a complex transponder plan due to frequency reuse. Flight models 5-9 have a maritime transponder package with seven channels from 1535 to 1542.5 MHz. Flight models carried the following extra frequency segment: 4192.5-4200 MHz, probably a TT&C frequency range.

The Intelsat 5, 5A, and 6 satellites make use of the 11/12 GHz region and downlink on 10.95-11.2 GHz (All ITU regions). Plans call for future development to expand these satellites' capabilities as follows:

10.95-11.2 GHz (All regions)
11.7-11.95 GHz (Region 2)
12.5-12.75 GHz (Regions 1 & 3)

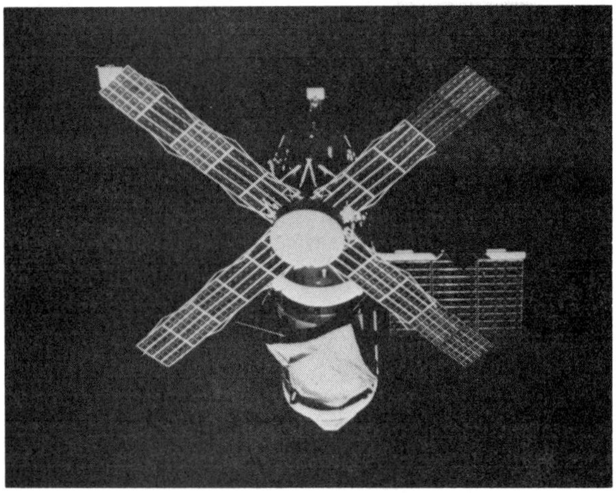

Intelsat V (Ford Aerospace)

INTELSAT V TRANSPONDERS

Global	WH	WZ	EH	EZ	Band Width	Center Freq (MHz)
---	1-2	1-2	1-2	1-2	77	3742.5
---	3-4	3-4	3-4	3-4	72	3825.0
---	5-6	5-6	5-6	5-6	72	3905.0
(7-8)or7-8*		7-8	7-8*	7-8	72	3995.0
(9) or 9		---	9	---	36	4055.0
10	---	---	---	---	36	4095.0
11	---	---	---	---	36	4135.0
12	---	---	---	---	41	4177.5

**On uplink transponders 7 and 8 are broken into 24 MHz segments each.*

14/11 GHz

WS	ES	Bandwidth (MHz)	Center Freq (MHz)
1-2	1-2	77	10992.5
5-6	5-6	72	11155.0
7-12	7-12	241	11577.5

TT&C beacons are as follows:

11.198/11.452 GHz Global beams
3947.5/3952.5 MHz Global beams
6175 MHz Command channel

The nominal bandwidth for the C-band portion of the Intelsat 5 satellites is 34 MHz; however, multiple bandwidths are employed. The chart below depicts the reuse of available frequencies used by the Intelsat 5 and 5A satellites. When a transponder with a bandwidth larger than 56 MHz is used, a multiple number is employed (e.g. 1-2). The same center frequency is used for each transponder in a set of four. Frequency reuse (4X) is accomplished by spatial and polarization isolation.

FREQUENCY BREAKDOWN

Freq Range	Tr No	Band-Width	Antenna Usage
3704-3781	1-2	77	W/E hemispheric W/E zonal beams
3789-3861	3-4	72	W/E hemispheric W/E zonal beams
3869-3941	5-6	72	W/E hemispheric W/E zonal beams
3949-4031	7-8	72	Global beams or W/E hemispheric W/E zonal beams
4037-4073	9	36	Global beams or W/E hemispheric
4077-4113	10	36	Global beams
4117-4153	11	36	Global beams
4157-4198	12	41	Global beams

14/11 GHz Range (All regions)

Freq Range	Tr No	Band-width	Antenna Usage
10954-11031	1-2	77	W/E spot beams
11119-11191	5-6	72	W/E spot beams
11457-11698	7-12	241	W/E spot beams

These satellites also use the standard Intelsat video channels.

HALF CHANNELS

Tr No	Half Chan	Center Freq	Tr No	Half Chan	Center Freq
1A	1	3715	7A	13	3965
1B	2	3734	7B	14	3984
2A	3	3755	8A	15	4005
2B	4	3774	8B	16	4024
3A	5	3795	9A	17	4045
3B	6	3814	9B	18	4054
4A	7	3835	10A	19	4085
4B	8	3854	10B	20	4104
5A	9	3875	11A	21	4125
5B	10	3894	12A	23	4165
6A	11	3915	12B	24	4184
6B	12	3934			

STW 1 and 2

China launched its first geostationary comsat on April 8, 1984; although officially designated PRC-15, this satellite is probably the ITU-listed STW 1 satellite located at 125° E. This C-band 6/4 GHz satellites has two transponders downlinks in the 4 GHz region with a total bandwidth of 36 MHz. With a 50 dBW EIRP at edge signal this satellite could perform a DBS function for the Chinese; designated STW 2, it will be located at 70° E and contain one 36 MHz-wide transponder somewhere around 4 GHz. Design lifetime for both satellites is three years.

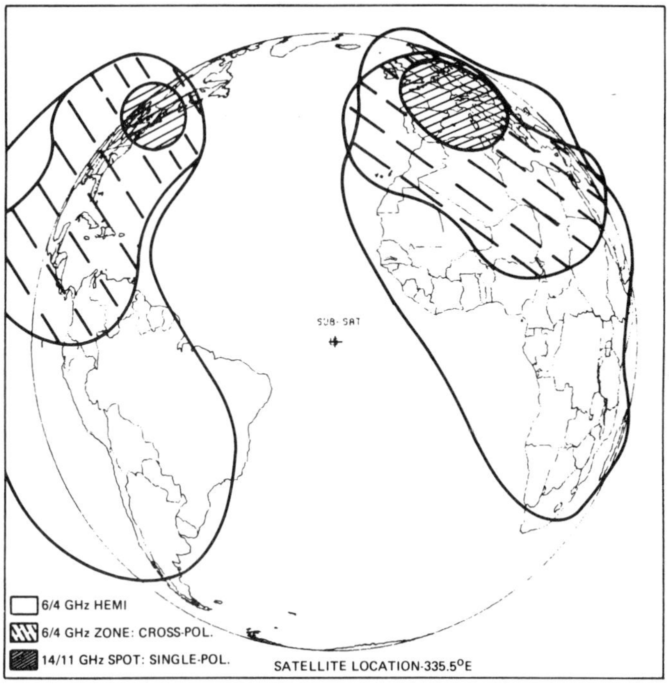

INTELSAT 5 ATLANTIC SATELLITE TRANSMIT CAPABILITIES
NOTE: The 14/11 GHz spot beams are steerable and may be moved to meet traffic requirements as they develop.

USSR Communications Satellites

For several years now, the Soviets have been quietly building a giant geostationary communications system consisting of many different satellites that serve several missions.

The Russian civilian and military communications satellites, like the U.S. comsats, are moving to higher frequencies. It is difficult to distinguish between military and civilian systems launched by the Russians because of the interoperability of their systems and the uncertainties associated with attempting to tie down a particular mission of a particular satellite in an ever-changing operational approach by the Soviets.

Some of the veil of Soviet secrecy has been lifted now that all nations file geostationary positions and frequencies with the International Frequency Registration Board of the International Telecommunications Union (ITU).

The Soviets named their global geostationary satellite positions, "Statsionar" (Russian for "stationary"). The Soviets have filed with the ITU the following Statsionar positions:

35° E	Statsionar 2
40° E	Statsionar 12
45° E	Statsionar 9
53° E	Statsionar 5
80° E	Statsionar 1/13
85° E	Statsionar 3
90° E	Statsionar 6
95° E	Statsionar 14
99° E	Statsionar T/T2
130° E	Statsionar 15
140° E	Statsionar 7
170° W	Statsionar 10
25° W	Statsionar 8
14° W	Statsionar 4
8.5° W	Statsionar 11

Each of these Statsionar positions can contain a variety of satellites and communications platforms riding on host spacecraft.

Gorizont

The first Gorizont ("Horizon") satellite to be launched at the end of 1978 posed something of a problem for Western analysts. The announced orbit, although nearly geosynchronous with a period of 1,420 minutes, was not geostationary due to an inclination of the orbit to the equator of 11.3°. Nor was the orbit, ranging from between 48,365 and 22,581 km, anywhere near circular. Its purpose was to "further improve and develop communication systems with the aid of artificial Earth satellites" and it was reported that it was "planned to use Gorizont satellites for relaying Soviet TV transmissions of the 22nd Summer Olympics in 1980."

There was some speculation that such an orbit was designed to give increased Northern Hemisphere coverage; however, subsequent Gorizonts were placed in the conventional near-geostationary orbits and it must be assumed that this unique orbit was the result of some attitude or propulsion error. These spacecraft have been observed both at 14° W (Statsionar-4) and 53° E (near to Statsionar-5). The R.A.E. table lists an object, 1978-118D, and suggests that this was in the 47° intermediate orbit, but no such fragment is listed in the NASA Satellite Situation Report as having decayed or still remaining in orbit.

Models of the Gorizont satellites were displayed at Le Bourget in 1979 and 1981. The cylindrical body and annular thermoregulator are similar to those of Ekran. The arrangement of the solar cell arrays differs in configuration from that of Ekran and amounts to some 2.3 square meters with further cells located around the circumference of the upper part of the spacecraft. Three-axis body stabilization is employed and attitude is determined from readings of three infrared and one visible light horizon sensors.

In addition to global-coverage horn antennas there are spot-beam receiving and transmitting antennas. The large circular dish, slightly more than 1 meter in diameter, supplies the 5 GHz downlink, as does the larger of the three "orange peel" shaped antennas. The two smaller antennas operate in the 6 GHz band used for uplink. These "orange-peel" antennas produce a beam that is wider in one direction than in the orthogonal and are adjusted prior to launch to change the coverage area from one satellite to another.

Gorizont satellites are equipped with a 40W repeater and five 15W repeaters. Channels carried on each are 6,7,8,9,10 and 11, all capable of simultaneous operation. The Gorizont satellites have revealed some flexibility in beam switching, the Atlantic satellite apparently having the capability to switch at least some of its transponders to European spot-beam pattern, in addition to global and northern hemispheric beams. In the spot-beam mode, an EIRP as high as 36 dBW (beam edge) has been estimated, some 7 dB above the global beam power. Again, these values would be achieved with a transponder RF output power in the region of 15° W.

The Gorizont satellites' primary function would appear to be the provision of international TV circuits for Intersputnik. The Atlantic satellite stationed at Statsionar-4 (14° W approximately) has carried up to five independent TV feeds simultaneously. Gorizont 2 began operation at this location in time to cover the Non-Aligned Nations Summit in Havana, Cuba, in September 1979. Initially, all six transponders operated, channel 8 being reserved for FDM/FM or TDMA telephony traffic, while channels 6,7,9,10 and 11 were available for TV. Feeds have been seen for a number of Intersputnik members, including Czechoslovakia, Poland, Hungary, and the German Democratic Republic as well as the U.S.S.R.

By the beginning of 1980, only the channel 8 and channel 10 transponders remained operational, but this situation was remedied in July of that year, when Gorizont 4 took over at the Statsionar-4 location with six channels, five of which were to provide Intersputnik and European TV coverage of the Moscow Olympic Games. Channel 8 remained dedicated to telephony.

Various transmitting antenna beams were in use over this period, but by October 1980 utilization had fallen to three channels--6,8 and 10, with TV on 6 and 10, channel 6 being selected for the European spot beam, as with Gorizont 2. In this mode, an energy-dispersal (ED) triangular waveform of 2 Hz, with a peak-to-peak deviation of 5 kHz, was added to the TV signal (nominal peak deviation 15 MHz). Otherwise, no ED is used in the Soviet systems. In early 1981, the satellite resumed six-channel operation, with channel 6 carrying Moscow Central TV's "channel 1" programs, channels 7 and 11 the "II Orbita" feed and channel 10 being reserved for Intersputnik.

Gorizont 3 in the Indian Ocean Statsionar-5 location has a similar TV usage of channels 6 and 10, the "Orbita II Vostok" variant of Moscow TV, with FDM/FM telephony additionally in channel 11.

All TV is 625 line, 25 frames/second with SECAM color. Even when taking a PAL-coded Eurovision source, the color is converted to SECAM before the feed is uplinked. The PWM "sound-in-vision" technique does not appear to have a place in Intersputnik exchanges. Standard audio subcarrier frequency is 7.5 MHz, though 8.2 MHz is also used. During Moscow Olympics coverage, further subcarriers were added for additional language feeds, down to a frequency of 5.5 MHz.

Gorizont satellites are currently occupying the following positions in the Statsionar system:

Statsionar 4	Gorizont 7
Statsionar 5	Gorizont 9
Statsionar 6	Gorizont 8
Statsionar 7	Gorizont 6

Gorizont signals can be received on U.S. TVRO setups since six Soviet transponders fall in the range of U.S. receivers:

1)	3658-3692	3675 MHz (6)
2)	3708-3742	3725 MHz (7)
3)	3758-3792	3775 MHz (8)
4)	3808-3842	3825 MHz (9)
5)	3858-3892	3875 MHz (10)
6)	3908-3942	3925 MHz (11)

The breakdown of programming on each Gorizont 7 channel is as follows:

1) 3675(6)	Soviet TV's first program
2) 3725(7)	Occasional video
3) 3775(8)	Single channel per carrier chan.
4) 3825(9)	Soviet TV's second program
5) 3875(10)	Soviet first program feeds
6) 3925(11)	Occasional video

When not transmitting video, active channels will display the Soviet 0167 test card.

Raduga

Late in December 1975, the Soviets launched "a new communications satellite" which they named Raduga ("Rainbow"). It carried equipment "to ensure uninterrupted round-the-clock telephone and telegraph radio communication in the centimetric waveband and simultaneous transmission of color and black-and-white Central Television programs to the network of Orbita stations."

In addition, the satellite carried "a three axis system of fine orientation towards the earth, a power supply system with sun-seeking guidance and tracking of solar cell batteries, an orbital correction system, a thermoregulation system, and a radio system for precision measurement of orbital parameters and satellite control"

No model of a Raduga has been publicly displayed nor pictures published. The Royal Aircraft Establishment's *Table of Artificial Earth Satellites* does not list any probable mass, dimension or configuration. An estimate of mass between 2000 and 4000 kg for a cylindrical body, 5m in length and 2m diameter, has been mentioned. The same report gives two solar panels and two antennas and specifies one TV channel and 10 telephony/data channels, each of which is capable of carrying 100 multiplexed telephone circuits.

Raduga satellites appear to have a primary mission of military communications using the GALS system to be described later.

The Statsionar-2 satellite carries the "Orbita-IV Vostok" version of Moscow TV in channel 10. The same type of audio pulses are transmitted, but in this case it is the 7.5 MHz subcarrier which is modulated with TV program audio, the pulse-width modulation (PWM) bearing an unrelated feed. Channel 6 is apparently occupied by a lightly loaded FDM/FM carrier, while the remaining channels transmit multiple carrier SCPC-type signals. The channel occupancy of the Statsionar-1/Statsionar-3 Raduga is not known.

Current Statsionar positions for Raduga satellites are:

Statsionar 2 (35° E)
Statsionar 9 (45° E)
Statsionar 3 (85° E)
Statsionar 8 (25° W)

It would appear that some time in the future, Statsionar 10 (170° W) will have a Raduga satellite occupying that slot.

The center frequencies are:

Transponder	Center Frequency	Transponder	Center Frequency
1	3425	7	3725
2	3475	8	3775
3	3525	9	3825
4	3575	10	3875
5	3625	11	3925
6	3675		

Ekran

The first Ekran ("Screen") satellite was described a "for further development of telecasting systems." The relay equipment ensured telecasting of Central Television's color and black-and-white programs to the "network of collectives using receiving devices situated in the inhabited localities of Siberia and the extreme North."

The large 96-element, cophased, helical antenna array, measuring 5.7 x 2.1m, was deployed as 4-by-6 matrices on four panels resembling a bedspring mattress. This provides a gain of 28 dB on the axis of the radiation pattern.

The familiar cylindrical body has been retained with the vernier correction engine but this was surrounded by an annular panel forming part of the thermoregulation system. Four solar cell panels were mounted in pairs at the end of booms extending at right angles to the axis of the satellite, each consisting of sub-assemblies arranged in 6-by-3 matrices.

Ekran satellites have, with one exception, been given the Statsionar T designation. The seventh Ekran, in 1981, was designated Statsionar 1 in the Tass announcement but this might have been a typographical error.

Two Ekran satellites are used at the Statsionar T location. Their successive operation ensures transmission of one color TV program and two radio broadcast program for 12 to 16 hours within a 24-hour period.

The Ekran satellites have operational lifetimes of some two or three years but future Ekrans are expected to have operational lives of up to five years.

A key feature of the television broadcasting system is the comparatively high power of the Ekran satellite's transmitter: about 200W as opposed to the 30-40W of the other communications satellites. This has permitted simpler design of the ground antenna and receiving equipment. Yagi arrays have replaced the large dish antennas required for the Orbita system. The receiver does not demand constant servicing, nor does the associated local relay station which distributes the received television program to the surrounding area using the normal television channels which can be received by standard television sets such as the "Rubin" or "Rekord."

The ground equipment of the Ekran system needs so little attention that it can even be set up in isolated villages in the taiga where a low-power relay of only 1W is all that is necessary. The Ekran receiving network had nearly 2,000 stations with its own relay equipment by the end of 1981 which brought territories inhabited by 20 million people into the U.S.S.R.'s main television network.

The 714 MHz downlink chosen for the Ekran system made possible the use of the simple and inexpensive receivers and Yagi antennas. Two types of collective-use receiving stations have been developed which operate reliably under conditions of Siberia and the extreme North. The first class, or professional, types are equipped with 32 or 16 element antennas giving a signal-to-noise ratio of 55 dB which is relayed to local TV centers and high power TV repeaters. These receiving stations, in combination with the repeater, have been given the identifier STV-100.

The second class, or set, types feed low power TV repeaters or cable distribution networks. They have four element antennas with a signal-to-noise ratio of not less than 48 dB and have the identifier STV-1.

The series production of ground stations, antennas and repeaters is carried out by enterprises of the Ministry of the Communications Equipment Industry. Receiving stations are installed and commissioned by organizations of the U.S.S.R. Ministry of Communications.

Transmission of television and radio broadcast programs to the Ekran satellites is conducted from a special ground transmitting complex with a transmitter output of 10 kW through a 12m diameter uplink antenna with a gain of 54 dB at 6 GHz.

The Ekran satellites at Statsionar-T (99° E)

cannot be observed from Florida, but the TV transmissions have been resolved experimentally in Malawi and South Africa, some 13 degrees off beam, a direction in which the EIRP is estimated as 30 dBW. It is understood that the Peking (China) authorities have made a formal complaint to Moscow over the high level of 714 MHz signals reaching China, and their potential for interference to other services.

The programs transmitted are a version of Orbita TV, between 1200 and 1800 Moscow time. Audio is believed to be via a 6.5 MHz subcarrier for compatibility with the intercarrier audio of Soviet domestic TV receivers. The signals, though of adequate strength for direct home reception within the target area, are also relayed by attended and unattended rebroadcast stations, in terrestrial AM TV format.

The Moskva System

The Ekran concept of a powerful transmitter and simple receiver is not always acceptable due to over-crowding of the frequency bands and the close positioning of satellites in the geosynchronous orbit. Restrictions caused by possible interference to the ground communications facilities of neighboring countries using the same frequency bands make it impossible to extend the zone of operation of the Ekran system to other regions of the U.S.S.R.

Progress in technology permitted the development of the Moskva system which, in conjunction with a network of microwave links and land lines, enables both first and second programs of Central Television to be relayed to a vast area.

Gorizont satellites operating with a 4 GHz downlink through a 40W transponder and narrow-beam antennas transmit to stations equipped with 2.5m dish antennas. One such antenna, receiving signals from a Gorizont satellite, was displayed in Vienna during Unispace 1982.

The receiver rack has a weight of 120kg and inputs 13.5W to the associated cable network. An uncooled parametric amplifier with noise temperature less than 100°K, mounted on the antenna, and a receiver are used to demodulate and amplify the television and radio broadcast signals.

The complete receiving unit is small; the components can all be located, for instance, at a post office, in a village club or at low power TV repeater stations, and the fixed antenna can be mounted on the roof or in the ground using simple supports.

Central Television and All-Union Radio

The Moscow-based services are supplied to five broadcasting zones, each of which comprises two or three time zones. Zone 1, comprising Moscow time plus eight, nine or ten hours, covers the most eastern part of the Soviet Union. This receives the first channel TV through the Orbita network and a Molniya 3 satellite.

It is intended that the second channel will become available before the end of the current five-year period through a new network to be called Double, using Gorizont-Statsionar 7. Molniya 3 will then be used exclusively for the far northern territories.

Zone 2, for Moscow time plus six or seven hours, receives first channel through Orbita, Ekran or Molniya 3. Here as well it is intended to add the second channel before the end of the current five-year plan period through the new Double network using Gorizont-Statsionar 6. As with Zone 1, the Moskva system will be used for channel 1 and the Orbita system for channel 2.

Zone 3, comprising time zones for Moscow time plus four or five hours, covers the central parts of the U.S.S.R. including Novosibirsk and Irkutsk. These receive channel 1 through the Ekran system and channel 2 through Orbita or Ekran.

Zone 4, for Moscow time plus two or three hours, receives first channel through Moskva and second channel through Orbita. The fifth zone, European U.S.S.R., on Moscow time plus one hour, has the same service as Zone 4 but from the Statsionar 4 rather than the Statsionar 5 location. The first radio program and Mayak (Beacon) program are available to all zones through the Orbita network.

Loutch

Probably designed to compete with the INTELSAT 5 and ECS satellites, these are presumably communications packages being flown on a host spacecraft.

As proposed, the Loutch packages have ten transponders in the international fixed-satellite service subbands: 10.95- 11.2 GHz and 11.45-11.7 GHz (also used by Intelsat & ECS). Each transponder has a nominal bandwidth of 34 MHz like the Gorizont C-band satellites, and 50 MHz separated center frequencies.

The following is a breakdown of the Loutch transponders:

Chan	**Uplink**(MHz)	**Downlink**(MHz)
1	14025	10975
2	14075	11025
3	14125	11075
4	14175	11125
5	14225	11175
6	14275	11475
7	14325	11525
8	14375	11575
9	14425	11625
10	14475	11675

Eight orbital positions were proposed in two groups for the Loutch system:

Loutch 1	14° W	Statsionar 4
Loutch 2	53° E	Statsionar 5
Loutch 3	90° E	Statsionar 6

Loutch 4	140° E	Statsionar 7
Loutch P1	25° W	Statsionar 8
Loutch P2	45° E	Statsionar 9
Loutch P3	85° E	Statsionar 3
Loutch P4	170° W	Statsionar 10

The significance of the "P" in the lower set of Loutches is now known: The upper set of Loutches correspond to Gorizont Statsionar positions and the lower set correspond to Raduga positions in the Statsionar system.

Similarly, we may determine that Statsionar 10 will have a Raduga satellite at that location through the use of Loutch and other special communications packages.

Volna

Volna is also a special communications package, much the way the Inmarisat package is part of the Intelsat 5 F5-9 satellites. Volna (Russian for "wave") is an eight system communications package that will use host satellites for maritime and aeronautical use. Some of the transponders filed with the ITU are for the 225-400 MHz UHF band, probably serving the same purpose as our Fleetsatcom Marisat/Leasat satellite constellations.

A breakdown of the Volna system is as follows:

Volna 2/4/6/8

Maritime service
Uplink 1636-1644 MHz
Downlink 1535-1542 MHz

Aeronautical service
Uplink 1645-1660 MHz
Downlink 1543-1558 MHz

Volna 1/3/5/7

--as above plus UHF transponders:

Uplink 335-399 MHz
Downlink 240-328 MHz

The Russians have filed the following orbital locations with the ITU for their Volna satellite constellation system: (*indicates a UHF transponder location)

Volna 1*	Statsionar 8	25° W
Volna 2	Statsionar 4	14° W
Volna 3*	Statsionar 9	45° E
Volna 4	Statsionar 5	53° E
Volna 5*	Statsionar 3	85° E
Volna 6	Statsionar 7	140° E
Volna 7*	Statsionar 10	170° W
Volna 8	Statsionar 6	90° E

The special UHF transponder Volnas will probably use the Radugas as their host satellites and the rest of the system will probably use the Gorizont satellites.

Gals

Russian for "tack," this system will consist of four communications packages that appear to be takeoffs on the U.S. military DSCS system. They will use the Raduga satellites as their host vehicle.

Strictly 8/7 GHz military communication packages, Gals will use:

Uplink 7.9-8.4 GHz; Downlink 7.25-7.75 GHz

Orbital locations for the Gals system is as follows:

Gals 1	Statsionar 8	25° W
Gals 2	Statsionar 9	45° E
Gals 3	Statsionar 3	85° E
Gals 4	Statsionar 10	170° W

Potok

According to the ITU they have a down link in the 4 GHz band and will be located at 14.6° W, 168° E, and 80° E. These do not correspond to any known Statsionar positions and therefore will probably be on their own satellite platforms.

SDRN
(Soviet "TDRSS-ski")

The Soviets have filed documents with the International Frequency Registration Board in Geneva, Switzerland, that give details of the SDRN geostationary satellite system.

This system, similar to the U.S. TDRSS system, will be used by the Soviets for communicating with their manned space stations and other spacecraft operating in low earth orbit.

Uplink from spacecraft to SDRN is 15.05 GHz while downlink to spacecraft from SDRN is 13.52 GHz; Downlinks from SDRN to ground stations at Moscow and Khabarovsk will be at 10.82, 11,32 and 13.7 GHz. Up link from ground stations to SDRN will be at 14.62 GHz.

Geostationary locations from SDRN are as follows: WSDRN-16° W, CSDRN-95° E, and ESDRN-160° W.

It may well be that the experimental SHF Cosmos 1366 was a development flight related to this proposed system and that the mysterious Cosmos 1426 and Cosmos 1516 were operating with Cosmos 1366 in the SDRN mode.

Molniya & Orbita Systems

The first Molniya 1 ("Lighting") satellite was launched on April 23, 1965, in a flight nearly parallel to that performed by Cosmos 41 the previous year. Molniya 1 was described as having the main task of relaying television programs, long distance bilateral, multichannel telephone, radiophoto, and telegraph communications.

Two days after the launch, the first television

broadcasts were relayed from Moscow to Vladivostok through Molniya 1. On the 27th of April, a return program was carried from Vladivostok to Moscow for further distribution by land line to all Soviet bloc country members of the Intervision system.

Molniya 2 and 3 series satellites appeared as time went by and, at one time, all three types were operational simultaneously. The Molniya 2 satellites were phased out and currently eight Molniya 1 and four Molniya 3 satellites are operational at any one time. Molniya provided the first routine color relays and forms the basis of a Washington-Moscow "hot-line."

From time to time scientific instruments onboard Molniya-1's transmitted TV cloud-cover pictures to earth. Two simultaneously operating Molniya 1 satellites were used in January and July 1971 during magnetic storms to investigate the dynamics of relativistic electrons in the outer belt inside and outside the plasmasphere. A Molniya-2 satellite gathered data on protons and alpha-particles on October 25, 1975.

Molniya 1

The Molniya 1 is a complex craft some what similar to the early lunar and planetary craft. The main body is a pressurized cylinder with conical ends and external cooling/heating coils, related fo the temperature regulating system, which are wrapped around the main cylinder. Small correcting rockets maintain the attitude of the spacecraft in the required position.

Atop the main cylindrical body are the special correction motor system and sun- seeking optical sensors. This subsystem is conical in shape and is surrounded by a ring of gas bottles containing fuel for the correction system.

At the opposite end of the main body are earth-seeking optical sensors. Extending from either side of the main body are the structures for two high-gain antenna systems -- steerable parabolic dishes used for the main communications tasks. A Molniya 1 displayed in the State Tsiolkovskiy Museum of the History of Kosmonautics at Kaluga had a four-helix array surround ing the horn. At the bottom of the main body, six panels radiate outward like petals on a daisy. These are covered with 500-700 watt solar cells to power the craft.

Inside the craft are the main communications receivers and transmitters, the buffer batteries, various sensors and telemetry systems, an onboard computer, and other necessary equipment for house keeping and control.

According to a Soviet account, the Molniya 1 sun sensor locks onto the sun to maximize the power for the solar cell system, and there is a gyrostabilizer to maintain this attitude. The earth seekers then lock onto the earth to point the parabolic antennas to maximize signal and strength. The directional parabolic antennas have a gain of approximately 18 dB. The two antennas are not used simultaneously; the second antenna is supplied for redundancy to extend useful life.

The communications equipment of the satellite includes three complete transceiver systems--one active and two as standby to extend operational life. Most of the equipment is solid state except for metal-ceramic triodes, klystrons, magnetrons, and traveling wave tubes. The useful life of the tubes is 40-50 thousand hours.

Although in the first models the input noise level was 2,000-3,000°K, tunnel diodes were soon to reduce this by a factor of 2 to 3. There are four traveling wave tubes used, three active and one spare. Television is transmitted at a power level of 40 watts, and data and telephony at 20 watts.

The capability of the payload includes a complete television channel with additional capability for television audio, multichannel telephony, VHF telegraphy (by multiplexing some of the telephone channels), and photofacsimile. A transponder RF output of 40W gives global beam coverage via a parabolic antenna.

Since the transfer of TV traffic to the higher frequency bands of Molniya 2 and subsequent Molniya 3 series, the Molniya 1 downlinks have carried other forms of modulation. Observations at Sheffield, England, since 1976 have revealed the presence of signals in the 974 to 998 MHz range, suggesting a transponder bandwidth on the order of 30 MHz, adequate for FM television, and extending perhaps from 970 to 1,000 MHz. Polarization is right-hand circular.

Channel Utilization

Current usage of the UHF channel is divided between two independent sets of Molniya-1 satellites. The four active spacecraft falling into the standard orbital plane groupings A through D carry a series of FM carriers, some of which are frequency-division multiplexed (FDM) between approximately 974 and 985 MHz (carriers are grouped around 980 MHz), for three hours each side of apogee on the Asian loop of the orbit over the Soviet Union. They are also used in their North American loop over Canada.

The second set of four satellites, those with orbital planes intermediate between the four standard groups, appears to operate only on the Asian loop and carries an estimated 12.5 kilobits-per-second, digital, frequency-shift-keyed (FSK) signal of 140 kHz deviation close to 998 MHz. No detailed analysis of modulation parameters or information content of these signals has been attempted.

The Molniya 1 satellites are dual-fed at changeover, the "incoming" and "out going" spacecraft simultaneously carrying the same traffic for a period of up to five minutes.

Since 1964 these Class 1 satellites have included Cosmos 41, Cosmos 1423, and Molniyas 1-70.

Molniya 2

On November 24, 1971, the U.S.S.R. launched its first Molniya 2 communications satellite. The same launch vehicle was used for the Molniya 1 launch so it may be presumed that there was no great increase in the size and weight of the satellite and that the main improvement lay in electronics. Solar panels were increased, adding about 50 percent to the power. With the change to the higher 4 and 6 gigahertz (GHz) frequencies, the earlier umbrella-like antennas with their clusters of three horns were replaced by arrays of five hours each.

The Class 2 Molniya satellites include Cosmos 837, Cosmos 853, and Molniyas M 2-1 to M 2-17.

Molniya 3

The first Molniya 3 was launched on November 2, 1974. To date, no model of the spacecraft has been put on public display. The principal difference between Molniya 3 and Molniya 2 satellites seems to be its color television relay and higher communications frequencies. Earlier Molniyas broadcast mainly black and white television with numerous color programs.

Three channels have been observed, corresponding to numbers 6, 8 and 10. Only two of the three appear to be occupied at any one time. Received signal levels are consistent with an RF power of some 15W to the global beam horn antenna, giving a beam edge EIRP in the region of 29 dBW. The constancy of signal strength with change of up and down path lengths during the active loop of the highly elliptical orbit suggests a closed-loop control of EIRP involving the uplink terminal.

The North American loop of the orbit is the one employed for internal TV distribution from Moscow, on channel 10, to the Orbita rebroadcast stations serving the far northern and eastern regions of the U.S.S.R. National TV programs are repeated at intervals appropriate to feed the range of time zones of the Soviet Union, though Moscow time (GMT+3 hours) is quoted throughout. Between the program transmissions an electronically generated test pattern is radiated, keeping the channel occupied with television signals almost 24 hours per day.

Program audio is carried by a two-channel time-division multiplex method employing two analog width-modulated pulses transmitted in the horizontal blanking interval of the modified TV waveform. A separate audio channel utilizes a 7.5 MHz subcarrier added to the baseband video signal. The program content of this channel is unrelated to the TV picture.

Also on the Molniya 3 North American loop, channel 8 is active in a single channel per carrier (SCPC) mode. Among the carriers in this transponder are believed to be those of the Washington, DC, to Moscow Direct Communication Link (DCL or "Hot-Line") with terminals at Fort Dietrick, MD, and Vladimir, 180 km east of Moscow (Details in next article).

Observations around change-over on this loop reveal that the uplink is not dual fed; the incoming satellite is activated approximately one minute before hand-over, at which time transponder noise can be seen in downlink channels 8 and 10. Acquisition of the incoming satellite is accomplished in less than one second, immediately after which the transponders on the outgoing satellite are deactivated.

On Molniya 3's Asian loop, the satellites carry multiple carrier FDM/FM and FM/SCPC traffic in the channel 6 and 8 transponders. Channel 10 TV signals have occasionally been seen on the Asian loop, transmitting test waveforms only. It is probable that, during the period of interest, the Raduga geostationary satellites have taken over TV distribution service to Orbita stations in those regions south of the limit of full-time visibility from the active portion of the North American loop. In this sense, the Molniya 3 and Raduga satellite systems are seen to be complementary.

Class 3 satellites include Cosmos 1175, Cosmos 1305, and Molniyas M 3-1 to M 3-30.

Washington-Moscow Hot Line

The United States and U.S.S.R. on September 30, 1971, as part of the initial Strategic Arms Limitation Talks (SALT) negotiations, signed accords on the prevention of accidental warfare which authorized the establishment of a new Washington-Moscow "hot-line" via satellite. The main objective of a satellite link would be increased reliability.

Although the terrestrial hotline, which was established as a result of the Cuban missile crisis, has never been a target of planned sabotage, it has been subject to occasional disruption with section of the cable blacked out by fire, pilfered, and once plowed by a Finnish farmer!

The hot-line system consists of two duplex telephone circuits equipped for secondary telegraphic multiplexing; one circuit is on the Molniya system and the other is on the Intelsat system. Encoded teletype messages go from the United States to Moscow in English via the Intelsat system and from Moscow to the United States in Russian via Molniya 3 satellites.

Originally, the Russians intended to use a station in the suburbs of Moscow for Intersat and a Molniya station at Vladimir. However, because of severe winter weather conditions in the Soviet Union, the Russians have constructed a second Intersat station approximately 50 miles from L'vov to ensure increased dependability. The United States has a Molniya station at Fort Dietrick in Frederick, MD, and an Intersat station at Etam, WV. The system began operation in the second half of 1976.

On January 16, 1978, the "hot-line" began using two independent satellite systems to transmit messages between Washington and Moscow. It was expected to provide a communications service of higher reliability than the cable link or the discontinued high-frequency radio system. The DCL would be less vulnerable than those systems since it eliminated dependence on third-country facilities. In addition it was not susceptible to the interruptions caused by atmospheric interference problems common to HF radio systems.

Circuits are tested hourly using a variety of sample messages such as nonpolitical passages from magazine articles and books. All messages are automatically encoded upon transmission and decoded upon receipt. Both United States and Soviet systems operate simultaneously so that if one system fails, the other continues to provide communications.

One notable use of the "hot-line" occurred during the Arab-Israeli Six Day War in 1967 when President Johnson advised the Soviet Union of U.S. ship and aircraft movements in the Mediterranean following an Israeli attack on the U.S.S. Liberty.

Future International Comsats

AMS 1 & 2

Israel plans to launch two hybrid satellites beginning in 1986; both satellites would be located at 15° E with uplinks/downlinks at:

5.925-6.425 GHz/ 3.7-4.2 GHz
14.0-14.5 GHz/ 11.7-12.2 GHz

ARINC AvStar Satellites

ARINC use of HF air-to-ground radio could be sharply reduced in the near future. Their request that the FCC approve a global aviation communication satellite system could result in deployment by late 1991 and completion by 1994 with a total of six satellites in geostationary orbit.

When the AvStars are all deployed, ARINC plans to sell or lease each satellite's transponders to Aviation Satellite Corporation. Each AvStar satellite will radiate two wide-area (global) beams and seven spot beams to serve high-density air routes, offering voice and data communications capacity equivalent to 400 voice channels.

ARINC interest in deploying an AvStar system was sparked by United Airlines, following its acquisition of Pan American World Airway's Pacific routes; however, Northwest Airlines has become the first carrier to subscribe to the service.

Motivation for an earlier-than-expected deployment of these satellites was due to the action of the USSR in shooting down Korean Airlines Flight 007 in the fall of 1983.

The orbiting relays will use 14 MHz of the L-band (1.5-1.6 GHz) reserved for aviation satellite services. ARINC's only other competitor, Inmarisat, has reserved the remaining 14 MHz of the 28 MHz-wide band for aeronautical satellite use.

The proposed system will provide air traffic and related services for the flight crew and will also provide automatic surveillance of aircraft, on-board navigation systems to periodically transmit aircraft position reports to ground stations.

Additionally, spacecraft would provide voice and data communications for en-route passengers, eventually underwriting much of ARINC's cost in the AvStar system.

If the FCC approves the system, the six satellites, each weighing 2866 lbs (1300 kg) and lifted aloft by a McDonnell Douglas Delta booster, would employ overlapping geosynchronous orbits to assure that two spacecraft would be available to all aircraft flying at latitudes less than about 75 degrees. The proposed sequence of satellite deployment would be at longitudes 58W, 114W, 63E, 22W, 132E and 173E.

ISI 1 & 2

International Satellite, Inc. plans to launch two satellites sometime in the future. These satellites would be located at 56° and 58° W. Uplinks would be at 14.0-14.5 GHz and downlinks would be at 10.7-10.95/ 11.7-12.2/ 12.5-12.75 GHz.

Orion 1/2/3

Beginning in 1986 Orion Satellite Corp. plans to launch three Ku-band satellites to be located at 37.5°/47°/50° W, respectively. Uplink for the three satellites is 14.0-14.5 GHz and downlink is 10.45-10.7 GHz and 10.95-11.2 GHz. Each satellite will contain twenty-two 36 MHz transponders. The design lifetime for these satellites is 7.5 years.

Pacific Area Satellite Systems A & B

These Japanese hybrid satellites will be located at 170° E. Expected for launch beginning in 1987, these satellites will have 5.925-6.425 GHz uplink and 3.7-4.2 GHz downlink for C-band, and 14.0-14.5 GHz uplink and 11.7-12.2 GHz downlink for the Ku-band. Design lifetime is seven years.

Videosat 1 & 2

Not much data is currently available about these French satellites which will occupy the 32° E and 37.5° W positions. Uplinks will be at 14.0-14.5 GHz and downlinks will be at 12.5-12.75 GHz with a design lifetime of seven years.

Other Satellites

Several other future satellite systems include:

Satellite	Country	Band	Year	Location
F-Sat 1/2	France	S,C,E	86/87	7°E/11°W
Apex	ESA	S,C,E	--	10°E
Italsat	Italy	E	86	13°E
NatSat 1/2	Nigeria	C	--	14/16°E
Sircal 1A/1B	Italy	U,X,K,E	87	16/22°E
SABS	Saudi Arabia	K	--	17°E
DFS-1/2	Germany	S,K,E	87	23.5/28.5°E
Zohreh 1/2/3/4	Iran	K	--	34/26/47/41°E
Paksat 1/2	Pakistan	K	86	38/41°E
GDL-3/4/5/6	Luxembourg	X,K	--	10.5/20°W/1/19°E
Satcol 2	Colombia	C/1/3 GHz	--	75°W
Satcol 1A/1B	Colombia	C/1/3 GHz	--	75.4°W
STSC-1	Cuba	C	88	83°W

Chapter Six

DIRECT BROADCAST SATELLITES

One of the more exciting developments is the Direct Broadcast Satellite (DBS), used to transmit programming directly to home satellite dishes. TVRO enthusiasts who now require 10-foot and larger dishes to receive present C-band signals will need dishes only 2 to 3 feet in diameter to receive DBS programming.

These satellites are designed to provide programming to rural areas and small cities not served by cable systems. Scrambled TV signals ensure that only those authorized to receive the signals can view them.

U.S. DBS systems were discussed during the ITU RARC-83 telecommunication conference. At present I do not have the results of that conference, but I do have information on systems that have been proposed to the FCC before the conference. These systems will be covered in this chapter as well as the European, Asian, and Pacific DBS systems that have been formalized by the ITU WARC-77 conference. Of course as more details become available, they will be presented in "Signals from Space" in *Monitoring Times* and future editions of this book.

DBS Europe

In 1977, the World Administrative Radio Conference (WARC) divided up the 11.7-12.2 GHz band into 40 channels to be shared by all European countries for DBS services.

Each European country has been allocated up to five channels in either the upper or lower portion of the band for DBS; these channels can be used for either one TV service with several subcarriers for stereo, or up to 20 different audio services.

Four orbital positions, each containing several satellites, have been reserved for European DBS satellites. For optimum usage of available spectrum space, DBS satellites will use both left- and right-hand circular polarization.

Each channel is allocated to a group of countries which share the frequency by using different orbital positions and polarities. The following list of channel assignments is for western and southern Europe.

France, San Marino, Turkey: 1,5,9,13,17
Ireland, Germany(West): 2,6,10,14,18

EUROPEAN DBS BAND PLAN
(Frequencies listed are video centered.)

Channel	Freq (GHz)	Channel	Freq (GHz)
1	11.727	21	12.111
2	11.746	22	12.130
3	11.765	23	12.149
4	11.785	24	12.168
5	11.804	25	12.187
6	11.823	26	12.206
7	11.842	27	12.226
8	11.861	28	12.245
9	11.880	29	12.264
10	11.900	30	12.283
11	11.919	31	12.302
12	11.938	32	12.322
13	11.957	33	12.341
14	11.976	34	12.360
15	11.996	35	12.379
16	12.015	36	12.398
17	12.034	37	12.417
18	12.053	38	12.437
19	12.072	39	12.456
20	12.091	40	12.475

Greece, Lichtenstein, Luxembourg,
Portugal: 3,7,11
Andorra, Austria, United Kingdom:
4,8,12,16,20
Belgium, Cyprus, Monaco: 21,25,29,33,37
Switzerland: 22,26,30,34,38
Denmark: 12,16,20,24,36
Finland: 2,6,10,22,26
Iceland, 21,23,24,27,29,31,33,35,37,39
Sweden: 4,8,30,34,40
Norway: 14,18,28,32,38
Italy: 24,28,32,36,40
Netherlands, Spain, Vatican: 23,27,31,35,39

The European DBS channels are further divided by polarization and orbital locations as follows:

37° W (Satellite unknown)

Left-hand circular polarization:

Andorra 4,8,12,16,20

Right-hand circular polarization:

San Marino	1,5,9,13,17
Lichtenstein	3,7,11,15,19
Monaco	21,25,29,33,37
Vatican	23,27,31,35,39

31° W Service by the UNISAT 1/2/3 sats.

Left-hand circular polarization:

Portugal	3,7,11,15,19
Iceland	21,25,29,33,37
Spain	23,27,31,35,39

Right-hand circular polarization:

Ireland	2,6,10,14,18
United Kingdom	4,8,12,16,20

19° W Several satellites are proposed for this slot including TV-SAT A3, Olympus 1, Lux-sat, TDF 1A/1B.

Left-hand circular polarization:

West Germany	2,6,10,14,18
Austria	4,8,12,16,20
Switzerland	22,26,30,35,38
Italy	24,28,32,36,40

Right-hand circular polarization:

France	1,5,9,13,17
Luxembourg	3,7,11,15,19
Belgium	21,25,29,33,37
Netherlands	23,27,31,35,39

5° E Two satellites appear to serve this slot: ESA's OTS and Sweden's Tele-X.

Left-hand circular polarization:

Finland	2,6,10
Norway	14,18
Sweden	4,8
Denmark	12,16,20
Nordic*	22,24,26,28,30,32, 36,40
Sweden	34
Norway	38

Right-hand circular polarization:

Turkey	1,5,9,13,17
Greece	3,7,11,15,19
Cyprus	21,25,29,33,37
Iceland**	23,27,31,35,39

Notes:

*Eight channels in a wide beam covering the Nordic countries: these are assigned to Finland (22,26); Sweden (30,40); Denmark (24,36); and Norway (28,32).

**Beam covers Iceland,the Azores, part of Greenland. Channels 27,35 are registered under Denmark.

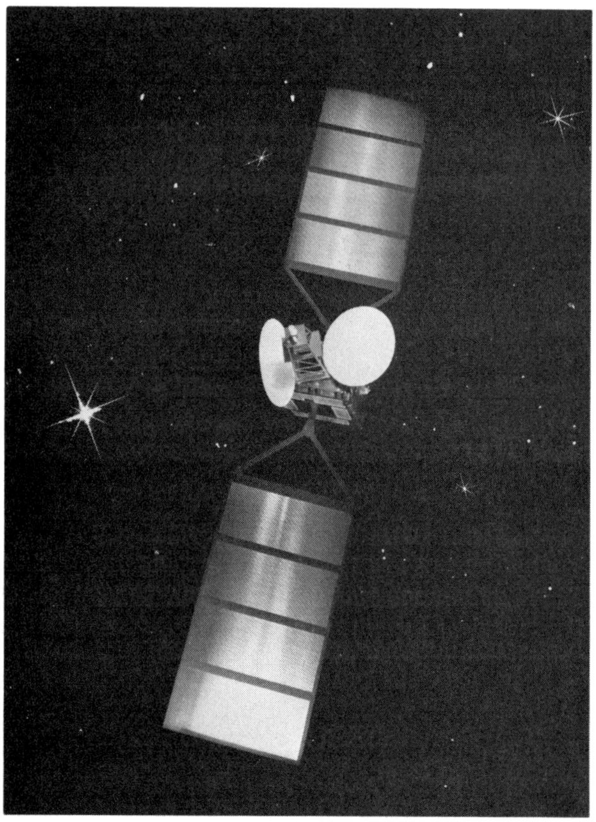

TDF-1 DBS satellite (Aerospatiale)

TDF 1A/2B and TV-SAT A3

This French/German DBS project will be co-located at 19° W. Each satellite will carry three channels of TV with two channels reserved for video and one channel reserved for audio service with a seven year lifetime. These satellites will be launched by ESA's Ariane rocket beginning in the 1985-1987 timeframe.

OLYMPUS 1 (L-Sat)

This ESA operated DBS satellite is a multipurpose TV platform schedules for launch in 1986 on an Ariane rocket. Olympus will carry five transponders for DBS operation and at least two of the channels will be for Italian DBS operation. This satellite will also carry a 30/20 GHz transponder package with a downlink around 3 GHz.

Olympus (L-Sat) DBS satellite (AEG-Telefunken)

TV DBS Satellite (AEG-Telefunken)

LUX-SAT

Radio-Tele-Luxembourg would like to transmit three channels of DBS programming on its satellite at 19° W. One channel would carry French language programming, another German, and the third channel would carry either English or Dutch programming. RTL's DBS would be a commercial venture supported by advertising.

UNISAT 1/2/3

United Satellite, Ltd. of the United Kingdom plans to launch three DBS spacecraft into orbit at the 31° W slot. These satellites will contain six transponders each and have a design lifetime of seven years.

Uplink 14.0-14.5 GHz
Downlink 10.95-11.2 GHz
11.7-12.2 GHz
12.5-12.75 GHz

TELE-X

Sweden, Norway and Finland's entry into the DBS market will be located at 5° E and transmit two channels of DBS video and two channels for data and occasional video transmissions. Tele-X will operate in the Ku-band.

USSR DBS

The Russians have been authorized five DBS orbital positions but have not filed their interest with the ITU for Ku-band at this time. Authorizations are 23° E, 44° E, 74° E, 110° E, and 145° E. The Ekran satellites located at 99° E are currently performing USSR DBS service on 702-726 MHz.

Pacific

There are two or three entries now in the Asian/Pacific area for DBS satellites. Japan, Australia, and possibly China all have systems proposed or that could be used for DBS broadcast.

BS 2A/2B

Japan launched its first DBS satel lites, BS 2A now located at 110° E on January 23, 1984. A second satellite is planned for launch sometime in 1985. Each satellite will carry two DBS transponders as follows:

Channel 1	11.91928 GHz Transmit power 100 watts
Channel 2	11.99600 GHz Transmit power 100 watts

These channels correspond to Europe's DBS channels 13 and 15. Japan is authorized channels 1,3,5,7,9,11,13,15 at the 110° E slot. These satellites also use the following TT&C frequencies: 11.70299

BS-2 DBS satellite (National Space Development Agency of Japan)

GHz at .1 watt and 2276.99 MHz at 1.3 watts.
Plans are to launch BS 3 in 1986.

U.S.

All U.S. DBS birds will use 17.3-17.8 GHz uplinks and 12.2-12.7 GHz downlinks.

Columbia Broadcasting System (CBS)

Satellites: Two in orbit, plus two spares
Channels: Three
Bandwidth: Two channels of 24 kHz needed for transmission of high definition TV.
Orbital Slots: 101° W & 148° W
Weight: 989 kg and 1,053 kg
Cost: $140 million for initial phase
Coverage: Four time zones
Services: CBS asked to be regulated as a hybrid, offering a range of services. It would provide for retransmission by local CBS-TV affiliates and cable operators, satellites-to-theater movie distribution, and home and business

AMERICAN DBS BAND PLAN
(Video)

Channel	Freq (GHz)	Channel	Freq (GHz)
1	12.22400	17	12.45728
2	12.23880	18	12.47186
3	12.25316	19	12.48644
4	12.26774	20	12.50102
5	12.28232	21	12.51560
6	12.29690	22	12.53018
7	12.31148	23	12.54476
8	12.32606	24	12.55934
9	12.34064	25	12.57392
10	12.35522	26	12.58850
11	12.36980	27	12.60308
12	12.38438	28	12.61766
13	12.39896	29	12.63224
14	12.41354	30	12.64682
15	12.42812	31	12.66140
16	12.44270	32	12.67598

subscription programming. All channels would deliver high definition television with stereophonic sound.
Design Lifteime: Seven years

Direct Broadcast Satellite Corp (DBSC)

Satellites: Two in orbit, plus two spares
Channels: Six primary (30 to 42 possible with spot beams in full operation)
Beamwidth: 22.5 MHz
Orbital Slots: 101° W & 148° W
Weight: 1,358 kg
Cost: $577 to $614 million
Coverage: Three time zones, 4 spot beams
Services: DBSC proposed to offer many more channels than any other satellite system; DBSC hopes to lease channels as a common carrier.
Design Lifetime: Seven years

Graphic Scanning

Satellites: Two in orbit, plus one spare
Channels: Two
Bandwidth: 18 MHz
Orbital Slots: 115° W, 143° W
Weight: Not available
Cost: Not available
Coverage: Two one-half CONUS beams plus spots to Alaska, Hawaii, Puerto Rico, and the Virgin Islands
Services: Graphic Scanning hopes to broadcast subscription TV including closed caption entertainment as well as electronic text services to subscribers. Each satellite would be independently programmed.

RCA Americom

Satellites: Four in orbit, plus two spares
Channels: Six
Bandwidth: 24 MHz (72 MHz for high definition TVRO)
Orbital Slots: 101° W, 119° W, 148° W, 157° W
Weight: 1,106 kg
Cost: $760 million
Coverage: Four U.S. time zones, Alaska and Hawaii
Services: RCA proposed to both lease channels and broadcasts directly to home and business. The company also plans to offer stereophonic sound and high definition television.

Satellite Television Corp (STC)

Satellites: Four in orbit, plus two spares
Channels: Three
Bandwidth: 16 MHz
Orbital Slots: 101° W, 119° W, 148° W, 157° W
Weight: 650 kg
Cost: $700 million for initial phase covering eastern time zones
Coverage: Four time zones
Services: The first of STC's three channels "Superstar," will operate 24 hours a day with a mix of major motion pictures, popular concerts, theater specials, and family entertainment. This channel will have the biggest appeal to the greatest number of subscribers.

Each of the other two channels, "Spectrum" and "Viewers Choice," will have about 15 hours of daily programming that will range from children's shows, film classics and sporting events to public affairs, cultural and education programs.

In addition to all these forms of information and entertainment, STC's service will feature pay-per-view capability; events such as major sports contests, entertainment spectaculars or even continuing education classes can be viewed for a nominal additional charge to the subscriber.

Optional viewing enhancements such as stereo sound, teletext, closed caption for the hearing impaired, and a second audio track for another language will also be offered.

United States Satellite Broadcasting (Hubbard)

Satellites: Two in orbit, plus one spare
Channels: Three
Bandwidth: 17.5
Orbital Slots: 119° W, 148° W
Weight: 546 kg
Cost: $228 to $230 million
Coverage: Four time zones and spot beams
Services: This system would carry the Dominion Satellite Network's programming to homes and businesses and would be supported by advertising. Programming would also be distributed by terrestrial broadcasters.
Design Lifetime: Seven years

Western Union

Satellites: Four in orbit, plus two spares
Channels: Four
Bandwidth: 18 MHz
Orbital Slots: 110° W, 119° W, 148° W, 157° W
Weight: 534 kg
Cost: $450 million
Coverage: Four time zones, Alaska, Hawaii, Virgin Islands
Services: Western Union asked to be regulated as a private, or discriminatory, carrier which would lease channels to selected customers.
Design Lifetime: Seven years

Chapter Seven

U.S. MANNED SPACE PROGRAM

Space Shuttle Transportation System

The Space Shuttle is America's newest and most versatile manned spacecraft. Unlike its predecessors -- Mercury, Gemini, and Apollo -- it is reusable, an aircraft-like spaceship which is designed for years of service and making space flight a relatively economical and routine event.

The Shuttle provides a flexibility never before achieved in space operations, allowing space to be treated as a resource rather than as a hostile environment.

Since the Shuttle is going to be with us for a number of years, this chapter should prove valuable to the monitor desiring to monitor Space Shuttle communications direct from the orbiter.

I would like to thank the Public Affairs Office of Rockwell International, prime contractor for the shuttle program, for their valued assistance in providing this information on the Space Shuttle communications system.

Communications

The S-band system is the primary means of communications between the orbiter and the ground tracking stations during the ascent, orbital, and entry phases of the flights.

There is one uplink from the ground tracking stations to the orbiter; phase-modulated (PM), it provides commands, voice and a coherent ranging signal to the orbiter.

Three downlinks from the orbiter to the ground tracking stations include one PM channel for voice, telemetry, two-way Doppler, and a coherent ranging signal. The other two are frequency modulated (FM) to transmit wideband data. One FM downlink provides operational real-time Space Shuttle main engine data during ascent, real-time video (television), or operations recorder dump data. The remaining FM downlink provides real-time development flight instrumentation (DFI) data; their downlink will be deleted in the future.

In addition to S-band, the orbiter has an ultra-high frequency (UHF) communication system which is used only for air-to-air and air-to-ground voice communications, and a Ku-band communications system for use in conjunction with the Tracking and Data Relay Satellite (TDRS) system.

The UHF is used after reentry during the approach and landing phase of the flight; this air-to-ground voice communication takes place between the orbiter and the landing site control tower, and the orbiter and chase aircraft which assist the orbiter flight crew in landing the orbiter. The air-to-air voice communication takes place between the orbiter and extra-vehicular activities (EVAs). In the pre-launch phase, the orbiter flight crew utilizes the launch umbilical for communications with the ground support and operations personnel until liftoff.

S-Band System
(1700 to 2300 MHz)

The uplink from the ground tracking stations will be phase-modulated on a center carrier frequency or either 2106.4 MHz (primary), or 2041.9 MHz (secondary) for the NASA Space Tracking and Data Acquisition Network (STDN) ground stations, and 1831.8 MHz (primary) or 1775.7 MHz (secondary) for the Department of Defense (DOD) Air Force Satellite Control Facility (SCF) ground stations.

The orbiter can transmit one PM downlink and two FM downlinks simultaneously. The PM downlink will be on a center carrier frequency of 2287.5 MHz (primary) or 2217.5 MHz (secondary) for the NASA STDN ground tracking stations. The S-band PM downlink transfers high-data-rate telemetry plus two voice channels (192 kbps - kilo bits per second); or low-data-rate telemetry, one voice channel (96 kbps), two-way doppler and two-way ranging.

The S-band FM downlink can originate from three FM transmitters on board the orbiter: two on 2250.0 MHz and one on 2205.0 MHz. The S-band FM downlinks transfer real-time engine data or television or operational instrumentation PCM (128 kbps) or playback of engine data or recorded high-rate data (192 kbps) and DFI (128 kbps).

All MCC-H (Mission Control Center--Houston) uplink commands are multiplexed with digital voice and transmitted on the S-band PM uplink. The S-band system removes the voice and transfers the commands to the on-board orbiter computers which route the commands to the intended systems.

The S-band PM uplink originates from an STDN remote station, NASA, or at a Dod SCF remote station. Each station has a choice of two uplink frequencies for data transfer under MCC-H control. The S-band uplink transfers the high data rate of two voice channels and commands (72 kbps) or low data rate of one voice channel and commands (32 kbps), two-way Doppler, and two-way tone ranging.

A choice of two S-banduplink and downlink frequencies prevents interference when two Space Shuttle orbiters are operating at the same time.

The two identical S-band transponders receive and transmit the PM uplink signal and downlink signal. Only one transponder operates at a given time; the other is a redundant backup. The selected transponder transfers the uplink commands and voice to the NSP (network signal processor), receives the downlink telemetry and voice from the NSP, retransmits on a coherent frequency for the two-way Doppler, and retransmits the two-way tone ranging signals.

The two on-board NSPs receiver commands and transmit telemetry data to the selected JS-band transponder. Only one NSP operates at a time, converting analog voice from the onboard audio central control unit (ACCU) to digital voice and multiplexing it with operational instrumentation telemetry from the pulse code modulation master unit (PCMMU). The selected NSP converts digital voice to analog voice for the ACCU and crew intercommunication and decodes commands for the on-board computers.

The PM uplink and downlink are normally coherent so that the frequency of the downlink is directly proportional to the uplink frequency. This is necessary for the two-way Doppler (change of frequency varying with velocity) data. By measuring the uplink frequency and knowing what downlink frequency to expect from the orbiter, the ground tracking station can measure the double Doppler shift that takes place and use that to calculate the radial velocity of the orbiter with respect to the ground station. These links are phase-modulated so that the S-band carrier center frequency will not be affected by the modulating wave. It would be impossible to obtain valid Doppler data if the S-band carrier center frequency were affected by the modulating technique.

The S-band also provides a subcarrier for two-way tone ranging. The ground will uplink ranging tones at 1.7 MHz and compute vehicle slant range from the time delay in the returned 1.7 MHz tones to determine orbiter range. Orbiter azimuth is determined from the ground tracking station antenna angles. A C-band skin tracking mode is also provided from the ground tracking stations to track the orbiter.

There are seven S-band antennas on the orbiter: four quadrant, two hemispherical, and one payload. All the antennas are covered with thermal protection systems.

The hemispheric antennas have a larger beam width, while the quadrant antennas have a higher antenna gain. The hemispheric antennas are so called because there are two of them, one on top of the orbiter and one on the bottom. The quadrant antennas are so called because there are four of them, two on each side of the orbiter, one on the upper half and one on the lower half of each side,which provides nearly total coverage in all directions.

The proper quadrant and hemispheric antenna to be used is selected automatically under on-board computer control, by ground command, or manually by the flight crew utilizing displays and controls on the flight deck.

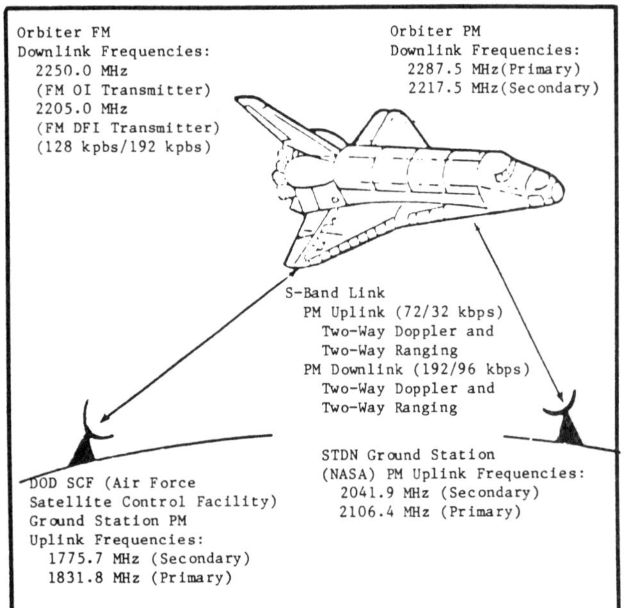

S-BAND RADIO FREQUENCIES

Audio System

It is through the audio distribution system that the flight crew can communicate with each other (intercom) as well as with ground support personnel. The system also provides for reception of caution and warning (C/W) aural tone signals in the event of an equipment malfunction or emergency, provides for reception of TACAN (tactical air navigation) identification signals from the ground tracking station and, in later flights, provides for a communications with attached payload (Spacelab), payload bay, and docking ring.

The audio distribution system interfaces with the S-band and Ku-band communications system via the NSP as well as UHF communication system for transmission and reception of air-to-ground and air-to-air voice signals. It also interfaces with the launch umbilical for voice communications prior to and up to liftoff.

During launch and reentry in the development flights,the flight crew members will be wearing ejection escape suits and will communicate with each other, and with the ground, through the helmets of their suits. When in orbit, they can, at their option, wear either the lightweight head set (LWHS) or a communications carrier assembly (CCA) headset. Regardless of the headset they use, they will also use a headset interface unit and a 1.21-meter (4 foot) crew communication umbilical (CCU).

The headset interface unit (HIU) is 6.35 centimeters (2-1/2 inches) by 12.7 centimeters (5 inches) and clips onto the belt of the flight suit. The unit has a volume control knob and a rocker-type switch with three positions: The transmit position allows access to intercom and external circuits; the intercom position allows access to intercom only; the center (springloaded) position is the off position.

The audio terminal units (ATUs) are control panels used by the crew members to select audio channel, volume, and mode of voice communication such as "PTT" (push to talk), "VOX" (voice activated), or "Hot" (live microphone).

Eight ATUs are available for use by the flight crew members. Two are located on the flight deck display and control panel forward station, three are located at the aft flight deck control and display panel station, two are located in the mid deck of the crew compartment, and one is located in the mid deck airlock for use during extra-vehicular activity (EVA).
Common to all ATUs are channel power switches, channel volume control thumbwheels, mode select knob, VOX sensitivity knob, page switch, and power switch. The channel power switches select access to any of five available channels for transmission or reception: two air-to-ground, one air-to-air, and two intercom.

The commander and pilot only have ATUs for reception of TACAN ground identification signals -- call letters in Morse code, repeated every 40 seconds. TACAN provides distance and bearing information in the band of frequencies from 962 to 1213 MHz.

All external UHF transmissions are sent to and received from the UHF transceiver via the ACCU. All three of the UHF frequencies -- 296.8 MHz (primary), 259.7 MHz (secondary), and 243.0 MHz (emergency) -- are pre-set in the UHF transmitter and cannot be altered by the flight crew.

During EVAs the UHF system is used for air-to-air voice communications with the flight crew member who is outside the orbiter using the UHF antenna in the airlock.

View of shuttle in orbit from SPAS satellite (Rockwell International)

Another UHF channel, 279.0 MHz, is reserved for one-watt space suit communications during the EVA.

Ku-Band System

The orbiter Ku-band system (15.250-17.250 MHz) provides a much higher gain signal with a smaller antenna than the S-band system, and higher data rates can be used.

One drawback of the Ku-band system is its narrow pencil beam which makes it difficult for the antennas on the TDRS to lock onto the signal. The S-band system will be used to lock the antenna into position first because it has a larger beamwidth. Once the S-band signal has locked the antenna into position, the Ku-band signal will be turned on.

The Ku-band antenna is gimbaled, which permits it to acquire the TDRS for communications acquisition or radar search for other space hardware. The Ku-band system is first given the general location of the space hardware from the orbiter computer; the antenna then makes a spiral scan of the area to pinpoint the target.

With communications acquisition, if the TDRS is not detected within the first eight degrees of spiral conical scan, the search is automatically expanded to

20 degrees. The entire TDRS search requires approximately three minutes. The scanning stops when an increase in the received signal is sensed.

Radar search for space hardware may use a wider spiral scan, up to 60 degrees. Objects may be detected by reflecting the radar beam off the surface of a target passive mode) or by using the radar to trigger a transponder beacon on the target (active mode).

Operations Sequence

Pre-launch

The crew, wearing ejection escape suits, communicate by voice through the launch umbilical. Voice communications through the S-band and UHF systems are checked.

Ascent

The flight crew,wearing ejection escape suits, communicate by voice via the S-band PM link to ground.

Orbit

The flight crew members remove ejection escape suits and put on light-weight headset interface units and communication umbilicals.

All of the S-band and communications links can be on at the same time when the orbiter is in contact with a ground tracking station.

Reentry and Landing

The flight crew again don the ejection escape suits. The UHF system is used during approach and landing for air-to-ground voice communications with the landing site control tower and for air-to-air voice communications with the chase aircraft.

Television

A color TV system is available in the orbiter crew compartment. One camera is for the flight deck and the other is for the mid deck. A black and white monitor is used to assist the camera operator in locating the appropriate field. A cue light on the TV camera shows when it is selected as the active camera; it will blink on/off if an over temperature condition is sensed by circuits within the camera.

NASA Tracking Network

The key elements of any shuttle mission are the capability to track the spacecraft, communicate with the astronauts, and obtain the telemetry data which informs ground controllers on the condition of the spacecraft and its astronauts.

The heart of this complex network is located at NASA's Goddard Space Flight Center in Greenbelt, Maryland, just outside Washington, D.C.; here the Spaceflight Tracking and Data Network (STDN) and the NASA communications network are located.

With the exception of very brief periods during the launch, flight, and recovery of space shuttles, Goddard serves as the heartbeat of the mission, receiving all telemetry, radar and air-to-ground communications and relaying that information to the Johnson Space Center in Houston, as well as to other NASA and participating Department of Defense facilities.

Most video (TV) transmission facilities used during the mission are provided and monitored for quality by Goddard personnel. Here the network operations managers and systems specialists keep the entire NASA/DOD network tuned for mission support.

Spaceflight Tracking and Data Network (STDN)

STDN is a highly-complex NASA worldwide network that provides reliable real-time communications with the Space Shuttle orbiter and crew; about 2,500 personnel are required to operate the network which consists of 15 ground stations equipped with 14, 30, 40 and 85 foot S-band antenna systems and C-band radar systems. These sites are augmented by 15 Department of Defense geographical locations providing C-band support and one Department of Defense 60 foot S-band antenna system.

In addition, there are six major computer interfaces providing real-time network computation support located at the Network Operations Control Center and at the Operations Support Computing Facility, both at Goddard; Western Space and Missile Center, California, Air Force Satellite Control Facility, Colorado; White Sands Missile Range, New Mexico; and the Eastern Space and Missile Center, Florida.

The network also has agreements with the governments of Australia, Spain, Senegal, Botswana, Chile, United Kingdom, and Bermuda to provide NASA tracking station support to the Space Shuttle program.

Space Shuttle Ground Support

North of Los Angeles the Vandenberg Space Launch Complex 6 ("Slick Six"), a giant concrete city of high technology, serves as the home of Space Shuttle's military missions.

For scientific and commercial spaceflights, the Merritt Island, Florida, S-band tracking station provides the approach data to the Launch Control Center at the Kennedy Space Center and Mission Control Center at Johnson during pre-launch testing and during the terminal countdown.

During the first minutes of launch and the ascent phase, the Merritt Island and Ponce de Leon, Florida, and Bermuda S-band stations provide appropriate tracking data. Other stations used during this phase of the mission include the C-band stations located at Bermuda; Wallops Island, Virginia; Grand Bahama; Grand Turk; Antigua; Cape Canaveral and Patrick Air Force Base, Florida. These tracking stations will provide both low speed and high speed tracking data to the Kennedy and Johnson Control Centers.

The Madrid, Spain; Indian Ocean Station Seychells; Orroral and Yarragadee, Australia; and Guam stations provide critical support to the Orbital Maneuvering System burns on the first revolution.

During the orbital phase, all the S-band and some of the C-band stations that see the Space Shuttle orbiter at least three degrees above the horizon will provide tracking, telemetry, air-ground, and command support to the Johnson Mission Control Center through Goddard.

During a nominal reentry and landing phase at Edwards AFB, the Goldstone and Buckhorn, California, S-band stations and C-band stations at the Pacific Missile Test Center, Vandenberg AFB, Edwards AFB, and Dryden Research Facility provide highly critical tracking, telemetry, command, and air-ground support to the orbiter and send appropriate data to the Johnson and Kennedy Control Center.

During a nominal reentry and landing phase for Runway 15 at the Kennedy Space Center, C-band stations at Nicholas Island off the California coast; Vandenberg AFB, White Sands, New Mexico; Stallion Station, Arizona; Scotts Peak, Arizona; and Mt. Lemmon, Arizona, will provide highly critical tracking data on the orbiter before it comes into view of the Eastern Test Range (ETR) C-band radars at Merritt Island and Patrick AFB, and the GSFC and Merritt Island S-band stations will provide highly critical telemetry, command and air-ground support as well as tracking data to Johnson and kennedy Centers.

Tracking and Data Relay Satellite Systems

The TDRS satellites relay data to and from the orbiter, unmanned earth-orbiting spacecraft, and the earth station at NASA's White Sands (New Mexico) Test Facility. TDRS offers simultaneous data relay services for up to 32 user spacecraft in orbits up to approximately 2693 nautical miles (3100 statute miles) through 85 to 100 percent of each spacecraft's orbit.

The TDRS operates as a repeater, relaying signals to and from the earth station and user spacecraft on three frequency bands: high capacity K-band, S- and C-band. The TDRS, however, will require much stronger signals to travel the longer distances from the orbiter to the TDRS and ground. A Ku-band system installed in the orbiter payload bay will provide the stronger signals.

The communications satellite measures more than 57 feet between the two extended solar panels, weighs some 5,000 pounds, and provides about 1,888 watts of electrical power. The spacecraft's umbrella-like S-band and Ku-band antennas measure 16 feet in diameter.

The first Tracking and Data Relay Satellite (TDRS 1) was deployed from the Space Shuttle Challenger on April 4, 1983. Problems developed during the Inertial Upper Stage boost phase after deployment from the orbit4r, placing TDRS 1 into an unsatisfactory elliptical rather than geosynchronous orbit.

Goddard, TRW Defense and Space Systems Group, and Space Communications Company

Shuttle Challenger returns to KSC after space mission (Rockwell International)

(Spacecom) developed plans to correct the spacecraft's orbit by a series of thruster firings and on June 19, 1983, the satellite achieved its planned geosynchronous orbit of 35,680 km at 67° W, northeast of Brazil. It was moved to its operational location, 41° W, after a series of communications tests during the STS 8 mission.

A second TDRS at 171° W (southwest of Hawaii) and a third on-orbit spare at 79° W will complement the system, providing communications coverage over at least 85 percent of low-altitude spacecraft orbits.

The TDRS was awarded in December 1976 by NASA to Western Union Space Communications Inc., a subsidiary of the Wester Union Corporation. TRW's Defense and Space Systems Group, Redondo Beach, California, builds the satellites and ground support equipment for Western Union.

NASA has leased the TDRS system and contract service from Spacecom for ten years.

Ground Spacecraft Tracking/Data Network (GSTDN)

TDRS is designed ultimately to replace all but a few of the GSTDN ground stations, but until the TDRS system is fully operational, GSTDN will remain in place as NASA's primary window for spacecraft communications.

NASCOM

The NASA Communications Network (NASCOM), managed by Goddard, provides the voice and data communications links connecting the network. During a flight, data flows from the shuttle orbiter to one of the TDRS satellites which transmits that information to the TDRSS ground station at White Sands, New Mexico; the data could also flow from the orbiter to one of the GSTDN stations. In either case, the data is retransmitted to a commercial communications satellite which sends the data to the

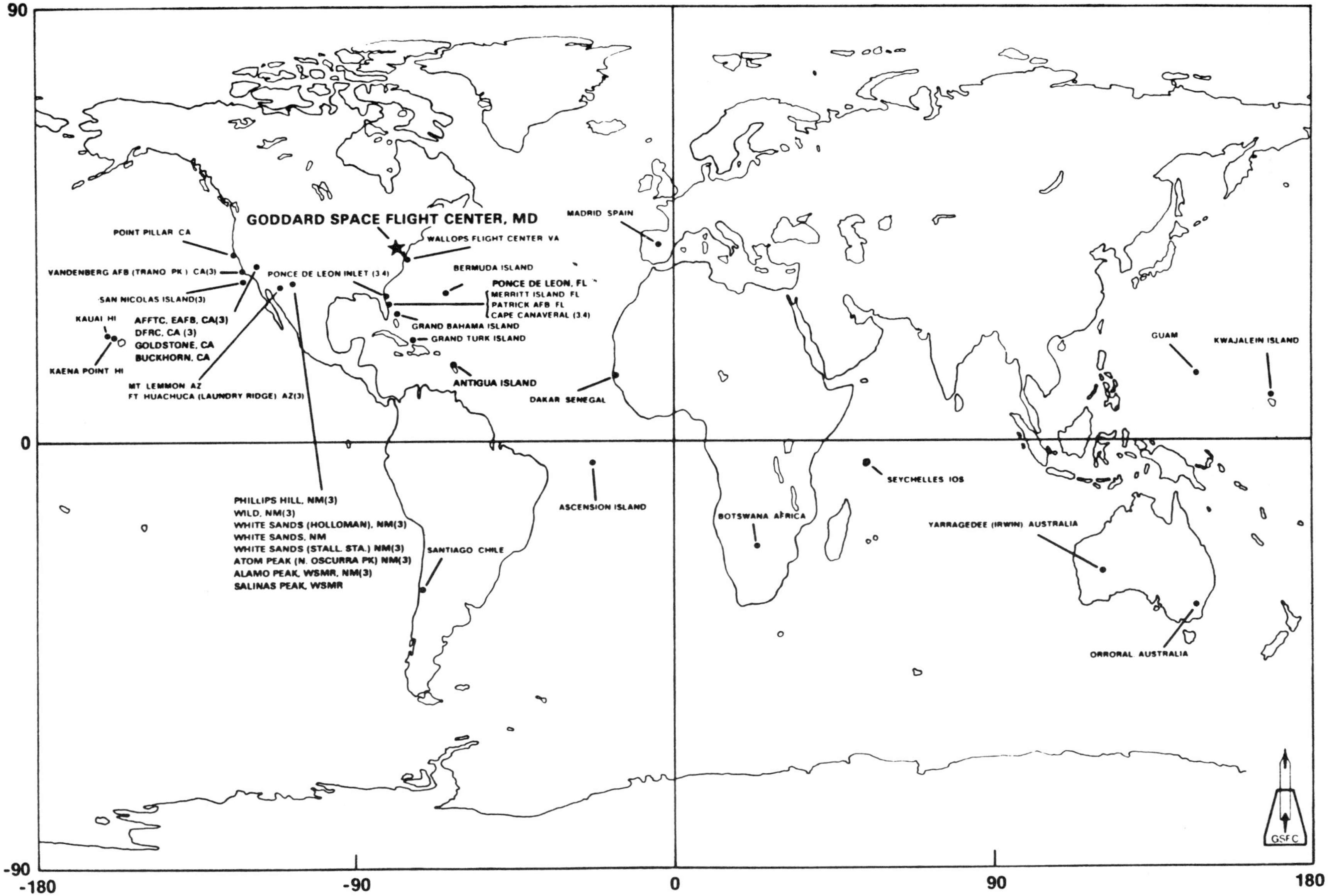

SPACEFLIGHT TRACKING AND DATA NETWORK (STDN)

Johnson Space Center. The Network Control Center at Goddard coordinates all the scheduling activities of this system.

Any TVRO home satellite dish owner is by now familiar with NASA's contract channel which provides shots of Mission Control at the Johnson Space Center as well as audio of air-ground communications. Full coverage of launches, landings and shuttle video are also carried on this channel. For the dyed-in-the-wool space junkie, this is the ultimate in monitoring.

Reports from listeners indicate that other missions have also been carried over the NASA contract channel, giving you the chance to watch your favorite satellite being launched from the Cape! Check for future missions on RCA Satcom 1R, transponder 9, at 139° W or one of the vacant channels on Satcom 5 at 143° W.

NASA TRACKING STATIONS

The following is a list of NASA tracking stations in the GSTDN system as of this writing.

Location	Capability
Ascension Island (ACN)	S-band, UHF A/G
Bermuda (BDA)	S-band, C-band, UHF A/G
Botswana (BOT)	UHF A/G
Buckhorn (BUC)	S-band, C-band
Dakar (DKR)	UHF A/G
Goldstone (GDS)	S-band, UHF A/G
Guam (GWM)	S-band, UHF A/G
Hawaii (HAW)	S-band, UHF A/G
Madrid (MAD)	S-band, UHF A/G
Merritt Island (MIL)	S-band, UHF A/G
Orroral (ORR)	S-band
Santiago (AGO)	S-band
Yarragadee (YAR)	UHF A/G

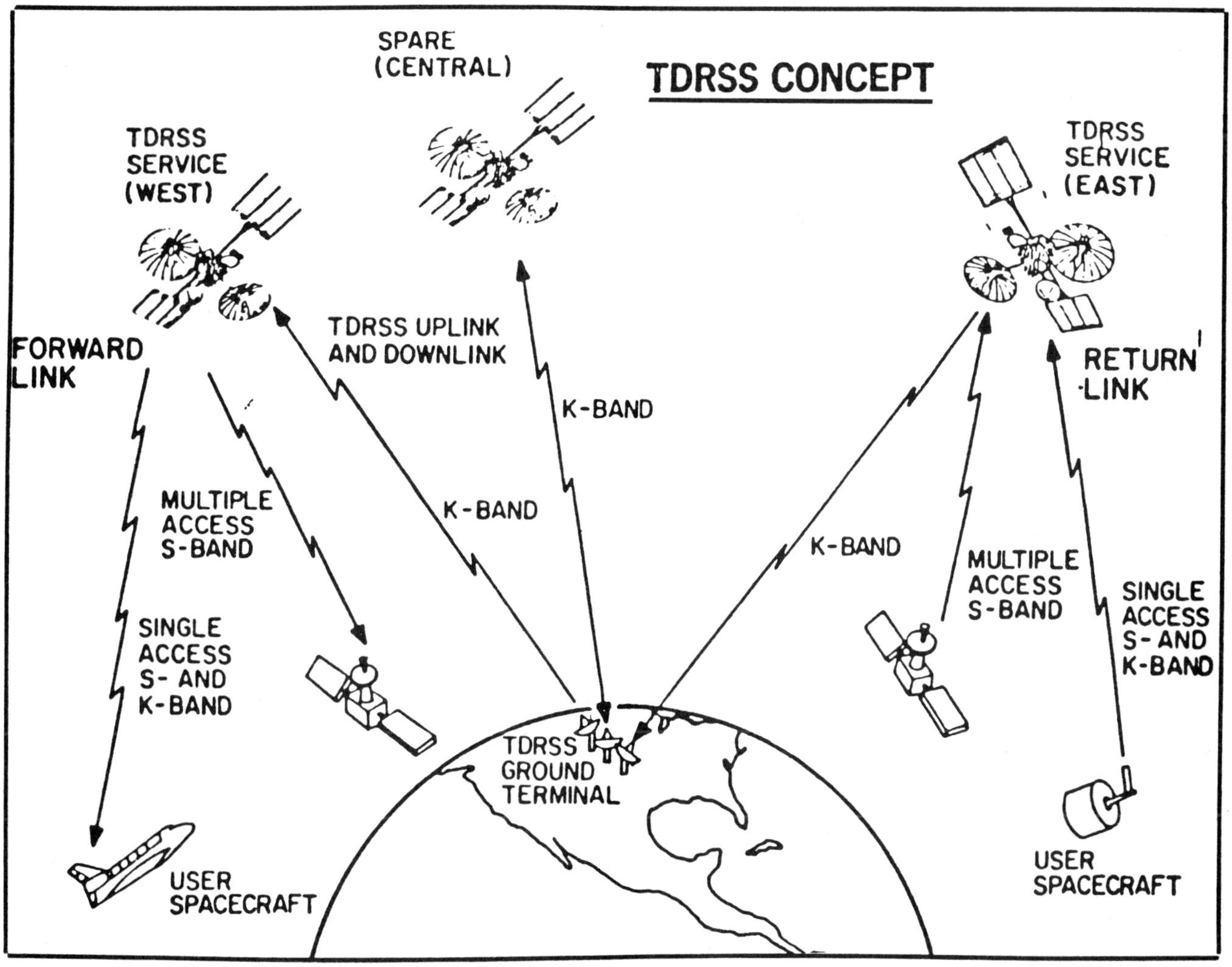

Space Shuttle Panorama

To better prepare listeners to receive Space Shuttle missions, this section has been extracted from Bob Grove's *The Best of Monitoring Times,* articles from *MT*, and Bob Grove's *Shortwave Directory*. Frequencies (kHz) recently monitored are underlined.

Eastern Space and Missile Center (ESMC)

Cape Canaveral, Florida
(Formerly ETR)

2356	6693	9132*	13635	18019	23479
2576	6701	9138	13735	18022	23485
2622	6708	9170	13742**	18051	23661
2638	6723	10147	13878	18196	23840
2678	6753	10159	14397	18310	23940
2716	6801	10215	14432	18331	24240
2744	6880	10230	14497	18354	24512
2764	6897 alt.	10270	14547	18434	24530
2800	6919	10301	14585	18700	24780
2820	6937	10305	14615	18769	24914
2836	7313	10310***	14650	18801	25130
3039	7332	10327	14896	18990	25161
3099	7355	10475	14937	19126	25198
3120	7387	10780Pri	14967	19143	25245
3187 alt.	7412	10850	15021	19303	25597
3219	7461	10880	15025	19371	26356
3365	7512	11104	15040	19390	26389
4704	7525	11183	15064	19928	26515
4714	7605	11205	15087	19961**	26684
4500	7675	11252	15051	19966	27720
4510	7697	11407**	15057	20186	27870
4520	7715**	11414	15067	20189	
4622 alt.					
4765	7742	11548	15471	20192*	
4825	7765	11621	15484	20195	
4855	7804	11634	15528	20198	
4860	7833	11984	15560	20266	
4864	7860	11988	15564	20272**	
4900	7910	12107	15575	20390 alt.	
4992	7919	12160	15610	20475	
5060	7985	12277	16167	20690	
5180	8779	12287	16216	21810	
5190 sec.	8788	13170	16246	22683	
5246	8964	13172	16420	22687	
5350	8972	13204	17470	22755	
5436	8981	13210	17490	22990	
5710	8991	13227	17552	23035	
5718	9006	13237	17554	23224	
5724	9022	13468	17668	23281	
5775	9043	13495	17975	23325	
5810	9115	13600*	18009	23413	
5822					

* Cape to Ascension
** Ascension to Cape
*** Antigua to Ascension

Western Space and Missile Center (WSMC)

USAF Pacific Missile Range *(Formerly SAMTEC)*
Edwards AFB/Vandenberg AFB, CA

3162
4486
4760
5700 sec.
5822
6820
6889
7705
9029
9212
10272
10352
10510
10660
10804
11510
13218 pri.
13756
14987
15793 Air
13900
15021
17428
19303
19640
20261

(Primary frequencies underlined)

Wallops Island, Virginia 3030

NASA Identifiers

Tactical Callsign	Identification
0	Patrick AFB, FL
1	Merritt Island, FL (AFE 71 - Malabar)
2	Jupiter, FL
3	Grand Bahama Is.
7	Grand Turk Is.
12	Ascension Is. (AFE 83)
13	Pretoria, S. Africa
89	Mahe, Seychelles
91	Antigua (AFE 86)
Abnormal One Zero	Vandenburg AFB, CA (AGD)
Abnormal Two Zero	Wheeler AFB, HI (AGD25)
Abnormal Four Zero	Kwajalein (ABK)
AGAR	EC-135 Advanced Range Instrumentation Aircraft 4050th Test Wing, Wright Patterson AFB
Ashley 10	Vandenberg AFB, CA
Ashley 12	Wheeler AFB, HI
Canaveral Control	Port Canaveral Harbor Control
Cape Leader	Contingency Emergency Base
Cape Radio	Cape Canaveral Communications (USAF)
Chase 1,2	T-38 Chase Aircraft
Dishpan	USS Redstone (Tracking Ship)
Eyesite	US Navy P-3 Orion Aircraft
Fisher	Cape Radio to Ships
Freedom	Booster Recovery Vessel (civilian contractor)
Gull Photo	USAF WC-130 Aircraft
Independence	Booster recovery vehicle (civilian contractor)
Jolly	USAF/ARRS SH-3 Helicopters
King	USAF/ARRS Aircraft
Liberty	Booster Recovery Vessel (civilian contractor)
Peapod	USAF AC-130 aircraft
Thinker 1	Senior Range Operations Officer
Variety	Aircraft

Patrick AFB Aircraft Frequencies (MHz)

126.4	Zone Control
141.3	Barracks 7
294.6	Barracks 7
294.6	Barracks 9
236.6	USAF Tower
335.8	Patrick Ground Control
348.4	Patrick Tower
340.9	Patrick Departure Control
393.0	Tower CK/SKID
264.8	Think 1
294.6	Variety 1 (USCG at SRB recovery area)
296.8	Space Shuttle CM
252.8	ARRS/Tng/TAC

Astronaut James "Ox" Van Hoften attempts to snag Syncom IV-3 (Leasat 3) during shuttle mission STS-51I (Rockwell Int'l)

NASA Space Telescope which was to have been launched by Shuttle (AEG Telefunken)

381.8	USCG Air/Ground
282.8	USCG/SAR
255.4	VFR Flight Service Station
344.6	PMSV Metro weather Patrick
372.2	PTD Base Ops
284.1	GBI Tower
358.3	Approach Control Patrick
369.2	GCA Discrete channel Patrick
297.2	"
378.8	"
372.8	"
383.0	PTD Patrick Ops
257.8	CK Reg/Mel
384.4	Melborne Tower
269.3	Miami ATC (Melborne sector)
269.4	Freeport Approach Control
255.4	UHF/DF
305.4	UHF/DF
282.0	USCG Secondary Air/Ground
263.0	Jacksonville ATC Daytona
385.6	Jacksonville ATC GNV
307.0	Orlando Approach Control
284.7	Orlando Departure Control
292.2	Avon Park Control

NASA Frequency List Cape Canveral, Florida

CT = Crawler Transport
ETR= Eastern Test Range
LC = Launch Complex
RCVY/HGR = Recovery Hangar
SLF= Shuttle Landing Facility
SRB= Solid Rocket Booster
VAB= Vehicle Assembly Building

All frequencies MHz except first entry

1610 (AM)	VIP Grandstand rebroadcast
117.800	Shuttle control rebroadcast
171.000	Utilities/Parking "Bravo Control"
156.1875	Telemetrics/EMI
173.6875	Security primary/Astronaut escort
162.6125	Launch support "Alpha Control"
173.6625	Range Safety/Prime emerg. egress
416.5000	Range Safety Officer (ETR)
170.4000	GSA Supply/Maintenance
170.1500	Base Communications/Press
170.3500	Public Affairs/Paging
148.4550	SRB RCVY/HGR AF Cranes
173.5625	Fire/Emergency Egress/Rescue
171.1500	General maintenance/fuels "NOVA"
173.1750	Security Tac
170.1750	Rail/Trucking Ops
158.9400	Civil Defense (Rcv 155.715)
173.7875	Fire/Backup emergency egress
155.3700	Florida Intercity Police
162.0125	Marine Ops
154.1600	Intercity Fire Department
148.4850	Timing/Photo/Status
163.5125	Cape SRO
163.4625	Cape Security
163.4875	"
163.5625	Cape Fire
142.5000	O&C Crane Ops
142.8600	"
143.0400	"
139.3000	Operational loan pool
139.9000	"
164.0000	Radiation monitoring
407.3250	LC 39 Convoy Command
407.4750	LC 39 VAB crane ops
408.1500	LC 29 Convoy purge
408.1750	LC 39 VAB crane ops
408.8000	LC 39 ops pool
409.0500	LC 39 VAB crane ops
409.1250	LC 39 ops pool
409.1750	LC 39 convoy cooling
121.7500	Chase aircraft/ground
126.3000	Air/ground control
126.6500	Weather recon
284.0000	Air/ground control
243.0000	Military emergency
121.5000	Civilian emergency
916.0000	Crawler moves
929.0000	"
939.0000	"
165.1875	Visual surveillance
165.6125	Unknown
165.8975	"
171.2625	Camera/recording production
173.5375	Unknown
416.1500	"

Edwards AFB Shuttle Support

121.8000	Lakebed traffic control
123.0500	Loma Linda Hospital
138.4500	A/G VHF command control
162.6250	Purge VHF net
164.1000	Convoy command VHF net
168.0000	Cooling VHF net
169.4000	PAO release (JSC)
169.6000	TV direction (JSC)
4680	Chase plane
7245	TV 1 long range
13012.5	TV 3 lakebed
1771.8310	TV 2 convoy (JSC)

Addition NASA Frequency Allocations (kHz)

3380	Nationwide tracking net
3385	"
3395	"
6982.5	"
14455	"
14452	Wallops Island, Virginia
20089	"
22745	"
2624	Houston, Texas
25590	"
27725	"
2017.5	Hampton, Virginia
2541.5	Moffett Field, California
27725	"
4683.5	Fairbanks, Alaska
5470.5	"
2207.4	New Orleans, Louisiana
2383.4	"
2431.4	Mobile, Alabama
27900	Eglin, Texas

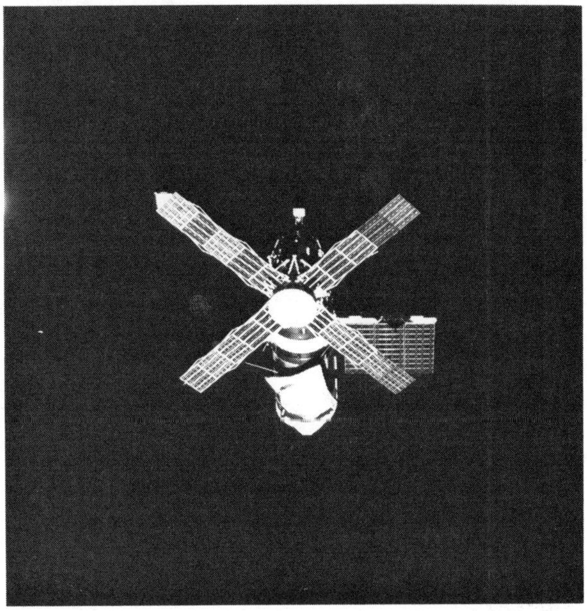

Overhead view of Skylab space station taken from the CSM during final "fly around" inspection (NASA)

Chapter Eight

U.S. MILITARY SPACE SYSTEMS

Edited by Bob Grove

Satellites have always had important military roles, but in recent times dependence on satellites has grown dramatically.

Early in the U.S. space program, the United States proved sensitive and defensive about Soviet charges that it was practicing aggression in space. The trauma of the Gary Powers U-2 capture and subsequent collapse of public information policies produced a very secretive attitude within some segments of the United States Government. We knew that information gathering was important to maintaining peace and was not aggressive by our standards.

No one knew what the Russians were up to within their closed society; our national security demanded that we examine all possible contingencies regarding the rapid expansion of Soviet strategic missile capabilities.

While legal interpretations of the U-2 incident showed that government aircraft overflights in the airspace of other nations were not necessarily espionage or illegal, the storm of protest it produced went far beyond abstract principles.

By contrast, flight through outer space could hardly be considered invasion of airspace. No logic would support the concept of sovereignty extending outward from earth to sweep limitless regions of the universe as the earth revolved on its axis. Nonetheless, in the absence of firm international law on such points in these early years, a cautious policy seemed appropriate.

Hence, in the fall of 1961, United States information policy on its military operations progressively tightened to withhold details of space hardware and operations plans; and on November 22, the first unspecified launch occurred. Subsequent launches drew such added attention from the world press that military information policies had to be tightened even more.

After the launch of Discoverer 38 in February 1962, no more names were announced for flights in that series. Information on recovery of all capsules returned from space by the military disap^^peared. Even the previously-widely-publicized Transit navigation satellites went under cover.

Thus, from March 1962 on it became U.S. policy to have no public names for its military space flights except when sharing unclassified experiments with other agencies.

Even so, the flights did not become truly secret. First, the launch is announced at the local launch site, naming the launch vehicle used; second, the orbital elements of most of these flights are published in the NASA *Goddard Satellite Situation Report*; third, the name of the launch vehicle and the orbital elements for all flights without exception are registered at the United Nations for the payloads and for all associated pieces of debris.

Even with this degree of openness, there is growing Soviet criticism about these military operations, including accusations that U.S. flights have gone beyond passive military support and have placed weapons of mass destruction into orbit. The low resolution pictures taken by NASA's TIROS weather satellites have been described by some Soviet writers as "spy in the sky" flights, and the shuttle program is considered strictly a military venture to spy on the Soviet Union.

Readers are referred to James Oberg's recent book, *The New Race for Space*. Oberg does an excellent job of documenting the Soviet fears about the Shuttle program and U.S. astronauts.

Information that will be presented in this chapter of Communications Satellites has been obtained from public, unclassified sources and on-the-air monitoring including the ITU, *Aviation Week and Space Technology*, NASA's *Satellite Situation Report*, *Spaceflight* magazine, RAE's *Table of Earth Satellites*, and numerous other public domain sources.

Who Manages the Spy Satellites?

The National Reconnaissance Office (NRO) established in 1960, is so highly classified that even its name is officially secret. One branch, the Special Projects Office, operates from the Air Force Space Division near Los Angeles, California.

Handling requests from the Central Intelligence Agency (CIA), Defense Intelligence Agency (DIA), the Department of State, and the joint military services, NRO dictates parameters to the Satellite Control Facility in Sunnyvale, California which, in turn, signals the recon satellites from any of eight tracking stations around the globe.

Photo Reconnaissance

Discoverer Series

On February 28, 1959, the U.S. launched from Vandenberg AFB its first photo recon mission, Discoverer 1, which orbited the earth until March 5, 1969. The satellite apparently burned up during reentry and never ejected a film return capsule.

There were 38 satellites in the Discoverer series, all launched from Vandenberg. The first 16 satellites in the series was launched by the Thor-Agena A rocket, 19.2 feet long and five feet in diameter. Fully loaded with fuel, the Agena weighed 8,000 pounds; the Discoverer payload (fuel consumed) weighed 1,300 pounds including the 245 pound reentry capsule.

With the flight of Discoverer 16, the U.S. Air Force introduced an improved Agena second stage called Agena B, 25 feet long. Capable of carrying an orbital payload of 6,000 pounds, the Agena B could be used with both the Thor and Atlas first stage rockets.

Discoverer 17, launched November 12, 1960, weighed 2,100 pounds and carried a capsule, directive antenna and a whip antenna. This is the first reference in ITU documents to some sort of transmitting system aboard the Discoverer flights. Earlier flights no doubt carried at least a radio beacon aboard the recovery capsule to aid military recovery crews in locating the film return capsules.

These satellites represent the first generation of recoverable photo recons for the United States.

SAMOS

SAMOS (Satellites and Missile Observation System) represented the first generation of radio transmission type photographic reconnaissance satellites; they were launched from the Western Test Range by the Atlas/Agena launch combination and carried photographic equipment developed by the Eastman-Kodak Company.

According to the September 12, 1960, issue of *Aviation Week*, SAMOS carried cameras with focal lengths as long as 40 inches.

SAMOS 1 This first attempt at putting a radio transmission type recon into space on October 11, 1960, was a failure.

SAMOS 2 On January 31, 1961, SAMOS 2 was launched from the WTR shortly after noon Pacific time and achieved orbit around 2:55 PM PST. Launched into a 549 x 472 km orbit, inclined 95 degrees, SAMOS 2 orbited the earth over 500 orbits.

The satellite used film with a resolution capability of 100 lines per millimeter from 300 miles; ground resolution was better than 20 feet and, with a 10 degree field of view of the lens, an area of over 2,500 miles could be photographed on a single picture. Barely a month in orbit, SAMOS 2's transmitter was cut off without explanation.

SAMOS 3 This was a launch failure due to the satellite not achieving orbit on September 9, 1961. This was the last SAMOS launch identified as such before the military secrecy order took effect.

SAMOS 4 An Unidentified launch attempt using the Atlas/Agena combination occurred on November 22, 1961.

SAMOS 5 The first unidentified launch into orbit, December 22, 1961, using the Atlas/Agena booster. Initial orbital parameters ranged from 747 km apogee to a 232 km perigee, inclined 90 degrees.

This satellite remained in orbit nearly eight months, presumably transmitting a fresh look at the Soviet Union after nearly a year of inactivity. Its 232 km perigee would improve the ground resolution twofold over the SAMOS 2, if the same camera and lens were used. The frequencies used for SAMOS photo transmission are still classified.

Second Generation Recons

Development of the second generation recoverable recon began in late 1960, with the new generation of recon making its debut in mid-1963. These satellites were launched by the Atlas/Agena D combination and would remain aloft only three to five days. Launch times from the Western Test Range normally occurred between 11:30 AM to 2:00 PM local time; by 1965, however, launches were being timed much more precisely and all of them occurred within a few minutes before local noon.

By 1962, Eastman Kodak had reduced the size and weight of the camera system sufficiently to permit the system to be launched by the Thor/Agena rocket system. The Atlas/Agena pair was reassigned to carry heavier cameras with longer focal length lenses used with the recoverable film pack satellites.

Cameras and the satellites' recoverable film capsules were built by General Electric's Missile and Space Division in Philadelphia. These satellites were apparently used to obtain high resolution photos of targets spotted by the radio-transmission type recons. These satellites had lower perigee points than the radio recons, giving them higher resolution.

KH-5

A new launcher, the Thrust-Augmented-Thor (TAT)/Agena D, was developed to loft more than 2,000 pounds of payload into orbit, nearly twice the useful payload that the Thor/Agena could lift. This permitted U.S. Satellite recons to carry more film and batteries for a longer useful life in orbit.

The first two launches, February 28, 1963 and March 18, 1963, were failures. The first successful launch in the new series occurred on May 18, 1963. During the rest of 1963, six more launches into orbit were conducted in this series of search-and-find satellites.

In 1964, launchings of the second generation radio-transmission-type recons occurred at roughly one a month intervals. While these satellites normally remained aloft three to four weeks before natural decay took place, some missions were inexplicably shorter. Occasionally, a second satellite would be launched while the former was already in orbit, suggesting that the first satellite either malfunctioned or ran out of film.

Sometimes there would be a brief gap in continuity, as if the planned launch has been delayed because of booster problems or payload problems.

During 1964, the U.S. orbited 12 radio-transmission-type recons by the TAT/ Agena D combination. These satellites were normally launched in the early afternoon, so that target areas would be illuminated by the sun at an angle, casting slight shadows on all objects.

Ground stations used to receive radio recon transmissions include:

New Boston (near Manchester, NH)
Vandenberg AFB, CA
Kaena Point, HI (island of Oahu)
NW Field, Guam
British Seychelles Island (Indian Ocean, NE of Madagascar)
Kodiak Island (Alaska)
A seventh station in an eastern African country was not revealed to avoid embarrassment to the host government.

Each station used a 60 foot dish to receive pictures from these orbiting spy satellites. The government also used six shipboard stations, each using a 30 foot dish. The first of these ships, the General H.H. Arnold, became operational in 1964.

Third Generation Recons

KH-7

In the summer of 1966, the third generation of radio transmission recons made its debut; launched by the Long-Tank-Thrust-Augmented Thor/Agena D boost complex, the first launch of this advanced skyborne TV spy station was August 9. This new photo intelligence gathering satellite carried not only a higher resolution camera but an infrared scanner as well, probably manufactured by HRB-Singer and TI, Inc.

Equipped with a new system called the SGLS (Space-Ground Link System), the spacecraft could transmit pictures to ground stations at a much faster rate.

Twenty-nine R/T satellites were launched from 1967-1970.

KH-8

The third generation high resolution recons are still in use. Launched on the Titan 3B/Agena D booster complex, the KH-8 provides high resolution imagery and carries mapping cameras and a multi-spectral camera four or six bands), all manufactured by Itek. Some 26.5 feet long with a diameter of about five feet, these satellites weigh three tons. Their image resolution is reportedly about six inches.

Launched from Vandenberg AFB into 347 x 125 km orbits inclined 96.4 degrees, the usual lifetime for these satellites is about 50-80 days.

In the October 6, 1980, issue of *Aviation Week*, defense personnel expressed concern that the U.S. only had four satellites of this type left in USAF inventory, held in reserve for serious military emergencies.

According to the same article, these vehicles have two recoverable reentry capsules, recovered by Lockheed C-130 aircraft as the pods descend near Hawaii. A follow-on system of high resolution recons will be launched via the space shuttle in the mid-1980's and will also have a six-inch resolution.

The KH-8 satellites will be replaced by the CIA's KH-11 digital imaging satellite. The KH-11 will also replace Big Bird, discussed next.

Fourth Generation Recon: Big Bird

KH-9

On June 5, 1971, A Titan 3D/Agena rocket boosted the first fourth generation photo surveillance satellite into orbit. Launched from Vandenberg, the satellite is known by many names: LASP (Low Altitude Surveillance Program), Project Code 467, and the most common, Big Bird.

These spacecraft carry a variety of intelligence gathering devices on board its Agena platform: a high resolution camera manufactured by Perkin-Elmer with a six-inch resolution, a search and find camera, a side looking radar scanner to see through clouds, infrared capability to allow night monitoring, and a 60-kilogram electronic ferret device which is usually cut loose and boosted into a higher ferret orbit for search and find missions.

These satellites perform both recoverable and radio-transmission photo-recon missions, carrying four to six recoverable film pods that are used during its 180 day average mission.

Measuring 50 feet in length and 10 feet in diameter, and weighing some 25,000 pounds, these monsters use a 20 foot erectable dish to transmit photographs back to seven U.S. Air Force ground stations on a frequency near 1800 MHz.

The normal orbital characteristics for Big Bird are 270 x 160 km, inclined 96.4 degrees. The Agena engine is used to prevent orbit decay and maintain 180 days in space. Built by Lockheed in Sunnyvale, California, Big Bird could be history by the time this book is in print.

The satellites were normally placed into sun-synchronous orbits, much like the Landsat earth resources satellites, allowing the satellite to pass over the same targets at the same time each day. The series of pictures thus obtained show changes occurring such as troop movements, flight tests, construction of new missile sites, and the number and types of missiles being installed. Film is processed on board and then scanned by a laser device; the signals are then converted into digital signals and transmitted back to earth.

As mentioned before, Big Birds are capable of performing a certain amount of ferret activity. Not all missions, however, carry the small 60-kg ferret satellites. This could possibly be because earlier ferrets had not yet failed in orbit.

The Big Bird satellites are believed to have played a role in tracking down and identifying several new Russian aircraft and ballistic missiles, including the mobile IRBM SS-20. This satellite apparently replaced two other types.

The following is a list of known Big Bird launches:

Satellite	Launch date
Big Bird 1	June 15, 1971
Big Bird 2	January 20, 1972
Big Bird 3	July 7, 1972
Big Bird 4	October 10, 1972
Big Bird 5	March 9, 1973
Big Bird 6	July 7, 1973
Big Bird 7	November 10, 1973
Big Bird 8	April 10, 1974
Big Bird 9	October 29, 1974
Big Bird 10	June 8, 1975
Big Bird 11	December 5, 1974
Big Bird 12	July 8, 1976
Big Bird 13	June 27, 1977
Big Bird 14	March 16, 1978
Big Bird 15	March 16, 1979
Big Bird 16	June 18, 1980
Big Bird 17	May 11, 1982
Big Bird 18	June 20, 1983

KH-11 Digital Image Recon

Also known as Project Code 1010, the CIA's KH-11 represents the best in U.S. photo recon technology. These satellites are normally launched from Vandenberg AFB into 530 x 250 km orbits but may dip down within 200 miles of the earth's surface for close-ups. Inclined 97 degrees, they are placed in orbit by the Titan 3D/Agena booster complex. They have a two year lifetime and orbit the earth every 92 minutes.

Fully maneuverable in orbit, the KH-11 was recently stationed above Brest at the Poland/USSR border.

About 65 feet long, seven feet across, and weighing about 30,000 pounds, these satellites are capable of operating for longer periods of time and return real-time digital images of photo recon targets back to earth as well as providing SIGINT function. Although the high resolution capability is not as good as the KH-9, these satellites return images to the Mission Ground Site at Ft. Belvoir near Washington, D.C., almost instantly, relayed by another satellite instead of having to wait for film return pods. The frequency used is not currently known.

Film capsules are dropped by parachute north of Hawaii where they are either plucked out of the air or float just under the ocean's surface, emitting a recovery beacon for frogmen.

From there the films go to the National Photographic Interpretation Center in the Washington Naval Yard. After development, specialists use computerized zoom capability to analyze details of tactical targets.

According to *Aviation Week*, a modified version of the KH-11, launched by either the Titan 3D or the Space Shuttle, will still transmit digital data with a resolution close to that of the film return recons. It is also believed that these later models will be retrieved by the Space Shuttle for refurbishment once their consumables are expended.

It is widely reported that the KH-11 uses the SDS relay satellites to send digital imagery back to U.S. ground stations while over strategic targets. This will be discussed further in the SDS section.

Future Developments

According to *Aviation Week*, two additional recon programs under study for the late 1980s would use digital imagery radar instead of cameras to see through clouds.

One will be stationed over Europe to pinpoint Warsaw pact armored units; the other (the Navy's "Clipper Bow" program) will use radar to spot Soviet ship movements.

The only new film return recon being developed would involve the placement of cameras in the space shuttle cargo bay. This system would be used over extremely high priority targets where the resolution provided by the digital image satellites is inadequate.

Military Weather Satellites

In the fall of 1966, the U.S. Air Force launched their first generation of military satellites, built by RCA ASTRO division, devoted to weather observation.

Launched on the Thor/Burner rocket vehicle, these satellites were put into sun synchronous orbits at a circular altitude of around 800 km. Seven satellites in the first generation of DMSP (Defense Meteorological Satellite Program) Block 4 were orbited from 1966 to 1969. They were spin-stabilized, cylindrical in shape and weighed 180 pounds.

All of the spacecraft in this series were equipped with two vidicon cameras, and the last three also carried an infrared sensor array.

Block 5A

The Block 5A spacecraft orbited in 1970 was the first three-axis-stabilized spacecraft flown for operational meteorological observation. Instead of vidicon cameras, these spacecraft used a line-scan sensor in the visual and infrared spectra.

This was also the first DMSP spacecraft with digital control. Three Block 5A satellites were successfully orbited in 1970 and 1971.

Block 5B/C

To meet military requirements for tactical land and shipboard sites, the direct readout capability was more than tripled and the recording capability was doubled in the Block 5B and 5C series of DMSP satellites.

Additional sensors were accommodated, including a precipitating electron sensor, an IR temperature sounder, and a cooled IR image channel. Reliability and mission effectiveness were enhanced. Digital telemetry was used for the first time in this series of satellites, and spacecraft weight nearly doubled that of the Block 5A.

Five Block 5B and two Block 5C satellites were orbited from 1971 to 1975. They carried similar equipment packages as listed:

Sensor AVE Package
- *Very High Resolution Visible
- *Very High Resolution IR
- *High Resolution Visible
- *High Resolution IR

Scanning Infrared Radiometer

Block 5D-1

The Block 5D spacecraft are the largest and most advanced of the DMSP series launched to date. Five satellites had been successfully orbited through 1982 since the initial launch in 1976.

The spacecraft contains a 300 pound payload of sensors, including a line-scanning radiometer, a vertical temperature/moisture sounder, and an auroral electron detector.

Block 5D-1 views the entire globe twice daily, taking visible light and infrared pictures with a resolution of up to ad o mile; in addition, global temperature soundings of the atmosphere are continuously gathered for transmission back to earth.

The Block 5D-1 is a new breed of "smart" spacecraft with guidance and navigation controlled by the satellite from the moment of launch to orbit achievement. Upon arrival on-orbit, the on-board computer precisely points the meteorological sensor payload package toward the earth's geo-center and performs operational functions with minimal ground control intervention.

Spacecraft control is through two miniature, onboard, reprogrammable central computers, particularly applicable to Shuttle deployment.

The Block 5D-1 satellites transmit on the following frequencies: 136.770, 137.500, 137.620, 137.770 and 1698.000 MHz.

Block 5D-2

On December 21, 1982, the first DMSP Block 5D-2 military weather satellite lifted off from Vandenberg AFB on an Atlas E rocket. This new series has a greater payload capability, additional downward-facing sensor mounting area, longer life, and increased flexibility over its Block 5D-1 predecessors.

Among the new sensors on the 5D-2 are:

* A topside ionospheric sounder to provide detailed global measurements of the electron distribution in the ionosphere
* A microwave imager to define the extent of service, identify precipitation areas, and measure soil moisture

  The Block 5D-2 spacecraft has a greater redundancy in selected critical systems and will have twice the predicted lifetime in orbit than the Block 5D-1--36 months rather than 18 months.

Additional spacecraft sensors for the Block 5D-2 include:

Acquire Meteorological Data in Visual and Infrared Spectra

* High Resolution (1.5 nm x 1.5 nm) global coverage
* Very High Resolution (0.3 nm x 0.3 nm) coverage of selected areas
* Near Constant Resolution vs Scan Angle

Meteorological Sensor for TIR service

* Visible and IR imager
* Microwave Temperature Sounder
* Gamma Detector
* Electron Spectrometer
* Other special meteorological sensors

The DMSP Block 5D-2 satellites are capable of providing sensor data on a real-time basis to local air weather service and Navy ground and shipboard terminals as well as storing the data in on-board recorders for later transmission to one of two command readout sites (Fairchild AFB, WA, or Loring AFB, ME)

These sites, as well as the Command Control Center (Offutt AFB, NB), are operated by the Strategic Air Command's 4000th Aerospace Applications Group. Recorded data is also recovered at the remote Hawaii Tracking Station of the Air Force Satellite Control Facility.

Data is relayed by satellite from both Command readout sites and the Hawaii Tracking Station to the Air Force Global Weather Central, at Offutt AFB, NB.

For a given geographic location, mobile ground terminals provide military commanders in the field with photographic-like prints of cloud cover four times a day. As the Block 5D-2 satellites pass overhead, the Mark IV and Mark III mobile ground terminals receive guided and labeled pictures of a section of the earth and its cloud cover; on every pass each satellite covers an area up to 4,500 nm long and 1,600 nm wide.

Mark IV technology now permits weather watchers to enlarge and print selected portions of infrared and visual weather data from both the military DMSP and civilian National Oceanic and Atmospheric Agency satellites.

As of this time the following frequencies are believed to be in use by the Block 5D-2 satellites: 136.700, 137.500, 137.620, 137.770, 1698.00, 1702.50 and 1707.00 MHz.

Listeners will be able to recognize the Block 5D-2 orbits based on the following information: 804 x 804 km circular orbit; 98.7 degrees inclination; and Atlas E launch vehicle from Vandenberg AFB.

Early Warning Satellites

These satellites are used by U.S. military forces to detect enemy missile launches and give U.S. strategic forces time to react. Most of these programs and their satellites are enclosed in a thick veil of secrecy.

MIDAS

This first U.S. program identified as performing an early warning mission by the United States also has the distinction of being the last to carry an announced name since the 1962 clamp down on space program secrecy.

Early MIDAS satellites went into near-circular orbits at roughly 3,200 km altitude. Subsequent launchings since 1962 of unidentified satellites into 3,200 km orbits--much too high for photo recon work--leave little doubt as to their intended function.

MIDAS (Missile Detection and Surveillance) satellites were built by the Lockheed Space Division. These satellites were equipped with instruments that could sense "thermal" infrared radiation that was emitted by the plume of ballistic missiles and other rockets.

The following is a summary of launches from Vandenberg AFB in the MIDAS series of satellites:

MIDAS 1 February 26, 1960; failed when the Agena second stage exploded.
MIDAS 2 May 24, 1960; put into a 480 km orbit initially from Vandenberg AFB. Failed after two days.
MIDAS 3 July 12, 1963, from Western Test Range; 3360 km orbit, inclined 91 degrees.
MIDAS 4 October 21, 1961, into the same orbit as MIDAS 3; last satellite in the series to be labeled
MIDAS 5 April 9, 1962
MIDAS 6 May 9, 1963
MIDAS 7 July 18, 1963
MIDAS 8 August 19, 1966
MIDAS 9 October 5, 1966

MIDAS 1-7 used the Atlas/Agena B combination launch vehicle; MIDAS 8 and 9 used the Atlas/Agena D launch vehicle.

Byeman (Code 949)

On August 6, 1968, using the Atlas/Agena D from Cape Canaveral, the USAF launched a new class advanced MIDAS satellite into a nearly synchronous orbit. Under the cover name of the Defense Support Program (DSP), this new generation of Early Warning satellite used a figure-eight ground track over western Asia to keep tabs on Soviet missile activity.

Built by TRW, these satellites carried SIGINT (Signal Intelligence) packages for ferret type operations. They resembled a ten-foot drum with a nose-like twelve-foot Schmidt infrared telescope.

Placed 35,700 km above the earth in near synchronous orbits inclined 10 degrees to the equator, the satellites etched a figure eight ground track over the higher latitudes of the Soviet Union, drifting from 10° north to 10° south latitude and anchored over Singapore.

Although this orbit allowed visibility of the northern reaches of the Soviet Union, during southern portions of the orbit the satellite was not very effective; thus, on April 13, 1969, another DSP Code 949 satellite was launched at 180 degrees opposite phase to allow monitors to view Russian and China constantly.

This satellite platform carried not just one sensor but several sensors packages; in addition to the infrared sensor used to detect ballistic missile launches and the SIGINT packages, the bird also carried Vela-type detectors for monitoring above-ground nuclear explosions.

Six of the Ballistic Missile Early Warning Satellites (BMWES), as the Code 949 satellites were also known, were launched.

The success of the twin DSP Code 949 satellites was considerable and these satellites served as prototypes for an even more advanced series of satellites that is still in use today, the DSP Code 647 satellites.

DSP Code 647 (IMEWS)

On May 5, 1971, "USAF Launch Operations 3811" was successfully lifted off from Cape Canaveral aboard a Titan III-C rocket. This was the first launch of the IMEWS (Integrated Missile Early Warning System) DSP Code 647 satellites. This satellite was parked into geostationary orbit over the Indian Ocean between the Maldives and Seychelles islands.

The second launch of the series was March 1, 1972, and was parked over the Panama Canal where it could detect submarine-launched missiles from the Atlantic or eastern Pacific Ocean. These satellites are the most important part of the U.S. Early Warning System; with three of these satellites, the U.S. can monitor all potential ICBM and SLBM launch areas for possible attack by enemy ballistic missiles.

According to one report, these satellites which are identical to the 949 spacecraft and have observed more than a thousand ballistic missile launches by the USSR, China, France--and the U.S.

Lauren Paine in his book *The Technology of Espionage* claimed that these IMEWS satellites included a large amount of SIGINT equipment. "It could pick up, record, and transmit emanations from all land communications systems, including radio and microwave telephonic transmissions. It was also rigged with other varieties of electronic surveillance equipment which were sensitive enough to do the work of hundreds of secret listening posts the United States intelligence community had in such places as Turkey, Iran, and West Germany. It was capable of duplicating everything those listening posts could do."

Information on hand at this time indicates that at least one of the IMEWS satellite's frequencies is centered around 250 MHz. Ground stations are located near Alice Springs in central Australia (Pine Gap; code name "Merino"); Woomera, Australia (code name "Casino"); Guam; at Menwith Hill (Harrogale), England; and Buckley Air National Guard Base in Aurora (Denver), Colorado. These sites use milcomsats to relay early warning data back to NORAD and SAC.

Satellite Early Warning System (SEWS)

This follow-on system to DoD's current IMEWS satellites, scheduled for deployment in the early 1990s, will carry better sensors currently being developed by DARPA/USAF.

Few details are now available about SEWS, but it will probably carry infrared mosaic arrays, VHSIC data-processing circuits, and advanced survivable features hardened against enemy attack.

The decision to go ahead with the SEWS system was due to be made in 1985.

DARPA is developing a parallel program called AMEWS (Advanced Missile Early Warning System), a comprehensive on-board data processing system to transmit messages directly to system users.

Nuclear Explosion Detection

Vela

Designed to detect and identify nuclear explosions, Vela satellites were the method used by the U.S. to verify violations of the 1963 test ban treaty with The russians.

There were 12 satellites launched in pairs in this series and they sometimes carried an ERS (Environmental Research Satellite).

Following launch from the Cape by Atlas/Agena D vehicles, these satellites were placed into 108,000 km orbits; each pair was phased 180 degrees to give full earth coverage. Uprated Vela satellites, starting with Vela 9/10, were launched using the Titan III-3 vehicle. Transmission life of the Vela satellites was around three years.

Vela satellites have been replaced by nuclear detection packages flown on the IMEWS satellites. On October 25, 1979, these Vela-IMEWS packages apparently detected a nuclear explosion in the atmosphere in the region of South Africa. The State Department later issued the following statement:

"The United States Government has an indication suggesting the possibility that a low yield nuclear explosion occurred on September 22, 1979, in an area of the Indian Ocean and South Atlantic including portions of the Antarctic continent, and the southern part of Africa. No corroborating evidence has been received to date. We are continuing to assess whether such an event took place."

Many theories about this event were proposed, but nothing was ever proven. This nuclear detection event remains a mystery today.

IONDS

A follow-on system to our current IMEWS-Vela type detectors, Integrated Operational Nudent (nuclear detection) Detection System (IONDS) will be carried by the NAVSTAR GPS satellites from 1988 onward. On-board sensors and GPS-furnished location data will provide rapid detection, location and reporting of nuclear detonations around the world.

During peacetime this system will be used to monitor test ban treaty violations and for intelligence gathering. During wartime the IONDS would provide attack damage assessments as well as the location of enemy warhead impacts so that the NCA could determine the most appropriate response to the attack. Secure links would relay data to terminals on the ground and on Air Force (SAC) command post aircraft.

The first IONDS has been built and has been turned over to Rockwell for integration in a GPS satellite.

Ferrets

Ferret satellites are classified top secret. They are electromagnetic recons used primarily to determine the location and characteristics of aggressors' radars.

The U.S. has a dual ferret program; the small ferrets are used to locate, log and determine enemy radar frequency bands, while larger ferrets are used for more detailed analysis of enemy transmissions.

During the early days of the program, most of these satellites were launched in pairs; single missions probably allowed the U.S. to focus its attention on very specific targets. Single ferrets were also launched as piggybacks to photo recon missions. It now appears that the third generation photo recons never carried any ferret type satellites.

A new geostationary class ferret made its appearance on March 6, 1978. Designed by the National Security Agency (NSA) and code named Rhyolite, the first satellite was placed over the horn of Africa to watch Russian ICBM activity at Tyuratam, Kazakhstan and Plesetsk.

The second Rhyolite was launched on April 7, 1978, and placed over Borneo in geostationary orbit to intercept Chinese and Soviet military communications and radar, and the end-of-mission aspects of Soviet missile tests. Two other Rhyolites were launched as in-orbit spares December 11, 1977, and April 7, 1978. All these satellites are believed to transmit near 1800 MHz.

Navy Ocean Surveillance

Project Whitecloud

The Navy Ocean Surveillance Satellites (NOSS) built by the Naval Research Laboratory are designed to monitor radio and radar transmissions from Soviet surface ships and submarines. These satellites are launched from Vandenberg AFB into near circular 1120 km orbits inclined 63 degrees by the Atlas F rocket.

Three spacecraft are deployed from a carrier vehicle into three parallel orbits with specific latitude/time/distance separation, indicating that interferometry techniques are being used to pinpoint the location of Soviet vessels.

The smaller NOSS satellites measure 3 x 8 x 1 feet. The eight foot side contains solar cells used to power the small satellite; the other sides support the array of antennas. According to *Aviation Week*, each satellite transmits its collected intelligence at slightly different frequencies: 1430.2, 1432.2 and 1434.2 MHz using 1 MHz bandwidth. Narrow-band telemetry is transmitted on 1427.23, 1427.43 and 1427.63 MHz.

NOSS transmissions have generated protests from some radio astronomers because of their proximity to the 1400-1427 MHz hydrogen line band allocated to radio astronomy.

These NOSS satellites generated national public attention in 1984 when it was discovered that the gift shop at NASA's Johnson Space Center was selling postal covers for $1.00 that provided accurate details and drawings of Project Whitecloud.

The drawing showed a mother spacecraft carrying three smaller spacecraft and, in addition, the words, "Built by the Naval Research Laboratory - carries radio-frequency antennas for detecting shipboard radar and communications signals" and "transmit data to parent spacecraft in order to cover a larger ocean surface."

Additional data presented on the cover told the purchaser that the satellite dispensed subsatellites into 700 mile high orbits, gave the mission name "Project Whitecloud" in bold letters, and gave information on the first NOSS launch in April of 1976 from Vandenberg.

According to NRL officials, Whitecloud is so secretive that NRL personnel are not even allowed to mention Project Whitecloud over the telephone.

A more advanced ocean surveillance satellite system proposed by the Navy, Clipper Bow, will use active radar to point out the exact location of Soviet ships. These new ROR/SATs would complement the current White Cloud satellites by permitting correlation of signal emissions received by Whitecloud with radar targets received by Clipper Bow to determine the class of vessel being watched.

As of this writing there appeared to be some problems with the future of Clipper Bow and it is not known if this program will ever be launched.

Navigation Satellites

Navigation satellites provide accurate position information anywhere in the world, day or night, and in any weather. NAVSATs was one of the first widespread applications of space technology for man's convenience and safety.

Transit

The idea of using satellites for navigation originated with the launch of Sputnik 1 on October 4, 1957. Scientists at Johns Hopkins University's Applied Physics Laboratory (APL) monitored that satellite and noticed that when the transmitted radio frequency was plotted on a graph, a characteristic curve of Doppler shift appeared. By studying this apparent change of radio frequency as the satellite passed overhead, they were able to show that the Doppler curve, when properly used, described the orbit of the satellite.

Further study showed that by monitoring a single pass of the satellite, all of its orbital elements could be determined; thus, navigation sets on earth could use the known satellite position and Doppler observations of the satellite to compute positions.

The Navy's Polaris submarines required accurate position information for launching missiles, since accuracy is dependent upon the certainty of position of the submarine at the time of launch. Thus, the Navy contracted with APL in 1958 to build navigation satellites based upon this new knowledge.

The first Transit launch was attempted on September 17, 1959, but it failed to orbit when the third stage of the launch vehicle did not ignite. On April 13, 1960, another launch was attempted and Transit 1B was placed in orbit, but not with the intended parameters.

The satellite was a small sphere with a diameter of 0.9 meters. Most of the mass of the satellite (121 kilograms) was devoted to chemical batteries (a few solar cells were also carried). The satellite used two stable oscillators and two dual frequency transmitters (162/216 MHz and 54/324 MHz) and, although the orbit was lower than planned, Transit 1B did demonstrate the operation of ultra-stable oscillators in space.

Transit 2A, launched on June 22, 1960, was the first satellite to test fly an electronic clock and also carried a Canadian cosmic radio noise experiment, thus giving it the distinction of being the first international satellite. Transit 3A, launched November 30, 1960, suffered a launch failure and is most noted for the international incident it created when some of the debris from the launch vehicle fell on Cuba and purportedly killed a cow.

Transit 3B also was affected by a launch vehicle malfunction which left it in a lower orbit than planned, but it was able to demonstrate the operation of a small magnetic memory system and the Army's Sector transponder for geodetic measurements in its one month lifetime.

Transits 4A and 4B, launched on June 29 and November 15, 1961, respectively, tested improved memory cores and oscillators. They were cylindrical rather than spherical and were the first satellites to use nuclear power sources (radioisotope thermal generators) in space.

The first prototype operational satellite, Transit 5A1, was launched on December 19, 1962, but the mission failed when the command receiver did not function properly. Transit 5A2 failed to orbit on April 6, 1963, while Transit 5A3, launched June 16, 1963, reached the proper orbit but experienced telemetry problems.

The next Transit series, 5B, represented the first satellites to rely entirely on nuclear power sources (radioisotope thermal generators, or RTG's). Transit 5BN1, launched September 28, 1963, tumbled in orbit and recovered in an inverted position such that its antenna pointed away from earth. On December 6, 1963, Transit 5BN2 was successfully launched and operated without failure.

Transit 5BN3 was lost in a launch failure on April 22, 1964, which caused another international incident when it was learned that a kilogram of plutonium (from the RTG) had been dispersed through the upper atmosphere over most of the southern hemisphere. As a result, Transit 5C reverted back to solar power. It was placed into orbit on June 4, 1964, and the Transit system was declared operational the next month.

Since June 19, 1964, Transit navigation satellites have continued to be launched, and the system has achieved a reliability record of 99.9 percent. Early problems with solar cell arrays which caused the spacecraft to cease functioning before its operational lifetime had been reached were solved and subsequent Transit satellites have operated for as long as ten years.

In July 1967, Transit was opened for use by the civilian sector, although some civilian oceanographic research ships had been equipped with Transit receivers during the previous year. The number of civilian users (American and international) of Transit is now on the order of four to five thousand and, as of July 1981, represented 93 percent of Transit use.

Transit Satellite Description

The Transit system relies on stable satellite frequency transmissions to provide high levels of accuracy. To control transmitter frequencies, the satellites use ultra-stable oscillators. Since the refraction of the satellites' transmissions by the ionosphere may cause inaccurate measurements, the transmitters operate on two frequencies so the user can measure the refraction and make requisite allowances in the calculations. The operational Transit satellites carry only one dual

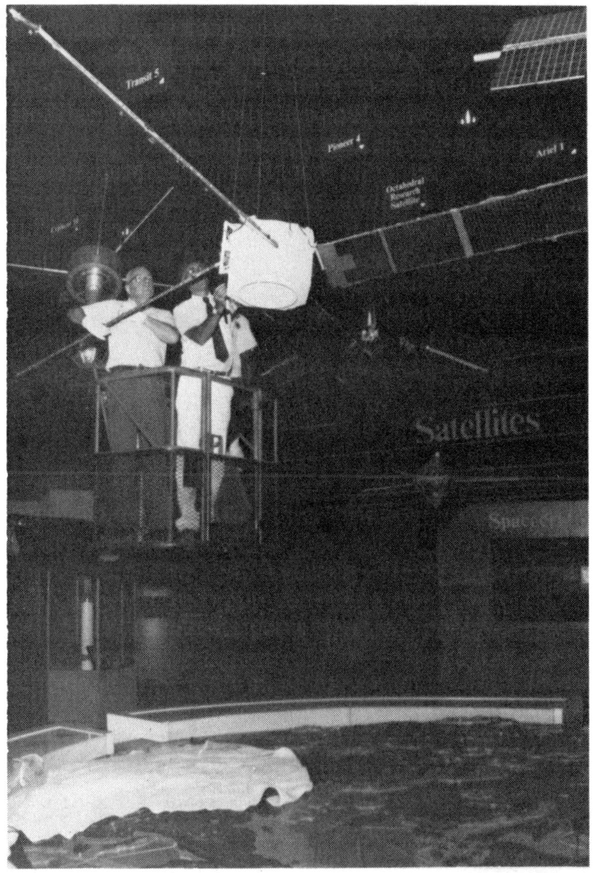

A Navy Transit satellite is removed from its home of more than 15 years at the Smithsonian Air and Space Museum in Washington, D.C. The satellite, called "Polar Bear," carries three experiments into space (Johns Hopkins photo)

frequency transmitter, are lighter than earlier models and can be launched by a Scout rocket.

The satellite is equipped with a memory core containing data on its orbital elements and the time of the day, and broadcasts this information at two-minute intervals. The memory can store 6,000 numbers, enough for 16 hours of operation. The information is updated in the satellite's memory core as it passes over "injection stations" located in Rosemount, MN, and Laguna Peak, CA, which also serve as tracking stations (together with stations at Prospect Harbor, ME, and Wahiawa, HI). The injection stations transmit commands up-dating orbital elements and time corrections to the satellite when it comes within range (approximately every 12 hours).

The total radiated power of the satellite is 1.5 watts, meaning that antennas on the receivers can be relatively small. Current generation Transit oscillators (5 MHz) have a stability one order of magnitude greater than those of the Transit 1B satellite, a drift rate of less than five parts in a hundred thousand million during a twenty minute pass. Current Transit satellites weigh approximately 60 kilograms and use solar panels for power.

To minimize atmospheric drag and provide a high Doppler shift, the satellites are placed in moderately high polar orbits at 1,000 kilometers with a period of 107 minutes. A minimum of four Transit satellites are needed, but even with this many available there are gaps during which data cannot be obtained. The average wait is 90-110 minutes (the length of time between usable passes increases closer to the equator), but there have been occasions when users have had to wait as long as eight hours to take readings. One of the goals of advanced navigation satellite programs is to reduce or eliminate these time delays.

Technology Improvement Programs for Navigation Satellites

The Navy has continued to improve navigation satellites and some of the developments include stronger transmitters, increased computer memory, improved reference clock stability, the capability to compensate for orbit disturbances, and attempts to harden the satellites against radiation. These improvements have been tested in the Timation/Navigation Technology Satellite series of launches and the Transit Improvement Program (TIP).

The first satellite in the TIP series was called Triad and was launched on September 2, 1972. It carried a new station-seeking system called DISCOS (Disturbance Compensation System) which compensated for the effects of atmospheric drag and solar radiation on the satellite's orbit. Triad also carried a new type of RTG for power.

TIP 2, launched October 12, 1975, carried electronics which were hardened against radiation, a computer rather than a hard-wired memory core, and a system for correcting frequency errors in the clock by ground command. The mission reverted back to solar energy for power and failed when the solar panels did not deploy. TIP 3 suffered the same fate after it was launched on September 1, 1976.

Despite the failures in the TIP series, the basic design for a new generation navigation satellite, Nova, was developed. The Nova satellites are the production version of TIP, and the first Nova launch was successfully accomplished on May 15, 1981. The solar powered Nova satellites are placed in orbits like those of Transit and weigh 368 pounds (166 kilograms) at launch.

Nova provides faster and more accurate position fixes than Transit, requires fewer position update transmissions from ground stations, and uses a 25-foot (7.5 meter) long gravity boom for stabilization. Transit user sets are compatible with the Nova satellites.

Navstar Global Positioning System

During the 1960's, the Navy and the Air Force were separately attempting to develop highly accurate, three-dimensional navigation satellite systems. These programs were combined in 1973 and renamed Navstar.

Navy Timation

The Navy program, begun in 1964, was called Timation and was concerned with developing highly-accurate, three-dimensional navigational clocks. The program was based on the principle that given velocity and the time required for a radio signal to be transmitted between two points, the distance or range between the two points can be computed. For this concept a precise, synchronized time of departure and measured time of arrival of the radio signal must be obtained. By synchronizing the signal transmission time to two precise clocks, one in a satellite and one at a ground-based receiver, the transit time could be measured and then multiplied by the exact speed of light to obtain the distance between the two.

The Navy Timation program envisioned a system of satellites in high-altitude circular orbits, testing the high precision clocks needed for this technique to work. The first two experimental passive ranging navigation satellites, Timation I and II, launched on May 31, 1967, and September 30, 1969, respectively, carried quartz crystal standards and were able to demonstrate the feasibility of the passive ranging technique. However, experimental data revealed that crystal clocks could not be made sufficiently stable for long periods of time due to radiation aging of the crystals.

Air Force System 621B

In 1964, while the Navy was starting the Timation program, the Air Force Space and Missile System Organization (SAMSO) began to look at the need for a highly accurate system to provide navigation information for aircraft. SAMSO concluded that satellites offered the best potential and, under a program called System 621B, conducted numerous studies between 1969 and 1972.

The Air Force's preliminary plan called for the three satellite constellations at 120 degree intervals over the equator, each with four satellites: one in geostationary orbit, the other three in slightly elliptic, inclined, 24-hour orbits. This system would not have provided coverage for the polar regions, however. No recorded flight tests of this system were made, but ground tests apparently were made at an inverted range at Holloman Air Force Base in New Mexico. (The inverted range concept is discussed further under the Navstar program description below.

Combined Program

In 1970, efforts began within the Department of Defense (DOD) to consolidate the development of navigation systems. In April 1973, the Navy Timation and Air Force 621B efforts were combined at the direction of the Deputy Secretary of Defense. The Air Force was designated the executive service for the joint program and was directed to prepare plans for a comprehensive system based on the formerly separate programs. The consolidated effort included the orbit configuration of the Navy Timation plan (which would provide worldwide coverage without foreign ground stations), while the signal was based on the Air Force 621B concept.

Timation III, which was renamed Navigation Technology Satellite 1 (NTS 1) after the programs were combined, was the first satellite in the Navstar development program. Launched on July 14, 1974, it tested two rubidium vapor atomic clocks (it carried two quartz standards as well) and operated at two different frequencies (335 MHz and 1580 MHz) to test dual frequency operation for improving accuracy. The satellite experienced tumbling in orbit caused by solar wind pressure against the solar cell arrays. This problem could not be corrected until 1975, but NTS 1 was able to achieve most of its goals.

NTS 2, launched June 23, 1977, tested a cesium beam atomic clock and had two L-band transmitters. This satellite was initially planned to serve as one of four satellites for use in testing Navstar, but a problem developed in a frequency synthesizer and the Navstar payload became inoperative.

There had been plans to launch an NTS 3 to test a hydrogen maser clock, but this flight has been eliminated. The hydrogen maser clock will probably fly on a Navstar satellite sometime in the future.

GPS - Navstar satellite

Navstar Technology

The Navstar satellites are being built by Rockwell International. Because their primary purpose is for military uses, the satellites have many features to protect them against hostile attack (whether through an actual attack on the satellite with an anti-satellite device or through jamming). Navstar is expected to provide ten times more accurate radionavigation in general, and four times more accurate weapon delivery, providing position accuracy within 16 meters (with an 18 satellite configuration which could be improved to ten meters with 24 satellites), velocity accuracy within 0.1 meter per second, and time accuracy within 100 nano seconds.

Navstar is a passive satellite system --the user does not need to query the satellite in order to obtain necessary positioning information. This type of system is advantageous because it can accommodate an unlimited number of users and does not risk giving away the position of the user in a military situation. Data transmissions from the satellite provide the user with the time, the precise orbital position of the satellite and positions of other satellites in the system.

The user can calculate distance by measuring the time it takes for the satellite's radio transmissions, traveling at the speed of light, to reach the receiver. Once distance from four satellites is known, position in three dimensions (latitude, longitude, altitude) can be calculated by triangulation, and velocity in three dimensions can be computed from the Doppler shift in the received signals.

Like the Transit satellites, Navstar operates on two frequencies. To increase resistance to jamming, the signals are pseudo-random noise (PRN) modulated and use spread spectrum technology. The first signal, called C/A (for coarse acquisition), transmits at 1575.42 MHz and will be the only frequency available for civilian users of the system.

The C/A signal is repeated every millisecond, enabling the ground receiver to align its internal clock with the satellite system's time reference. If four satellites are accessed, 30 meter accuracy can be obtained using just the C/A signal, although DOD has decided to permit an accuracy of only 200 meters for now (see section D).

Military users will be able to access a second frequency sequence operating as 1227.6 MHz which carries both the C/A signal and a P (for precise) signal with a repetition period of several days. Using the P code sequence, accuracies of ten meters can be obtained with a 24 satellite constellation and 16 meters with 18 satellites.

Navstar uses atomic clocks to provide the precise time information needed for this system. Each satellite now carries three rubidium clocks and one cesium clock. As noted earlier, problems with the rubidium clocks on early Navstar satellites have been solved, and hydrogen maser clocks are being developed. The satellites use solar panels for power supply and are three-axis stabilized. The satellites weight 760 kilograms at launch (455 kilograms in orbit) and are launched by an Atlas F vehicle. They will be launched by the space shuttle in the future.

A ground segment is needed to track and control the satellites in orbit and to update data in the satellites' memory banks. Orbital data and clock corrections are sent up to the satellites once a day via command and control data links which operate at 2227.5 MHz (downlink) and 1783.74 MHz (uplink). The ground segment headquarters is temporarily at Vandenberg Air Force Base, CA, with three additional tracking stations in Hawaii, Alaska and Guam. As noted earlier, plans to build a Navstar Master Control Station at Fortuna Air Force Station in North Dakota were dropped, and the final control station is now planned for SCOC in Colorado.

Communications Satellites

As with all users of the spectrum, the emphasis on communications has moved higher in frequency--and in elevation; now, seventy percent of all military communications are conducted via satellite.

Let's take a brief look at the development of satellite communications and an intensive look at tactical applications.

Moon Bounce

The first voice message to be transmitted via a space relay used the moon as a passive reflector. In 1954 the Naval Research Laboratory (NRL) transmitted over an earth-moon-earth link successfully, and in 1955 established the first transcontinental transmission via the moon. By 1959 a 400 megawatt duplex RTTY link between Washington D.C. and Hawaii became operational on frequencies of 400 and 450 MHz and by 1960 amateur radio operators were duplicating the feat with a fraction of the power.

Project Echo

The Echo 1 sphere, an artificial echo-producing body was placed in a 1600 km orbit following launch by NASA in August, 1960. It was a mylar-aluminum balloon 100 feet in diameter weighing 176 pounds. Echo^O2, launched in January 1964, was larger and more rigid, weighing 500 pounds; it was placed in a 1280 by 960 km near-polar orbit.

The Courier satellite built by Ford Aerospace & Communications Corporation for the world's first repeater type satellite (Ford Aerospace).

Both satellites were used by the United States and other countries for communications tests and remained in space until May 1968 and June 1969, respectively. NASA conducted its communication tests around 107.9 MHz.

Project West Ford

Project West Ford, sponsored by the U.S. Air Force, was an investigation using an orbital belt of dipoles as passive reflectors. On May 10, 1963, a MIDAS launch successfully dispersed the dipoles in a polar orbit at about 3,700 km altitude.

Designed for operation in the 8,000 MHz range, ground stations in Massachusetts and California measured the dipole belt characteristics and usefulness in point-to-point communications. Speech was transmitted in digitized form and received intelligibly, verifying theoretical predictions. The individual dipole orbits decayed with time and after five years the belt virtually ceased to exist.

Project Score

In December 1958, at Atlas 10-B vehicle put into an elliptical orbit a 150 pound payload capable of both delayed and real time transmissions of voice and teletype. The system operated successfully for 12 days, with an output power of eight watts at 150, 132.405, 132.435, 107.940 and 107.970 MHz.

Project Courier

In October 1960, a Thor-Able Star vehicle put Courier 1B, a DOD satellite, into a 1120 by 960 km orbit. Courier 1B operated for 17 days, with an output power of 3 watts at 2 GHz; it could receive and store 300,000 teletype words during each overhead passage and retransmit them to four ground stations. Tests of delayed and real time voice, teletype and facsimile were conducted between Fort Monmouth, New Jersey, and Puerto Rico.

Project Telstar

Telstar 1, a joint AT&T/NASA project, was launched on June 10, 1962, by a Delta vehicle and placed in an elliptical orbit 5604 km x 950 km. The satellite weighed 170 pounds and was operable on a communication duty cycle of 10%. Telstar 2 was launched in May 1963.

The Telstars provided tests for telephone, television, facsimile, and data among the United States, Great Britain, France, Italy, and Japan until March 1965. The output power was 3 watts with uplinks at 6 GHz and downlinks as follows:

136.0 MHz (telemetry); 4080 MHz (beacon)
Communications experiments:

--simplex	4170 MHz	2.25 watts
--duplex	4165 MHz	1 watt
	4175 MHz	1 watt

Project Relay

In December 1962, Relay 1, weighing 172 pounds, was launched by NASA into an elliptical 7380 km x 1310 km orbit. In January 1964, Relay 2 was placed into a similar orbit. The output powers were 10 watts with 1700 MHz uplinks and 4200 MHz downlinks.

Relay 1 operated successfully until February 1965, supporting more than 2000 communications tests among the United States, Europe and South America; Relay 2 was used by U.S. earth stations until September 1965.

Project Syncom

This NASA project included three satellites. Syncom 1 failed to achieve proper orbit. Syncom 2 was placed in an inclined synchronous orbit in July 1963 but was used for testing. Syncom 3 was placed in a synchronous equatorial orbit near the International Dateline in August 1964, and was used for TV transmissions of the Tokyo Olympic Games to the U.S. in the fall of 1964.

In April 1965, after its program of tests was completed, NASA transferred operation of both satellites to DOD. Neither is currently operational.

DOD Tactical Satellite Communications Activities

In 1965 the three services initiated studies on tactical/mobile satellite communications.

On July 1, 1967, LES 5 (Lincoln Laboratory Experimental Satellite), a UHF satellite repeater weighing 225 pounds, was placed into high orbit by a multiple payload Titan III-C. The UHF system was

Facsimile photos of President Kennedy and Governor General Doctor Namdi Azikiwe of Nigeria were exchanged by satellite.

Syncom 3

Antenna installed at Pt. Mugu, California which received TV transmission of the 1964 Tokyo Olympic Games via Syncom 3

Photo taken from TV monitor during telecast of 1964 Tokyo Olympic Games

Syncom IV-1 rises from the cargo bay of the earth-orbiting Discovery on its way to service for the U.S. Navy (Rockwell Int'l).

tested by SAC aircraft as well as by Army and Navy mobile units. This launch also deployed a gravity-gradient experiment (Dodge), an electronically despun array test satellite (DATS), and the third launch of IDCSP satellites.

In September 1968, LES 6 was launched in further support of the tactical communications study program. This was followed in February 1969 with the launch of TACSAT 1, a high powered experimental tactical communications satellite for use by all military services, to assess the role of satellites in tactical situations.

TACSAT 1 used the following frequencies:

Purpose/Uplink (MHz)	Downlink (MHz)
SHF Communications	
7977.5-7987.5	7252.5-7262.5
SHF Beacon	
---	7298.5
UHF Communications	
302.5-312.5 MHz	249.3875-249.8125
UHF Beacon	
---	254.1

The TACSATCOM program included LES 6 and TACSAT 1, which were both geostationary satellites. LES 6 is a small, single-band UHF satellite, whereas TACSAT 1 was the largest and most powerful communications satellite launched up to that time, with ability to operate in both UHF and SHF bands.

These two equatorial satellites were aligned with the central United States for preliminary tests, then repositioned for overseas testing and limited operation.

The TACSATCOM program primarily served the mobile user in tactical situations; some 100 research and development terminals, mostly using the UHF band, ranged from manpack size to those mounted on 1-1/4 ton trucks. A number of airborne terminals were also tested in the program. TACSAT 1 ceased operations in December 1972.

LES 8/9

On March 15, 1976, a Titan III-C rocket lifted off from the Cape carrying the last two LES satellites and two SOLRAD satellites. The two LES satellites were used to test design concepts for the new Defense Satellite Communications System Phase III (DSCS III) series and satellite crosslinking and operations in the 36 GHz band.

LES 8/9 also tested the use of nuclear power as an alternate source of energy; each satellite carried radioisotope thermoelectric generators aboard rather than solar panels that could be easily damaged by Soviet anti-satellite weapons. Nuclear power also extends the satellites' lifetime since no battery supply is needed during eclipse periods from the sun.

These satellites also tested crosslinking experiments on 225 Hz as well as 36.9-38 GHz to determine the feasibility of using only two satellites for global coverage instead of the usual three or four.

The UHF portion of the LES 8/9 satellites are used in the U.S. Army Satellite Satellite Communications Agency UHF satellite system; one wideband channel for the USASATCOMA system is subdivided as follows:

Downlink	Uplink	Channel
249.350 MHz	303.150 MHz	1
249.375	303.175	2
249.400	303.200	3
249.425	303.225	4
249.450	303.250	5
249.475	303.275	6
249.500	303.300	7
249.525	303.325	8
249.550	303.350	9
249.575	303.375	10
249.600	303.400	11
249.625	303.425	12
249.650	303.450	13
249.675	303.475	14
249.700	303.500	15
249.725	303.525	16
249.750	303.550	17
249.775	303.575	18
249.800	303.600	19
249.825	303.625	20
249.850	303.650	21

The current orbital locations of the LES 8/9 satellites are 108 and 104 degrees west, respectively. Communications monitored on these satellites have been encrypted and no clear voice has been reported to date.

Marisat (Gapfiller)

When Gapfiller was launched in April 1976, Secretary of the Navy William Middendorf II filed the first message sent from a shore station to the satellite which retransmitted the message to ships at sea:

"The transmission of this message marks the achievement of a major milestone as we move toward the realization of a real-time worldwide command and control system. Operations on the Atlantic Gapfiller represent the first major step toward improved links between our fleets and shore commands."

Thus, the satellite communications age came full time to the U.S. Navy. As part of its agreement with COMSAT General, three MARISAT satellites were launched into geostationary orbit from Cape Canaveral aboard a McDonnell Douglas 2914 Delta between February and October 1976. These satellites have a diameter of 85 inches, an overall height of 150 inches, and an in-orbit weight of 727 pounds. Hughes Aircraft, under contract to COMSAT General, built these three multifrequency COMSATs, the world's first space maritime communications system, placed over the Atlantic, Pacific and Indian oceans.

Each satellite is equipped with three transponders: The L-band transponders (1.5/1.6 GHz) are used for commercial ship-to-shore communications, while the C band (4/6 GHz) frequencies are used for commercial coast-to-coast communications links in the United States.

Satellite tracking, telemetry and command functions are conducted over C-band frequencies. Commercial maritime services--voice, teletype, facsimile, and high speed data--are used by over 2,000 vessels including Jacques Costeau's Calypso which left the ATS 1/ATS 3 satellites in favor of the MARISAT satellite system.

On October 1, 1984, the Navy again signed a contract with COMSAT to extend its used of the MARISAT satellites for another year, the ninth year the Navy has used the Gapfiller system.

The following is a system breakdown of the MARISAT (GAPFILLER) satellite series.

Satellite	Ocean	Longitude	Launch Date
MARISAT F1	Atlantic	15° W	2/19/1976
MARISAT F2	Pacific	176° E	10/14/1976
MARISAT F3	Indian	73° E	6/9/1976

The following is a frequency breakdown of the MARISAT (GAPFILLER) satellite series:

Commercial Services:

L-band downlink: 1537-1541 MHz
C-band downlink: 4195-4199 MHz
TT&C: 3945.5/3954.5 MHz

UHF Gapfiller:

Downlinks/Uplinks
254.150/307.750 MHz
Channel 1 - 25 kHz bandwidth
257.550/311.150 MHz
Channel 2 - 25 kHz bandwidth
Channel 3 consists of a wideband 500 kHz channel as follows:

Downlink/Uplink		
248.850/ 302.450 MHz	Ch.	1
248.875/ 302.475		2
248.900/ 302.500		3
248.925/ 302.525		4
248.950/ 302.550		5
248.975/ 302.575		6
249.000/ 302.600		7
249.025/ 302.625		8
249.050/ 302.650		9
249.075/ 302.675		10
249.100/ 302.700		11
249.125/ 302.725		12
249.150/ 302.750		13
249.175/ 302.775		14

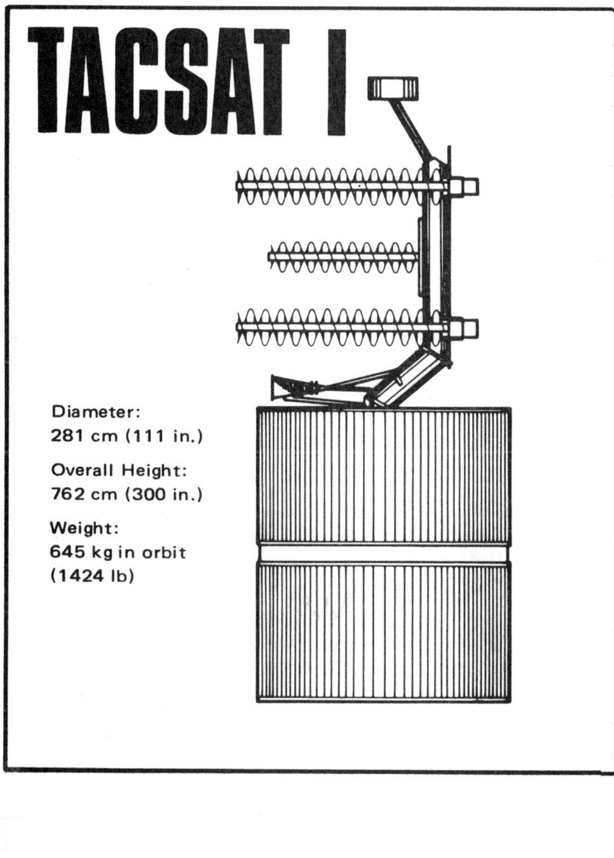

249.200/ 302.800	15
249.225/ 302.825	16
249.250/ 302.850	17
249.275/ 302.875	18
249.300/ 302.900	19
249.325/ 302.925	20
249.350/ 302.950	21

AFSATCOM

The AFSATCOM (Air Force Satellite Communications Network) is the primary means of controlling strategic forces and disseminating emergency action messages (EAM) by the U.S. Air Force and the Department of Defense. Initiated in 1978, the system will be replaced by MILSTAR in the 1990s.

The AFSATCOM system consists of airborne and ground based terminals which provide low-capacity, secure, two-way teletype communications between National Command Authority (NCA), Joint Chiefs of Staff (JCS) and nuclear-capable forces deployed worldwide.

Pacific MARISAT launch

The main element of the system consists of dedicated UHF transponders carried on a variety of other host satellites, including DSCS phase 3 and SDS satellites (each of which carries a one-channel AFSATCOM transponder), the FLEETSATCOM satellites (each carries a 12-channel, 5 kHz narrowband transponder and one 500-kHz wideband transponder for presidential directives), as well as several highly-classified satellites.

AFSATCOM terminals are carried on a variety of platforms including the E-4B airborne command post, RC-135 reconnaissance aircraft, SAC strategic bombers, TACAMO aircraft, various ground based command posts, and ICBM command centers. Navy and Army command posts and, eventually, some SAC tanker aircraft, will also carry AFSATCOM terminals.

SDS

The Satellite Data System consists of a network of three satellites launched from Vandenberg AFB into high Molniya-type 40,000 x 450 km orbits, inclined 63 degrees.

These satellites provide two-way data/voice communications between remote tracking stations and the Central Satellite Test Center at Sunnyvale, California. These stations are also used to control various surveillance satellites of the Defense Department such as the KH-11 and KH-9.

It is believed that these satellites use the 1700-1900 MHz range as well as 243.6925-243.7325 MHz and 243.8525-243.9225 MHz to relay vital real-time data over the pole to ground stations operated by the U.S. Air Force.

The SDS is also an important part of the AFSATCOM network, providing U.S. strategic forces with that crucial trans-polar link that equatorial geostationary satellites cannot provide. The AFSATCOM transponder associated with the SDS might use the 243.7075-243.7325 MHz range.

Listeners should be aware that these satellites are not found along the geostationary equatorial belt like the FLEETSATCOMs, but will appear north of the listener when the satellite is turned on and operating.

FLEETSATCOM

"I take great pleasure in inaugurating...the first satellite of the fleet satellite communications system. The gap is filled." With that message by Admiral James L. Holloway III, then Chief of Naval Operations in 1978, the Fleetsatcom system became operational.

Since the first satellite was launched, there have been four additional launches of Fleetsatcom satellites. These satellites make up the current backbone of U.S. Navy communications worldwide.

The main body of the 2300 pound satellite, a hexagonal structure 7.5 feet wide and four feet high,

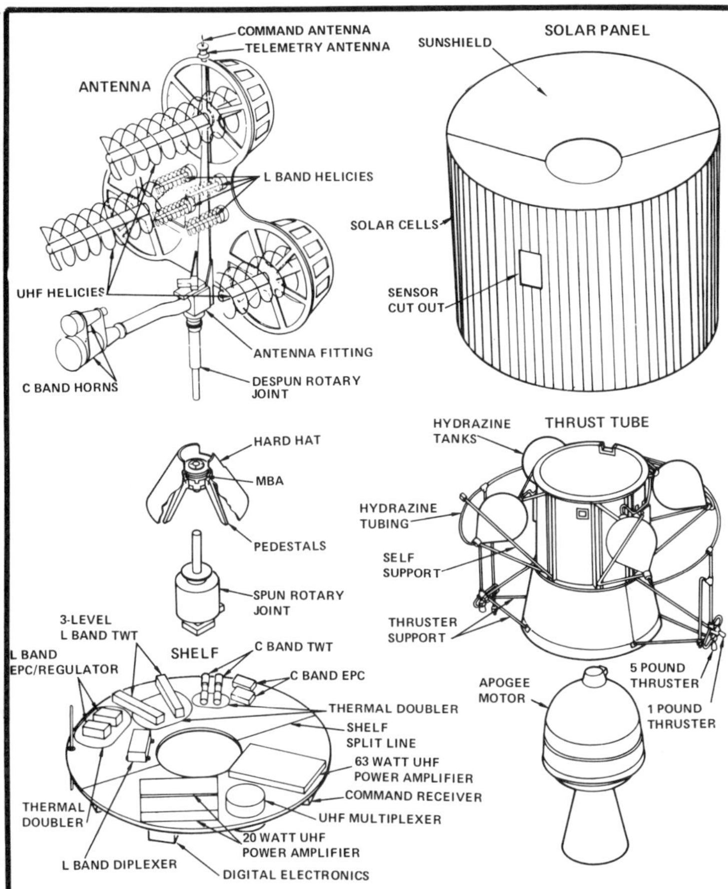

Exploded view of MARISAT spacecraft

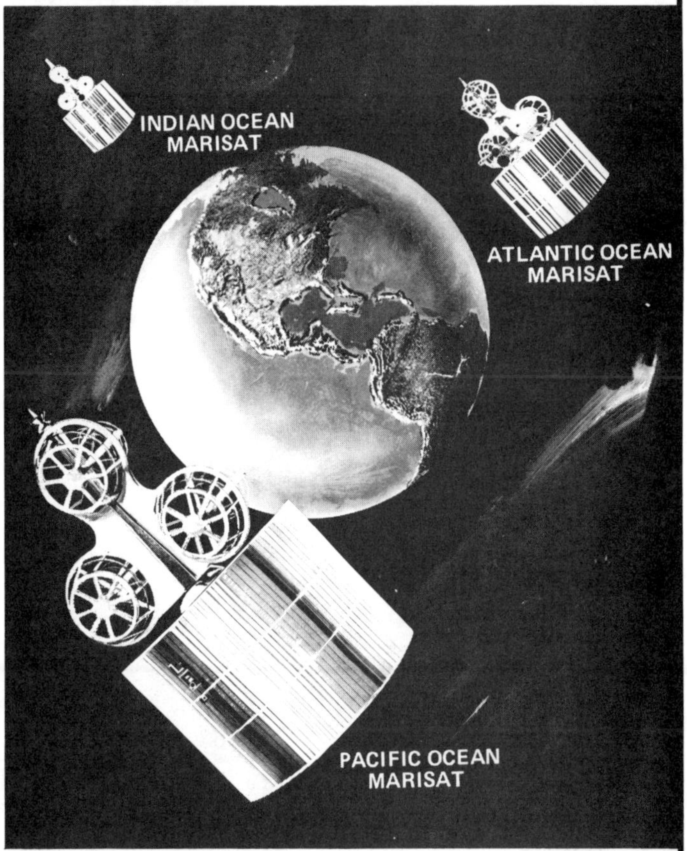

MARISAT satellite system

contains the payload and spacecraft modules. Two solar arrays attached to booms extending from the spacecraft module continually rotate to face the sun. These solar arrays produce a minimum 1,259 watts of power for at least five years, charging nickel-cadmium batteries that power the spacecraft during solar eclipse periods.

The satellites use three-axis stabilization to keep the antennas pointed at the earth. Hydrazine jets alter the satellites' longitude for station-keeping purposes.

The payload section contains three antenna systems, transponder equipment the electronics required to support 23 communications channels. The UHF receive antenna is an 18-turn helix, 12.5 feet long, offset from the satellite payload module.

The UHF transmit antenna is a deployable dish 16 feet in diameter. A separate conical helix antenna sits atop a central mast and serves the satellite's S-band tracking, telemetry and control functions on 2202.5, 2252.5 and 2262.5 MHz.

Each satellite contains the communications package outlined in Table 1. Table III is a synopsis of Fleetsatcom launches to date.

When Fleet F2 was originally launched in 1979, it was located at 23° West. With the launch of F3 in 1980, F2 was moved to the Indian Ocean relay slot.

Fleet F5 was launched as a spare to be available when F1, the oldest satellite in the constellation, died.

Table I

Ch	Communications Package	User
1	1 25-kHz fleet broadcast ch	US Navy
2	1 500-kHz wideband channel (transponder will employ division for 21-25 kHz accesses)	National Command Authority
3-11	9 25-kHz fleet relay ch	US Navy
12-23	12 5-kHz narrowband ch	USAF

Unfortunately for the Navy, the General Dynamics Atlas- Centaur nose cone shroud covering the satellite collapsed inward during launch, destroying the primary antenna. The satellite continues to transmit telemetry, but is essentially useless due to the accident.

After eight delays from its original launch date of May 22, 1986, the sixth Fleet Satellite Communications satellite was launched aboard the Atlas Centaur rocket from Cape Canaveral Complex 36 on December 4, 1986 at 9:30:00.17 p.m.

The spacecraft was injected into a transfer orbit of 35948 km by 162 km 28 minutes after launch. On December 6 the satellite was placed into geosynchronous orbit at 105°. Similar to earlier Fleetsatcoms, F7 also carries an extremely high frequency (EHF) communications package which will be used to test ground stations for the Milstar

Table II: FLEETSATCOM FREQUENCY PLANS

The breakdown of each of the 23 transponder downlinks is as follows (all frequencies in MHz):

	Alpha	Bravo	Charlie	Delta
Channel 1:Fleet Broadcast				
	250.450	250.550	250.650	250.350
Channel 2: 500 kHz wideband				
	260.350-260.850	262.450-261.950	262.050-262.550	263.550-264.050
Channels 3-11: Navy relay channels				
ch.3	251.950	252.050	252.150	251.850
ch.4	253.650	253.750	253.850	253.550
ch.5	255.350	255.450	255.550	255.250
ch.6	256.950	257.050	257.150	256.850
ch.7	258.450	258.550	258.650	258.350
ch.8	265.350	265.450	265.550	265.250
ch.9	266.850	266.950	267.050	266.750
ch.10	268.250	268.350	268.450	268.150
ch.11	269.750	269.850	269.950	269.650
Channels 12-23: AFSATCOM narrow band channels				
ch.12	243.945	244.045	244.145	243.845
ch.13	243.955	244.055	244.155	243.855
ch.14	243.960	244.060	244.160	243.860
ch.15	243.965	244.065	244.165	243.865
ch.16	243.970	244.070	244.170	143.870
ch.17	243.975	244.075	244.175	243.875
ch.18	243.980	244.080	244.180	243.880
ch.19	243.985	244.085	244.185	243.885
ch.20	243.990	244.090	244.190	243.890
ch.21	243.995	244.095	244.195	243.895
ch.22	244.000	244.100	244.200	243.900
ch.23	244.010	244.110	244.210	243.910

Table III: CURRENT FLEETSATCOM SPACECRAFT

Designator/Name	Intl Desig	Launch Date	Orbital Location	Bandplan
F1/Pacific East relay	78-16A	2/9/78	100oW	Alpha
F2/Indian Ocean relay	79-38A	5/4/79	75oE	Charlie
F3/Atlantic Ocean relay	80-04A	1/17/80	23oW	Bravo
F4/Pacific West relay	80-87A	10/30/80	172oE	Bravo
F5/In-orbit spare*	81-73A	8/6/81	44oW	Charlie
F6/In-orbit space	---	3/26/87	Lost, booster failure	
F7/In orbit spare	86-96A	12/4/86	105oW	Delta
F8/In orbit space	---	on hold	unknown	unknown

*Note: Satellite inoperative. The Atlas Centaur nose cone shroud collapsed during launch destroying the primary antenna.

advanced military communications satellite system.

The EHF package, built by the MIT Lincoln Laboratories, operates on 20 watts of power with an uplink at 40 GHz and a downlink at 20 GHz to accommodate low and high rate frequency hopping.

The Fleetsatcom satellites are the spaceborne portion of a worldwide DOD network to enable communications between naval aircraft, ships, submarines, ground stations, Strategic Air Command elements, and Presidential command networks. The Fleetsatcom program is managed by the Naval Warfare Systems Command.

The Fleetsatcom structure weighed 5,061 pounds at liftoff and has a span of 43.4 feet between the tips of its solar panels. The main body is 7.5 feet wide and 22.8 feet high.

The main body consists of three hexagonal modules, the payload, spacecraft and EHF package. The solar arrays extend from the spacecraft module, which contains the hydrazine reaction control system thruster and propellant tanks, Sun and Earth sensors and reaction wheel.

The payload module contains three antenna systems, transponders for 23 channels and communications electronics. The offset mast is the UHF (240-270 MHz) transmitting antenna. A separate conical-helix antenna on the central mast is the S-band tracking, telemetry and control antenna.

The super-high-frequency antenna horn protrudes through a hole in the UHF antenna mesh. The EHF antenna consists of a 5 degree steerable-spot beam and an Earth coverage aperture within look-through cutouts in the center portion of the UHF transmitting antenna.

Power for the spacecraft is produced by a 2,200 watt solar array; three 24-cell nickel-cadmium batteries power the spacecraft when it is in the earth's shadow.

Fleetsatcom F7 replaced Fleetsatcom F6, lost 51 seconds after liftoff from Cape Canaveral on March 26, 1987. The $83 million satellite was destroyed by ground controllers after its Atlas Centaur rocket appeared to be careening out of control. The Fleetsatcom was launched into an area of electrical activity and could have been hit as many as four times by lightning, the apparent cause of the failure.

Fleetsatcoms F7 and F8 are all scheduled to be inserted into parking orbits to replace current operational Fleetsatcoms as required. The new satellites all have a design lifetime of ten years; all four operational spacecraft now in orbit have lifetimes of five years and have exceeded those lifetimes.

FLEETSATCOM

It is believed, based on intercepts and preliminary monitoring, that F7 is carrying a new bandplan Fleetsatcom Delta which corresponds to Leasat Whiskey.

A breakdown of current Fleetsatcom spacecraft and their bandplans is presented in the accompanying tables.

Table IV

Ch	Communications package	User
1	1 25-kHz fleet broadcast ch	US Navy
2	1 500-kHz wideband transp (transponder will employ frequency division for 21 25-kHz accesses)	National Command Author-ity
3-8	6-25 kHz relay transponders	US Navy
9-13	5-5 kHz narrowband chans	USAF

Table VI

Leasat 2	Atlantic Ocean	Leasat Whiskey band plan
Leasat 1	East Pacific	Leasat Yankee band plan-unconfirmed
Leasat 4	West Pacific	Leasat Xray band plan-unconfirmed
Leasat 5	Ground spare	Leasat band plan unknown

Table V

Satellite	Object No.	Launch date	Shuttle Position	Orbital Mission
Syncom IV-2 (Leasat 2)	84-93C	8/31/1984	STS-41D	15° W
Syncom IV-1 (Leasat 1)	84-113C	11/ 8/1984	STS-51A	105° W
Syncom IV-3 (Leasat 3)	85-28A	4/12/1985	STS-51D	72° E
Syncom IV-4 (Leasat 4)	85-76D	8/27/1985	STS-51I	178° E
Syncom IV-5 (Leasat 5)				Ground spare

LEASAT

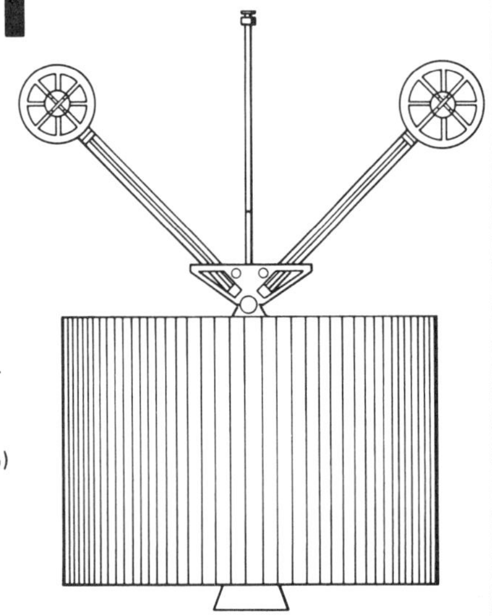

DIAMETER
422 cm (168 in.)
OVERALL HEIGHT
616 cm (243 in.)
WEIGHT
7938 kg (17,500 lb)
aboard Shuttle
1315 kg (2,900 lb)
beginning of life

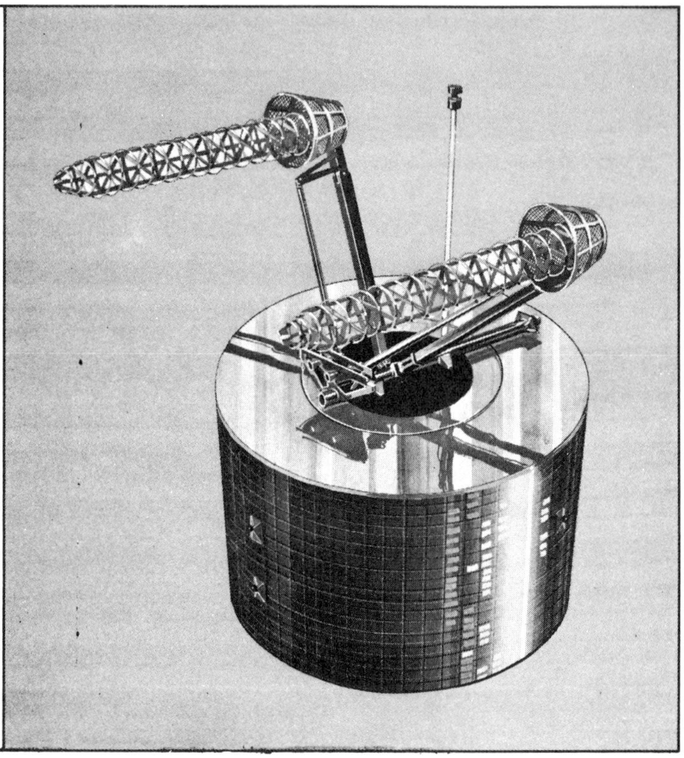

LEASAT

Assuming that the Fleetsatcom system would die from old age at the end of their five year lifetimes, Congress approved the lease on an interim series of communication satellites. Manufactured by Hughes Aerospace, these Leasats (Leased Satellite) provide communications to cover the period between Fleetsatcom satellite failures and Milstar satellite system launch.

The Space Shuttle has now launched four of the Leasat (Syncom) satellites as of this writing, and the satellites have been plagued with major problems almost from the beginning.

Leasats have fewer channels than the Fleets, but with DAMA (demand-assign, multiple access), excellent supplemental channel capacity is realized.

Leasats have a diameter of 168 inches, and overall length of 243 inches, and a weight of 2,900 lbs. each.

Each Leasat communications payload is listed in Table IV. Table V is a synopsis of Leasat launches to date.

Based on information available now Table VI is a lineup of band plans that are currently in use and those of future use on Syncom satellites.

Aviation Week reported that Leasat 2 suffered a wideband (channel 2) failure. Hughes engineers were working to correct the problem. The satellite's remaining six relay, five narrowband and one fleet broadcast channels are functioning properly.

Leasat 4 failed completely on September 6, 1985, after reaching geosynchronous orbit. Hughes Communications of Los Angeles, owner of the crippled Leasat 4, blamed the failure on a faulty cable between the satellite's UHF radio transmission system and its broadcast antenna. The satellite is now considered an $85 million loss and Hughes officials have abandoned hope of reviving Leasat 4.

Leasat 3 was the satellite that was recently repaired by "Ox" Van Hoften on the STS-51I mission. The satellite is currently under Hughes ground control. Telemetry data from Leasat 3 continues to confirm good satellite health. The liquid propulsion systems are intact and the solid propellant perigee-kick-motor temperatures appear to be rising toward acceptable levels. Hughes officials give a 50-50 chance of Leasat 3 being fully revived.

Leasat 1, over the continental U.S., continues to be the only fully operational Leasat in service. This satellite puts in a hefty signal, audible throughout the U.S.

Leasat 5 will replace Leasat 4 as the PacWest relay satellite. West coast monitors should be alert for a new set of milsatcom frequencies that will appear shortly after launch. More than likely the X-ray band plan will be in use.

Defense Satellite Communications System

The Defense Satellite Communications System (DSCS) is an integral portion of the global Defense Communications System, designed to provide vital communications services to the United States and Allied Forces throughout the world by means of satellites. The DSCS has been implemented in phases and the current series, Phase III, has just been introduced.

The DSCS program is designed to link military headquarters, field commanders, and logistic centers throughout the world. Major satellite terminals are located in the United States, Okinawa, Hawaii, Guam,

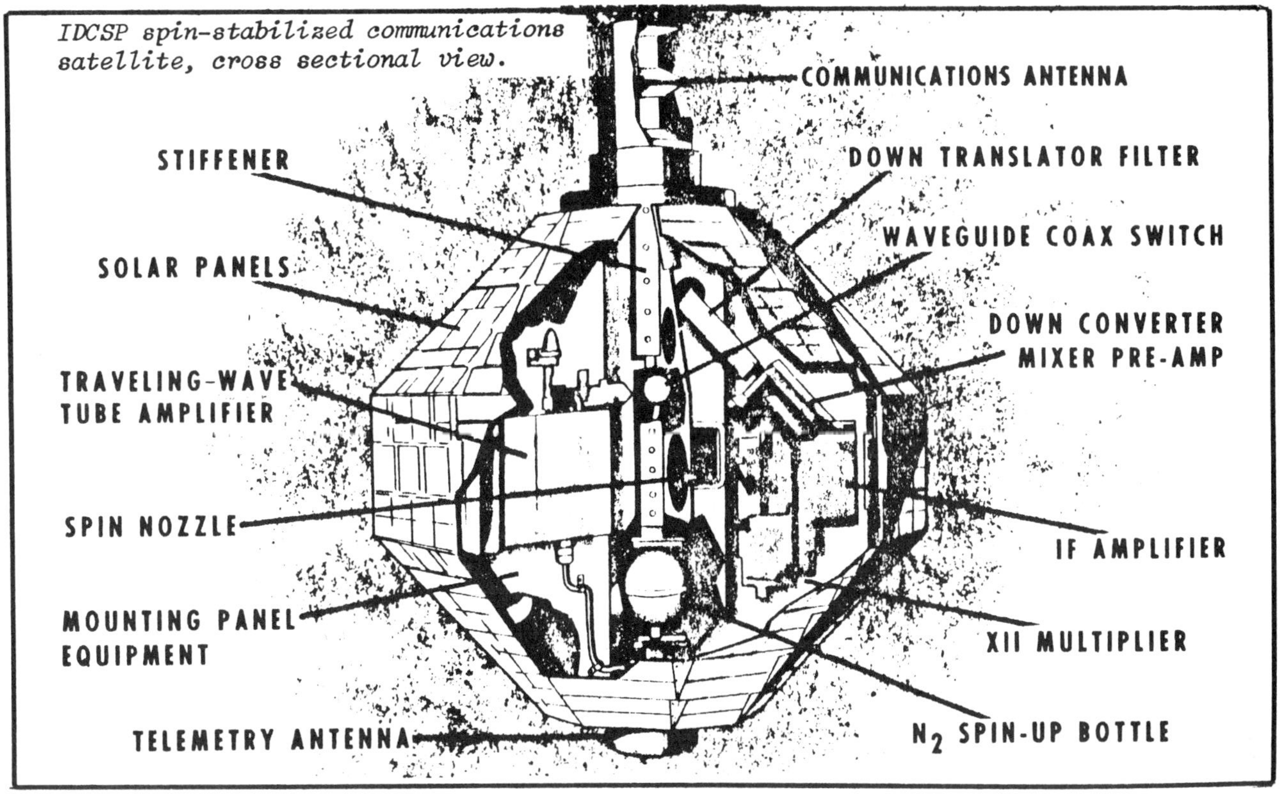

IDCSP spin-stabilized communications satellite, cross sectional view.

Australia, Korea, Thailand, West Germany, and Turkey.

Phase I

The first seven Initial Defense Communications Satellite Program (IDCSP) satellites were orbited on June 16, 1966. The next launch January 18, 1967, carried the next eight satellites in the orbital system; the third launch July 1, 1967, carried three IDCSP satellites into orbit; the final launch June 13, 1968, brought the total operational system to 26 satellites.

The IDCSP was a joint program with each of the military services having specific responsibilities under the direction of the Defense Communication Agency (DCA). The Air Force, as project manager, developed and launched the military COMSATs with the Navy responsible for the shipboard ter minals. The Army's project manager, the Satellite Communications Agency (SATCOMA), developed the ground terminals and conducted the test program.

Launched by a Titan III-C from Cape Canaveral, each IDCSP satellite, approximately 36 inches in diameter and weighing about 100 pounds, orbited between 22 and 23 hours, drifting about 1.3 degrees per hour. The mutual visibility time for a ground terminal through a given satellite varied from a minimum six hours to five days.

The following frequency breakdown is for all IDCSP satellites:

Channel	Downlink(MHz)	Uplink(MHz)
1	7267.0250	7985.7450
2	7271.7125	7990.4325
3	7277.9625	7966.6825
4	7285.7550	8004.4950

Beacon signals (modulated with 110 to 130 kHz tones) generated by each satellite were used by communications ground stations for satellite acquisition and tracking, and for individual satellite identification.

The telemetry system included a telemetry generator, transmitter, sun angle sensor, antenna, and instrumentation sensors that measured critical satellite parameters. Outputs from the sensors, along with individual satellite identifications, were processed as 56 analog telemetry signals and 18 binary signals.

Each of the launch groups had its own 401 MHz telemetry frequency chosen from the following pool:

401.0125	401.0375
401.0625	401.0875
401.1125	401.1375
401.1625	401.1875
401.2125	401.2375

DSCS Phase II

Starting in 1972, the IDCSP satellites were phased out; the newer DSCS Phase II satellites have the potential to carry up to 1,300 full-d with the central United States for preliminary tests, then repositioned for overseas testing and limited operation.

The TACSATCOM program primarily served the mobile user in tactical situations; some 100 research and development geostationary orbit and dispenses them with the correct velocity and spacing.

The satellite's platform carries two earth coverage antennas and two steerable "spot beam" antennas, making the spacecraft continuously visible to earth terminals located within about 5,000 nautical miles of the subsatellite point.

The first two DSCS Phase II spacecraft were launched in November 1971; since that time, 15 DSCS satellites have been placed into orbit. The last launch on October 30, 1982, included the first of the new DSCS Phase III satellites.

TRW developed the DSCS II operational network which consists of four active and two in-orbit spare satellites. To provide for all communication contingencies, the DSCS II satellites contain an on-board propulsion system so that the satellites may be repositioned to new locations in the Clarke equatorial belt.

These satellites also tested crosslinking experiments on 225 Hz as well as 36.9-38 GHz to determine the feasibility of using only two satellites for global coverage instead of the usual three or four.

The UHF portion voice, data, and teletype transmissions among fixed, large terminal users, although some transportable terminals are used by ground mobile forces and the SAC E4B airborne command post. Other users include NATO and the British Ministry of Defense.

The following is a breakdown of the DSCS II communications transponders:

Ch 1: 7975-8100 MHz uplink earth coverage
7250-7375 MHz downlink earth coverage
7250.1 MHz beacon, earth coverage
Ch 2: 8125-8175 MHz uplink narrow beam
7400-7450 MHz uplink narrow beam
Ch 3: 8215-8400 MHz uplink narrow beam
7490-7675 MHz downlink narrow beam
7675.1 MHz beacon, narrow beam
Ch 4: 7900-7950 MHz uplink earth coverage
7700-7750 MHz downlink narrow beam

DSCS Phase III

This newest generation of DSCS satellites is currently being deployed by the U.S. military. Twelve satellites are planned to be launched from either the Space Shuttle or the Titan III-C rocket. Built by General Electric, DSCS III satellites are expected to have a seven year life-time.

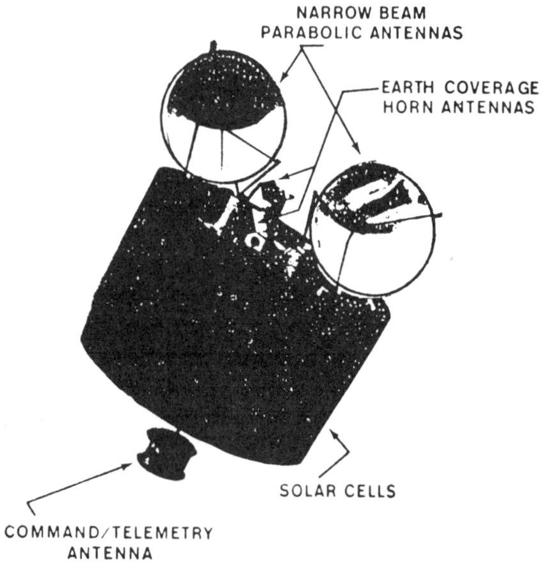

Phase II synchronous satellite, exterior view

These satellites will have many refinements not previously available including pseudo-random noise and spread spectrum techniques to resist jamming; additionally, multibeam antennas can be used to null interfering signals. A steerable dish antenna can be pointed by ground controllers towards different users to help reduce jamming directed toward the satellite.

Users of the DSCS III satellites will employ spread spectrum modems and encryption devices to secure all transmissions through the satellites' communication channels. Six SHF communications channels handle digital voice, data and teletype traffic. In addition to the SHF transponders, DSCS III will also carry an AFSATCOM 225-260 MHz UHF transponder.

The six SHF transponders are are follows:

Channel	
Channel 1	7250-7310 MHz
2	7330-7390
3	7410-7495
4	7515-7575
5	7595-7655
6	7675-7725

The following geostationary orbital positions have been designated for DSCS satellites:

60° E
175° E
13° W
135° W

By the late 1980s, MILSTAR will take over many DSCS users, but DSCS III will continue to pass much of the tactical traffic as well as retain such users as the National Command Authority and the World-Wide Military Command and Control System (WWMCCS).

NATO

NATO Phase I

In 1967 NATO decided to participate in the DSCS Phase I test program, initially leasing and finally purchasing two 15-foot diameter MASCOT satellite ground terminals built by Philco-Ford. These terminals were used to train personnel for the advanced system (NATO SATCOM Phase II) and to demonstrate the feasibility of SATCOM communications for NATO.

NATO Phase II

In 1967 a Memorandum of Understanding between the United States and NATO was signed whereby the U.S. would build and launch two Skynet 1-type satellites for NATO. The U.S. Air Force Space and Missile System Organization (SAMSO) provided overall management for the program.

System studies indicated that two minor changes in the Skynet 1 satellite would provide a better system for NATO. The first modification was to change the equal power division between the 20 and 2 MHz channels of the transponder to a 6:1 ratio, respectively.

The second change was to shift the antenna aiming point from the subsatellite point on the equator to between 4O° and 45°N latitude since all NATO earth terminals would be north of the equator.

The NATO SATCOM Phase II included launching the two modified Skynet-type satellites and building a satellite ground system. NATO II-A was successfully launched into synchronous orbit March 1970 and ceased operation April 21, 1972. NATO II-B was successfully launched February 1971 and ceased operation May 25, 1974.

The operational requirements dictated the used of twelve 42-foot-dish fixed earth terminals near the capital cities of Belgium, Germany, United States, United Kingdom, Norway, Turkey, Italy, Canada, the Netherlands, Denmark, Greece and Portugal.

The frequencies used by the NATO II series satellites were as follows:

Uplink:

2-MHz Channel	20-MHz Channel	Beacon
7976.02 - 7978.02	7985.12 - 8005.12	-----

Downlink

7257.3 - 7259.3	7266.5 - 7286.4	7299.5

The NATO II satellites used a 375-400 MHz subsystem for telemetry tracking and control.

NATO Phase III

The NATO III communications satellite system consists of three satellites in geostationary orbit above the Atlantic Ocean; orders have been placed for a fourth U.S.-built satellite with an option for a fifth, extending the present system's life-time into the late 1980s.

The NATO III space system provides rapid and secure communications among NATO member countries through a network of ground and shipborne stations. The program, funded entirely by NATO, is interoperable with the U.S. DSCS satellites.

The initial satellites in this series were built and launched under management of the Air Force Systems Command Space Division by NASA's Delta 2914 Space Booster from Cape Canaveral; the additional satellites will be launched from the Cape aboard Delta 3914s.

NATO III satellites are 86 inches in diameter, approximately 119 inches high including antennas, and weigh 849 pounds on orbit. The cylindrical surface of the satellite is covered by solar cells to provide electrical power to operate the satellite.

An electrically driven motor controlled by the attitude and antenna control subsystem keeps that three horn antennas--one narrow-beam transmit, one wide-beam transmit, and one common receive--pointed at their proper targets on earth. A 64-element crossed-dipole antenna array is provided for command and control of the spacecraft.

The communications transponder is a three-

The NATO-IIID communications satellite is shown in the final assembly area at Ford Aerospace & Communications Corporation's Western Development Laboratories Division, Palo Alto, CA (Ford Aersopace).

channel, frequency-translating repeater operating in the 7 to 8 GHz range. Signals are received by the wide beam receiving antenna, then amplified, shifted, in frequency, further amplified, and retransmitted back to earth through the appropriate antenna.

An X-band beacon is provided for ground station acquisition, tracking and identification of individual satellites. Command and control of the satellite is provided by an on-board S-band receiver.

The NATO III-A spacecraft was launched in April 1976, and the NATO III-B in January 1977. The III-B spacecraft has been used by the United States Defense Satellite Communications System (DSCS). The NATO III-C satellite was launched November 18, 1978. Orbital locations for the three oper ational satellites are: 18°, 22.5°, 50° W.

The operational frequencies used by the NATO III satellites are:

7245 to 7275 MHz
7272 to 7332 MHz
7347 to 7445 MHz
7339.5 MHz TT & C

Skynet

The Skynet satellite communications program was developed by the United Kingdom to meet its military needs. Studies began in 1962 when the U.K. built five earth terminals, one shipboard terminal (HMS Wakeful), and one land mobile terminal.

The U.S. built and launched two satellites intended to provide a five-year lifetime: Skynet 1A was successfully launched in November 1966 and it functioned as intended until January 1972 when the second of two traveling wave tubes failed; Skynet 1B was launched in August 1970 but it was not placed into synchronous orbit because of the failure of the apogee kick motor.

The follow-on Skynet 2 program used two higher-power satellites: Skynet 2A, launched January 1974, did not achieve orbit because of a second stage misfire; Skynet 2B was successfully launched in December 1974 and placed in geostationary orbit over the Indian Ocean.

Long distance strategic point-to-point digital communications and selected tactical communications with mobiles required two independent satellite bands:

Uplink:		
2-MHz Channel	20-MHz Channel	Beacon
7976.02-7978.02	7985.12-8005.12	--
Downlink:		
7257.3-7259.3	7266.4-7286.4	7299.5

Like the NATO II satellites, Skynet 1 and 2 used a UHF array in the 375 to 400 MHz range for TT & C with redundant UHF transponders and command/telemetry processing equipment.

Skynet 3

This British project never got off the ground; The British decided instead to participate in the U.S. DSCS and NATO projects.

Skynet 4

Planned for launch in the mid-1980s, the Skynet 4 project provides for two three- axis stabilized satellites, each having a design lifetime of seven years. These satellites will be launched from the Space Shuttle and a PAM-D payload-assist rocket.

These satellites will carry a four channel SHF (7/8 GHz) repeater and two independent high-power, narrow-band channels suitable for communications with UHF mobile terminals. Uplink for these two UHF channels will be around 300 MHz and downlink around 250 MHz. Skynet 4 will also carry an EHF research and development package in the 43 to 45 GHz range.

Polar Bear testing facility (LTV Aerospace and

Future Military Comsat Systems

Lasercom

A space-based laser communications system described as having "mind-boggling potential" is being developed by the Air Force Space Division, Los Angeles, California, to meet future Department of Defense requirements for rapid, secure, world-wide communications: Lasercom, the Laser Communications System. When operational, the system would be capable of transmitting in slightly over one second the entire text of the Encyclopedia Britannica!

The Lasercom system uses a modulated laser light beam to carry information rather than the radio frequency waves now used by radio and television.

The shorter wavelength of the laser system, measured in millionths of a meter, makes it possible to send larger volumes of data than conventional RF systems and to focus the laser light that carries the data into narrow beams.

Present RF satellite communications systems can handle about 24,000 voice channels or eight color television channels at a time; the suitcase-sized Lasercom can simultaneously handle about 40,000 voice channels and 12 color TV channels. If only used for a telephone relay it could carry over 250,000 conversations at once.

A Lasercom system being developed by McDonnell-Douglas Corporation can use terminals that are smaller, lighter, and require less power than present RF terminals. Lasercom will eventually be used on board space satellites with airborne, ship and ground facilities.

Seen from earth the laser beam from a Lasercom satellite would only be as bright as a star or about one-twentieth as bright as a 60-watt light bulb.

From a satellite orbiting 22,000 miles from the earth, the Lasercom light beam is less than one-half mile in diameter when it reaches the earth's surface; compared with the 300-mile-wide signal broadcast from current radio link satellites, the narrow Lasercom beam makes enemy eavesdropping and jamming nearly impossible.

Test models of the Lasercom system have been flown on board aircraft from air to ground checks and have been used between ground stations. A series of tests, using full scale prototypes was completed in December 1980 at the White Sands test facility in New Mexico.

Plans call for the Lasercom system to be test flown in the mid-1980s on a host satellite furnished by the Air Force Space Division's Space Test Program office.

Milstar

DOD's next generation of highly survivable comsats, Milstar, will operate in the EHF region using 44 GHz uplinks and 20 GHz downlinks. Launches from the Space Shuttle are expected to begin in 1987.

The narrow beam EHF signal provides virtually

unlimited capacity and will be extremely difficult to monitor or jam; components' weight will also be greatly reduced.

Milstar will consist of a network of eight satellites, five in geostationary orbit (four active and one in-orbit spare) and three polar-orbiting satellites. Milstar will also have an AFSATCOM UHF capability to allow current users of UHF to gradually phase in to Milstar.

These satellites will carry ECM and encryption equipment as well as cross-links between satellites (developed by Lasercom) to ensure that critical communications can be maintained.

A Polar Bear in Orbit

Museums are where you find old, historical things, put there never to be used again...until now. On November 14, 1986, a U.S. Navy Transit navigation satellite that hung in the Smithsonian Air and Space museum for about eight years was launched into orbit! The satellite was placed in the Smithsonian in the mid-1960s straight from assembly in response to a Smithsonian request for an operational spacecraft. This satellite is now called Polar Bear (Polar Beacon Experiment and Auroral Research).

Polar Bear was conceived as a follow-on to the satellite mission called Hilat (High Latitude), a satellite which carried two similar experiments to Polar Bear's and was launched in June 1983. As with Hilat, the program office requested a spare Transit from the Navy, but no satellite was available.

A trade was arranged between the Smithsonian, who was known to have a partially-tested Transit, and the John Hopkins University Applied Physics Laboratory in Baltimore, MD. The lab had a Transit that didn't qualify for spaceflight.

The 275-pound Polar Bear was on display from 1976-1984. While displaying the satellite, the Smithsonian museum stored the satellite electronics. When tested at John Hopkins' lab, all the electronics worked and only one solar cell out of thousands on the solar panels was rated marginal.

Miscellaneous Military Satellites

Listed below is a pot pourri of the DOD's numerous research and minor military satellites with known frequencies.

Satellite Name	Int'l No.	Period	Mission (minutes)	Frequencies (MHz)
TRAAC	1961AH2	105.8	Transit Research & Attitude Control	136.65/ 324.0
Anna 1B	1962BM1	107.9	Geodetic Satellite	54.0/ 162.0 216.0 324.0
ERS9 (TRS 4)	1963-30B	107.8	Radiation research	136.890
ERS 13 (TRS 6)	1964-40C	2,364	Radiation research	136.771
Radsat SN-43	1964-83C	106.1	Radiation research	136.563
ERS 17 (ORS 3)	1065-58C	2,608	Environmntal research	136.768
EGRS 7 (SECOR)	1966-77B	167.5	Geodetic satellite	136.800
ERS-15 (ORS-1)	1966-77C	167.6	Five coldwelding exp.	136.440
EGRS- (SECOR 8)	1966-89B	167.6	Geodetic satellite	136.830
OV5-3 (ERS 20)	1967-40D	2,829	Radsat	136.260
EGRS 8 (SECOR 9)	1967-65A	172.1	Geodetic satellite	136.840
EGRS 13 (SECOR 13)	1969-37B	107.2	Geodetic satellite	136.800
OV5-6 (ERS-26)	1969-46B	3,115	Solar Flare monitor	136.380
---	1969-82E	103.49	Unknown	137.410
SERT 2	1970-9A	106.0	Ion engine test	136.920 136.230 136.928
TOPO 1	1970-25B	106.9	Army SECOR type sat.	136.840
SSU precursors	1971-11A	104.9	Project Whitecloud Precursor	138.800
SSU precursors	1971-110C	104.9	" " "	137.080
SSU precursors	1971-110D	104.9	" " "	136.320
SSU precursors	1971-110E	104.89	" " "	137.050
SOLRAD 11B	1976-23D	7,333	Solar study sat.	136.530
P76-5	1976-47A	105.5	Comm propagation exp.	137.648 378.609 390.082 401.550 401.5515 413.028 424.501 435.974 447.447 458.920 1239.084 2891.171
Scatha (P78-2)	1979-07A	1416	Scientific satellite	2222.5 2241.5 2250.5 5765

The satellite was launched into a 1000-km polar orbit aboard an Air Force Scout G-1 vehicle from Vandenberg Air Force Base, California. Polar bear carries three joint-sponsored experiments that will be used for improving communications over the earth's polar regions by minimizing the natural interference of the polar environment. It has an expected lifetime of three years.

One experiment, the Auroral Imaging Remote Sensor (AIRS), sponsored by the Air Force Geophysics Lab, Hanscom air Force Base, Massachusetts, and built at the Applied Physics Lab (APL), photographs the aurora borealis (northern lights). AIRS is a multiwavelength ultraviolet and visible-light scanning device that can produce images (night and day) or the auroral oval that encircles the region above the poles.

The Beacon experiment, developed by the Defense Nuclear Agency, consists of three transmitters operating at VHF, UHF and L-band frequencies. This experiment measures electron-generated scintillation in the earth's ionosphere above the polar regions and samples the electron composition of these regions. The L-band transmitter also provides the satellite's telemetry link to the ground stations located in Sondre Strom Fjord, Greenland; Churchill, Canada; and Tromso, Norway.

The last experiment, an APL-designed Magnetometer Data Processor experiment, uses a three-axis vector magnetometer together with a digital data processor. This experiment provides a high-rate, high-resolution sampling of the earth's magnetic field over the poles in addition to supplying controllers with information on the spacecraft's orientation or position in space.

Using the existing satellite saved more than $2 million; Another $6-8 million was saved by using flight-qualified spare parts in the buildup of the Scout vehicle.

The Polar Bear is stabilized in space by a 60-foot-long gravity-gradient boom for roll and pitch, and by a momentum wheel (similar to a gyroscope) for yaw control. Initial tracking was provided by the North American Air Defense (NORAD) Command and processed at the Johns Hopkins' APL. Thirty days later the satellite was turned over to the Navy Astronautics Group (NAVASTROGRU) in California for operational tracking, command and control.

Although the frequencies have not been published and the Air Force would not release them, a listener found the VHF beacon on approximately 149.981 MHz; the UHF beacon is probably on 399.962 MHz. The L-band is probably around 1575 MHz.

Commercial Satellites

Much day-to-day command and support traffic is presently based on channels leased from commercial satellite operators. The DOD is also looking at the possibility of using direct-broadcast satellite systems for distributing non-critical messages. Users would require a simple receive-only terminal, the cost of which could be kept to a few hundred dollars.

Chapter Nine

SPACE SURVEILLANCE

NORAD

Just over two decades after the first man-made space object rocketed into earth orbit, the art of keeping track of satellites has evolved from two scientists with a desk calculator to the sophisticated space tracking system employed by the North American Aerospace Defense Command.

The nerve center of NORAD's spacetrack mission is the Space Defense Operations Center, or SPADOC, located deep inside Cheyenne Mountain in the Colorado Rockies. There, computers keep a constant record of the movements of thousands of man-made space objects circling the earth. These computers receive a steady flow of data from a network of radars and other satellite-watching devices located around the world.

SPADOC: Watching Space from Underground

Housed in a steel building within a chamber carved out of the solid granite core of the mountain near Colorado Springs, the SPADOC Computation Center maintains a computerized catalog of orbiting satellite payloads and pieces of space debris. The center charts the present position of these satellites, plots future orbital paths and, for the larger objects, forecasts the general time and location where they might re-enter the earth's atmosphere.

In the center's catalog, dating back to 1957 when the Soviets opened the space age with the launch of Sputnik 1, are more than 16,000 objects; more than 4,700 are still in space ranging from small fragments the size of a grapefruit to satellite payloads weighing several tons.

These satellites appear to SPADOC crews on 'scopes as computer-generated lines against a world background map. The displays show the present course of the satellite and the paths it will follow in the future.

While SPADOC is primarily interested in satellite payloads, it also keeps track of space debris--spent rocket bodies, launch hardware and fragments that result from in-space breakups of larger objects. In fact, the majority of the objects now in space are pieces of debris--"space junk."

Space Computation Center at NORAD (NORAD)

Some of the space vehicles and debris cataloged by NORAD are in deep space where atmospheric drag is considerably less than near the earth's surface; these objects will stay in orbit for hundreds or even thousands of years.

Objects no longer in space are logged at the center as having "decayed"; this means they have been recovered, burned up during atmospheric reentry, or have impacted on the earth, moon or other planets.

SPADATS: Network of Sensors

To perform its mission of keeping track of these man-made objects, NORAD uses a network of sensors to accumulate observational data. This sensor network consists of radars, telescopes, cameras, and radio receiving equipment.

Known as the Space Detection and Tracking System (SPADATS) the network makes some 30,000 satellite observations each day. All these observations are transmitted to the Computation Center computers for updating the master data base and correlation.

Although the system is keeping watch on thousands of satellites, it cannot track *all* man-made objects in space; deep space probes go beyond the range of most sensors, and there are gaps in the coverage around the earth.

Optical sensors are limited to hours of darkness and clear weather while radar beams are too narrow to search large areas of deep space at one time. Some pieces of orbiting debris are too small to track accurately.

Sorting Space Traffic

Each time a piece of orbiting hardware is detected by a sensor, the information is sent to the SPADOC Computation Center computers where the observation is compared with predicted locations of cataloged satellites. If the observation information cannot be verified or matched with a known satellite, then it may indicate a newcomer in space and the approximate orbit is determined by the center's technicians.

As the object continues to circle the earth, additional observations are made, orbital characteristics are established, and the new satellite is entered in the catalog.

Often, it takes several hours to accumulate enough information to form an accurate mathematical description--an orbital element--for each satellite. The set of figure includes the period (time it take for one revolution around the earth), the inclination to the equator in degrees, the perigee (closest point of the satellite to the earth's surface), and the apogee (high point of the satellite's orbit).

What Goes Up...

...Must come down. In addition to keeping track of them when they are launched and orbit the Earth, the SPADOC Computation Center follows satellites when they reenter the atmosphere.

The most celebrated reentry of a man-made object occurred July 11, 1979, when the American spacecraft Skylab impacted without harm to life or property in Australia.

Although man-made objects reenter from orbit on the average of more than one per day, only a very small percentage actually land on earth. The heat and force of atmospheric reentry causes most of them to break up and burn. Of those that do survive, most impact in the oceans (75% of the earth's surface) or uninhabited areas.

One of the most important military uses of trajectory information on a reentering satellite is to alert the Missile Warning Center. Without this prior knowledge, a returning space object could appear to be an incoming warhead and provide a false indication of missile attack.

Another reason for keeping track of a falling spacecraft stems from the 1967 United Nations space treaty that states each country is responsible for damage caused by its returning satellites; consequently, the SPADOC Computation Center closely monitors every space object large enough to survive reentry and impact on earth.

There are many factors that make it impossible to predict precisely where and when a satellite will come down: gaps in the tracking network prevent continuous surveillance; atmospheric drag and solar radiation influence both the speed and course of a returning object.

Depending upon when a satellite was last observed by one of the tracking sensors, and considering its speed of about 300 miles a minute, the area of atmosphere where the reentry is predicted to occur could be as much as 100 miles wide and 1,000 miles or more long.

Historically, 95 out of 100 objects have decayed within the NORAD prediction "confidence window"-- ∓20% of the time from the last observation to the predicted decay time.

Recovery of fallen space debris is not a NORAD responsibility; however, the command does make its information available to civil and international agencies concerned with such matters.

Computers play a large part in NORAD. Today nearly 100 computers are in use in the mountain to keep track of millions of bits of information received each day. (NORAD)

A Joint Effort

Like other elements of NORAD--a bi-national, combined command of United States and Canadian forces--the Space Detection and Tracking System involves people from several branches of the armed services. It has sensors operated by the U.S. Air Force, Navy, Army and Canadian Forces at such widespread locations as Korea, Alaska, new Zealand, and Italy.

Among these sensing devices are space-watching cameras with telescopes that can "lock on" a satellite more than 20,000 miles away. New "electronically agile" phased-array radars can maintain tracks on more than 20 new or 200 known satellites at the same time.

One of the phased-array radars is on Shemya Island, Alaska, near the end of the Aleutian chain. This radar looks northwest to the Russian land mass 450 miles away for intelligence-gathering on Soviet ballistic missile tests. This far-north sentinel can also warn of ballistic missile attack on North America as well as keep a watch on satellites.

Phased-array construction allows a radar to scan large expanses of space in fractions of a second. Energy fed into the individual antenna elements of the radar's face is electronically steered--there are no moving mechanical parts to limit the speed of a radar scan.

In a normal surveillance mode, the Shemya radar is designed to have a 99 percent probability of seeing an object the size of a baseball at 2,000 nautical miles. In the tracking mode, it can handle as many as 200 satellites to ranges of more than 1,000 miles.

New phased-array radars that watch for missiles as well as satellites are at Concrete, North Dakota, and Eglin Air Force Base near Fort Walton Beach, Florida.

Also contributing to the spacetrack mission are radars of the Ballistic Missile Early Warning System at Clear, Alaska; Thule, Greenland; and Fylingdales Moor, England.

Conventional radars are also part of the sensor network. An installation at Pirinclik, Turkey, has two huge detection radar antennas which resemble big curved screens standing on edge, and a dish-shaped tracking radar.

The immobile detection antennas transmit radar energy into space in the shape of a large fan. When a satellite penetrates the beam, the tracking antenna is alerted and locks its narrow beam of energy on target, following the space object across the sky to establish data on its orbit.

Space Cameras

The most sensitive and precise satellite-tracking instruments currently reporting information to NORAD are optical devices known as Baker-Nunn cameras. Named after their designers, the cameras are operated jointly by the U.S. Air Force and the Canadian Forces.

The three-ton, ten-foot-tall telescopic cameras can photograph light reflected from an object the size of a basketball out to approximately 20,000 miles in space. Normally, the cameras are directed to observe satellites known to be in orbit beyond the range of other sensors and for which precise data are needed.

The Baker-Nunn cameras must be operated in twilight or darkness when the space object is illuminated by the sun against the star field in the darkness of space.

If the camera is stationary, the moving satellite appears as a streak of light on the time-exposed film with the stars as bright dots. The camera can be swiveled on its gimbal ring to match the speed of the satellite, and then the object appears on the film as a dot and the stars as streaks.

The film is processed at the camera sites and the information is transmitted to the NORAD SPADOC Computation Center, where the satellite's position can be determined with great precision.

The U.S. Air Force Baker-Nunn cameras are at Pulmosan, Korea; Mt. John, New Zealand; San Vito, Italy; and Edwards Air Force Base, California. The Canadian Forces have a camera and another optical device at St. Margarets, New Brunswick.

Satellite watchers in the space computational center have tracked more than 15,500 man-made objects in earth orbit including Skylab. The center maintains a computerized catalog of orbiting space objects, charts their current positions, plots future orbital paths, and forecasts the times and general locations where they will reenter the earth's amtosphere.

Hawaii Facility

Another installation that uses telescopes and cameras is the Maui Optical Tracking and Identification Facility (MOTIF). Located atop Mount Haleakala at an altitude of 10,000 feet, MOTIF has visible-light and long-wave infrared sensors for collecting data on both near-earth and deep-space satellites.

Equipment at the facility includes dual 48-inch telescopes, a 22-inch acquisition telescope, two photomultiplier tubes, low-light TV cameras, a long-wave infrared array, and associated video and data-processing devices.

The telescopes detect and track the reflected light from the satellites. By noting the precise location of the telescope mount at precise instants of time, azimuth and the elevation of the telescope can be determined and the accurate position of the satellite can be calculated and transmitted to NORAD in near real-time.

Like the Baker-Nunn cameras, MOTIF's visible-light sensors must be operated in darkness and the satellite must be illuminated by the sun or other light source. The infrared sensor, however, can be employed whenever a satellite of interest is radiating enough energy to be detected.

Limiting factors for both the Maui and Baker-Nunn sites are that skies must be clear and winds must be less than 30 mph. The Baker-Nunn cameras have limited search ability. The MOTIF, however, does search deep space for objects that have not been cataloged by NORAD when it is not being used to provide positional or signature information on known satellites.

U.S. Navy Involvement

Another important contributor to the Space Defense Operations Center is the U.S. Naval Space Surveillance System, an electronic "fence" stretching 3,000 miles across the southern United States from Georgia to California. This energy field reaches 6,000 miles into space.

The system has been functioning since 1959 and consists of three powerful transmitter stations and six receivers. The transmitters are at Gila River, Arizona; Lake Kickapoo, Texas; and Jordan Lake, Alabama. The receivers are at San Diego, California; Elephant Butte, New Mexico; Silver Lake, Mississippi, Red River, Arkansas; and Fort Stewart and Hawkinsville, Georgia.

Each transmitter station sends out a continuous wave of radio energy in a fan-shaped pattern with a very narrow north-south dimension and a wide east-west spread. This creates the "fence"--a vertical east-west fan of radio energy extending thousands of miles into space.When a satellite passes through the beam, energy is reflected back to the receivers.

The receiver stations measure the reflected satellite signal and send their data to the system's operations center and headquarters at Dahlgren, Virginia, where the signals are processed by high-speed computers and the position of the satellite is precisely determined.

This information is then transmitted to the SPADOC Computation Center where it is used--along with data from the other sensors--to determine where the satellite has been, and where it will be on successive revolutions.

Other Contributing Sensors

Although not assigned to NORAD, other sensors contribute nearly full time tracking support to the SPADOC Computation Center as part of the Space Detection and Tracking System. These include a U.S. Army radar at Kwajalein Atoll in the Pacific; U.S. Air Force radars on Ascension Island in the Atlantic; Antigua Island in the Caribbean and at Kaena Point, Hawaii; and a deep-space tracking radar operated by the Massachusetts Institute of Technology Lincoln Laboratory at Westford, Massachusetts.

Improving Space Surveillance

A need for more timely information on the steadily growing deep-space satellite population has resulted in plans to improve NORAD's capability to detect and track high altitude satellites.

A system using telescopes, video cameras and computers is being developed to detect and track space objects from 3,000 to 25,000 miles from earth.

The Ground-based Electro-Optical Deep Space Surveillance (GEODSS) is being tested at an experimental site at White Sands Missile Range, New Mexico; it is the forerunner of a five-station worldwide network planned for nighttime surveillance of deep space. Like other NORAD optical sensors, the GEODSS equipment must be operated in clear weather at twilight or night when the sun illuminates satellites within its view.

The new system will have observatories girding the earth at intervals to insure worldwide coverage. The first three sites are operational at White Sands; Taegu, South Korea; and Maui, Hawaii.

Each installation will have three telescope sensor systems linked to video cameras. Two of the three will be 40-inch-aperture main telescopes for primary deep-space tracking and will be able to search large areas of the sky for faint, slow-moving objects. The other 15-inch telescope is intended for wide searches of lower altitudes, where objects are traveling at faster speeds.

The television cameras will feed their space pictures into a computer, which drives a display scope. Stars are automatically filtered from the night sky backdrop by the computer, and the satellites appear on the display screen as streaks of light.

Positional and identification data can be transmitted to the SPADOC Computation Center in seconds, thus overcoming the time-consuming film processing required by the Baker-Nunn system.

Communications Network

The lifelines of the Space Surveillance system and NORAD are the communications circuits--they must be fast and reliable. "When we pick up a telephone," a NORAD official declared, "we expect to talk to someone at the other end--right now."

NORAD's Cheyenne Mountain is at the hub of a vital communications system which utilizes satellites, microwave routes, buried cables, and a short-wave backup system. There are two microwave antennas on the mountain and two coaxial cables are buried deep from many miles from the facility. Any one of the antennas or cables can carry the entire communications load for the complex.

Most of the communications traffic is data, flowing from computers at surveillance and warning installations directly to computers in Cheyenne Mountain. The sensors, communications and computer systems form the largest and most complex command and control network in the free world.

With its communications hot lines, the NORAD underground operations center can contact the Pentagon or White House, Canadian Forces headquarters in Ottawa, the Strategic Air Command, other control posts of the aerospace defense system, and major military centers around the world.

These are the lines that also carry vital intelligence to the NORAD commander in chief and his advisors at their positions in the three-level Command Post.

On the lower level are the command director and his assistant along with data display and communications technicians; these people keep the Command Post functioning around the clock. They monitor the airspace situation, maintain information on air defense forces, and recall the commander in chief and other members of the battle staff when necessary. They also display on 'scopes and screens information depicting the current air defense status, as well as missile warning and space surveillance activities.

Overlooking the display screens from the upper level is the U.S. civil defense National Warning Center. It is located in the command post to have access to warning information at the same time it is available to NORAD. The center would sound the alarm over civilian alerting circuits.

The keystone of the control operation is a specially-designed command and control system based on the speed and reliability of computers that receive, process, store, and display air defense information.

If the NORAD commander calls for a particular display in the Command Post, a technician punches up the request; the system queries a computer, translates computer material into symbology, and displays it to the battle staff.

Data can be shown on console 'scopes or one of two screens, each 12 by 16 feet, that form part of the forward wall of the Command Post. Without leaving his or her chair, the display technician can build a picture of North America that will show tracks of unidentified aircraft approaching the continent. For more detailed information on any track, data can be called up from the memory core of the computer.

By pushing more buttons, the operator can command a computer to display the paths of orbiting satellites, positions of foreign submarines off continental shores, or the status of weapons available to NORAD for defense against bomber attack. Both the present and future orbital paths of a satellite can be projected on the screen; superimposed on a world background map, a satellite's ground trace tells where the space vehicle will be for several hours in the future.

Information displayed in the Command Post can come from a Distant Early Warning Line site on the Greenland ice cap or a Ballistic Missile Early Warning System radar in Alaska, for example. This information flows through the computer and onto the display consoles without ever being processed by human hands. This is done in "real-time" so the battle staff can watch the aerial track at almost the same time it is being tracked by a NORAD sensor.

In the event an attack comes, the NORAD commander in chief and his staff would provide attack warning to the National Command Authority and manage the aerospace defense battle over the United States and Canada from inside Cheyenne Mountain.

NORAD uses the AFSATCOM transponders, DSCS satellites and terrestrial systems for communications. They maintain a backup HF network with transmitter locations at Cheyenne Mountain; North Bend, Oregon; and Hamilton, California. USB radio checks are made approximately on the hour; callsigns change often and are similar to SAC ground station callsigns. The following frequencies (kHz) have been noted in use:

5297	9023
9793	11441
14894 Primary	20855

Chapter Ten

SOVIET MANNED SPACE PROGRAM

Introduction

On the evening of April 9, 1980, Geoffrey Perry, a stocky, silver haired schoolmaster at the Kettering Boys School, switched on his unmodified $100 Eddystone communications receiver. Perry was on vacation in his trailer, parked near the village of Bude in Cornwall, England.

He turned the dials of his receiver expertly and listened. In less than a minute, he heard what he had been waiting for: a series of beeps that sounded like a radioteletype signal. The signal sounded distant at first, then stronger as time wore on.

Perry told his wife to keep listening as he walked out to a telephone booth. He was calling Reuters news service--collect, of course--and told them that the Russians had just launched a manned flight into orbit.

Shortly after his phone call, a news bulletin flashed over the Reuters news wire announcing the flight of Soyuz 35, the first for the Russians in over a year. Moscow acknowledged the flight several hours later.

For Geoffrey Perry, it was another in a string of coups stretching back to the early 1960s when he first started monitoring Russian spacecraft.

In the nearly 30 years since the launch of their first satellite, individuals such as Geoffrey Perry have monitored and studied the Russian space program. These dedicated individuals have reported many successes in lifting the Soviet veil of secrecy that surrounds their space program.

One can liken the Russian space program to a best selling detective novel: The plot unfolds slowly, details are sketchy, clues are hard to obtain, but the results can be most satisfying when the mystery is finally solved. Listening to the Soviet space program is one of the most exciting and challenging aspects of the monitoring hobby today.

Some Background

Since October 4, 1957, when Sputnik 1 was placed in orbit by the Soviet Union, the question of Soviet-American rivalry in space and its broader implications have been matters of public concern, waxing and waning with the irregular occurrence of new space spectaculars. Sometimes the question raised by space events have been matters of serious economic and military concern; other times they have been domestic and international political issues.

The Soviet quest for military supremacy is no secret and has now expanded into space. With the development and employment of an orbital antisatellite (ASAT) weapon over a decade ago, the Soviet Union clearly signaled its recognition of space as an arena in which to conduct a war.

In the past ten years, the USSR has been launching more than 75 spacecraft per year, a rate four times that of the United States. Over the past few years, the rate has been close to 100 launches annually.

The annual payload weight placed into orbit by the Soviet Union is even more impressive: 660,000 pounds--ten times that placed into orbit by the United States. While this level of effort reflects the importance the Soviets attach to their space programs, it also reflects some technological weaknesses in the areas of longevity and flexibility.

Soviet military and military-related space programs range from extended manned missions to meteorological, communications, navigational reconnaissance, surveillance, targeting, and antisatellite missions.

The magnitude of the Soviet space program comes into even clearer focus with the realization that the USSR currently has under development a heavy-lift space launch system, comparable to the U.S. SATURN 5, that will be able to place payloads at least six to seven times the weight of those that the U.S. Space Shuttle places into orbit.

Since the USSR's successful launch of Sputnik 1, the Soviets have pursued a vigorous, deliberate and methodical program to exploit space for military purposes. The initial Soviet use of space for military applications occurred in 1961 with the launch of their first photographic reconnaissance satellite.

Throughout the 1960s and 1970s, Soviet military space programs expanded in scope and number. By the advent of the space shuttle era, these programs had developed into mature, integrated systems, for the direct and indirect support of Soviet military forces, and with the capability to threaten low-altitude U.S. and allied satellites.

The military, political, and economic value of

Soviet space programs dictates that general control and direction of the Soviet space effort rest with the Politburo. The fundamental decisions for military space systems research, design, development, testing, and production at the national level are made in the Soviet Defense Council.

Actual program management is overseen by the Military Industrial Commission which reports directly to the Council of Ministers. All five components of the Armed Forces take part in the development and operation of the Soviet space program. The Strategic Rocket Forces provide the launch and tracking support.

The USSR has invested heavily in three extensive space assembly and launch complexes at Tyuratam, Plesetsk, and Kapustin Yar. Each of these complexes is protected from external attack by surrounding networks of air defense, including manned interceptor bases and surface-to-air missile complexes.

New launch facilities are being built at Tyuratam to handle the newest generation of Soviet heavy-lift space boosters which include the one in the SATURN 5 class, another somewhat smaller expendable booster and a reusable system similar to the U.S. shuttle.

Additional developments include newer spacecraft with greater flexibility; upgrading of command, control and tracking sites; upgrading of ships dedicated to supporting space activities; and expansion of production, research, development and test facilities.

On any given day, 70-110 Soviet satellites are in orbit, more than half of which serve military purposes solely. Some 85 percent of all Soviet space launches are exclusively military or joint military/civilian missions.

The resources required to sustain this effort are enormous. The Soviet space program in the past has relied heavily on modified ballistic missiles for space system boosters. Eleven of 14 launch systems successfully flown by the Soviets have used stages from ballistic missiles.

The Soviets evidently intend to use their various space launch vehicles for at least another decade. This current family of vehicles, combined with the new generation of spacecraft systems, indicates that the Soviets' will be able, by the mid-to-late 1980s, to increase significantly their space programs both in number and weight of payload.

Current evidence indicates that the Soviets are developing a space transportation system (STS) that, in many respects, copies the design of the U.S. Space Shuttle. The new heavy-lift vehicle (G booster) is based on the core rockets of the STS, and a new medium-lift vehicle that is evidently designed for high launch rates is also under advanced development. Significant new launch support facilities at Tyuratam are nearing completion to support the aforementioned programs.

New spacecraft under development include a space shuttle that differs from the U.S. shuttle only in the respect that the main engines are not on the orbiter. Recent tests have been conducted on a small space plane (Kosmolyot) that could be a test vehicle or a scale version of a military space plane. The Soviets have also continued testing their space tug that would be used in space to move equipment and supplies from one orbit to another.

The Soviet shuttle could be first tested in the mid-to-late 1980s. The space plane scale model has already been tested three times, and the space tug probably will reach operational status late in the 1980s.

The Soviets continue to pursue their manned space programs maintaining in orbit the Salyut space station. This station is manned during most of the year, enabling the Russians to perform a variety of functions from space including military research and development, and manual augmentation of their photo reconnaissance and surveillance efforts.

A larger and permanently manned space station, expected during the next decade, will significantly increase their in-orbit operations capabilities. This station (possibly called "Kosmograd," Russian for "Star City") could be used as a stepping stone to interplanetary exploration and the establishment of bases on other planets. The Soviets, however, are more than likely to use such a station to perform command and control, reconnaissance and targeting functions.

The Russians have steadily increased their space photographic and electronic reconnaissance effort since the early 1960s. Each year over 50 of these satellites are launched to support military forces on a daily basis. They provide target identification, location and characteristics, order of battle, force monitoring, crisis monitoring, situation assessment, geodetic information for ICBM targeting, and mapping.

The Soviet military space program also reflects an ever-increasing use of space for worldwide surveillance and warning. The Soviets have brought a number of U.S. and Allied military forces under surveillance by these space based systems.

The surveillance satellites include an ICBM launch detection system and ocean surveillance systems. Two of the RORSAT satellites with nuclear-powered generators caused worldwide alerts during their breakup and reentries to earth in 1978 and early 1983.

Soviet efforts in the surveillance field are expected to lead to a multi-satellite detection, surveillance and attack-warning system against strategic and non-strategic ballistic missiles, and possibly bombers as well.

The Soviets are increasing their use of space systems for command, control and communications. Since the mid-1960s when the first Soviet Molniya communications satellite was launched, the USSR has continued to improve and expand its communications satellite programs to support its political leadership and its military, diplomatic and intelligence missions.

The Soviets are emphasizing the development of communications networks using satellite systems that will be placed in geostationary orbits. Currently, the geostationary satellites--Gorizont, Raduga, and Ekran--occupy orbits with the announced function of fulfilling part of the Soviet domestic and international communications requirements. These satellites could also provide military communications to ground, sea and air elements of the Soviet armed forces.

The Soviets have embarked on an ambitious expansion of their communication satellite program that will add measurably to their global command, control and communications capability. Over the next ten years, the Soviets will develop and deploy an even more advanced series of communications satellites, some of which might relay transmissions from manned orbital command and control platforms.

The majority of Soviet military space programs has been specifically designed to support terrestrial military operations; however, the development of an antisatellite system has extended Soviet military use of space to a capability for direct space warfare operations. It is important to re-emphasize the size of the Soviet program, demonstrated by a launch rate this is four to five times that of the United States, and the size of the Soviet annual payload weight placed into orbit--660,000 pounds--ten times that of the United States.

Soviet Launch Sites

The Soviet Union has three space launch complexes, just as does the United States. Even the functions of these three locations share a similarity with the United States sites.

Tyuratam

The largest and most versatile of the Soviet launch sites is near the rail stop village of Tyuratam in Kazakhstan at 45.6° N latitude, 63.4° E longitude. The Russians call it the Baykonur Cosmodrome, although it is about 370 kilometers southwest of that station stop on a different railway line.

It may have been thought that by giving contradictory information about the cosmodrome, the Russians would maintain some element of doubt in the western world, since the town of Baykonur is on the correct ground trace of the early Soviet flights which were at 65° inclination.

To this day the Russians pinpoint the launchpad for manned flights as being at 47.3° N and 65.5° E, which is patently false in light of the conclusive public evidence of initial revolution ground traces and known launch times.

Tyuratam was the site from which the first Soviet ICBM's were fired, all the early Sputniks, all manned flights, all lunar and planetary flights, the earlier communications satellites, all the fractional orbit bombardment system (FOBS), and the military interceptor flights. It is also the area from which all heavy payloads are put up by the Proton "D" launch vehicles. Evidence now indicates that Tyuratam will be the launch site for the largest Soviet launch vehicle (G booster), still under development.

In effect, Tyuratam is the "Kennedy Space Center" of the Soviet Space program.

The SL-4 space launch vehicle is the workhorse of the current Soviet manned space program. (DOD)

Plesetsk

The second of the Soviet launch sites is near the town of Plesetsk on the railway from Moscow to Archangel at 62.8° N latitude, 40.1° E longitude in European Russia. After many years of official silence the Russian publication *Pravda* in June of 1983 acknowledged the existence of Plesetsk.

It is finding increasingly heavy use, primarily as an operational site, in contrast to the experimental or specialized nature of the Tyuratam flights.

Plesetsk is, in effect, the "Vandenberg Air Force Base" of the Soviet Union. From here are launched many of the navigation satellites, weather satellites, and the majority of the military satellites.

Since mid-1977, most of the Molniya-class inclined-orbit communications satellites which previously were launched from Tyuratam have also been launched from Plesetsk. With its northern location, Plesetsk is used for missions which require coverage of extensive parts, of earth, since even flights launched due east for maximum payload capacity cover most of the inhabited earth.

Plesetsk has been discussed in the western press as a missile launching area. Its later space role presumably was known to western governments, but the first public disclosure of this space cosmodrome came for the Kettering Grammar School in England. Geoffrey Perry published the first clue in April 1966 shortly after the first Plesetsk launch in March, 1966. He pinpointed the location in November 1966 when flights at different inclinations had established a nodal point of crossing ground traces.

As additional kinds of missions were launched from the Plesetsk area, their orbital patterns of orbital inclinations suggested launch pads scattered over a considerable geographic area. Landsat pictures confirmed to the public that Plesetsk was spread over tens of kilometers, although it was not quite as large as the Baykonur complex.

When weather conditions are just right, an occasional Plesetsk launch has been visible from Sweden and Finland as the rocket rises above the horizon. The launch of the Meteor 2-7 satellite on May 14, 1981, was observed from the roof of the Department of Physics at the University of Umea at 2153 UTC. A student took a color photograph which was published in Jan. 18, 1982, issue of *Aviation Week and Space Technology*.

Kapustin Yar

The third Soviet launch site is near Kapustin Yar on the Volga River below the city of Volgograd at 48.4° N latitude, 45.8° E longitude, also in European Russia. This site was also finally recognized by the Soviets in a June, 1983, issue of *Pravda*. Early launches from Kapustin Yar consisted of small Vertikal rockets, and short and medium range missiles.

It was not until 1962 that small payloads were placed in orbit from the Kapustin Yar site using the smallest of the Soviet space launch vehicles, and only in 1973 did they start space launches from Kapustin Yar using the intermediate size launch vehicle.

The use of smaller vehicles at Kapustin Yar for launching vertical probes make this site seem to parallel the U.S. Wallops Island, Virginia and White Sands, New Mexico test areas.

Manned Space Program

Despite setbacks in their earth orbital program, from docking failures to space tragedies which have taken the lives of four cosmonauts, the Soviets' enthusiasm for manned space flight has not waned; and even though they have not succeeded in putting cosmonauts on the Moon, they talk about sending people to Mars in the near future.

They have extended the duration of manned space flight to 237 days, enabling Soviet physicians to conclude that, although there are medical problems with long duration spaceflight, none will prohibit flights of greater length.

This aspect of the Soviet space program should prove most interesting to the average monitor enthusiast. Since the Soviets make use of both HF-AM and VHF-FM, operations are readily monitored with inexpensive equipment in the radio shack.

Vostok

On April 12, 1961, Major Yuriy A. Gagarin became the first man to orbit Earth. His ship, Vostok 1 (code name "Kedr"), made a single orbit from Tyuratam and was recovered in Kazakhstan. The electrifying news produced the same kind of shock waves in the world as the launch of Sputnik 1, despite the advance notice implied by the Korabl launches earlier in 1960 and 1961.

There were a total of six Vostok launches, a summary of which is given below.

VOSTOK NO. 1

Launch date:	4/12/1961
Apogee/Perigee:	327km/181km
Inclination:	65.0
Booster:	A-1
Code name:	Kedr (Cedar)
Cosmonaut:	Gagarin

Frequencies used: 9109, 19995, 20006 kHz/ 143.625 MHz

VOSTOK NO. 2

Launch date:	8/6/1961
Apogee/Perigee:	257km/178km
Inclination:	64.9
Booster:	A-1
Code Name:	Oryel (Eagle)
Cosmonaut:	Titov
Frequencies:	15765, 19995,

20006 kHz/ 143.625 MHz

VOSTOK NO. 3

Launch Date:	8/11/1962
Apogee/Perigee:	251km/183km
Inclination:	65.0
Booster:	A-1
Code name:	Sokol (Falcon)
Cosmonaut:	Nikolayev

Frequencies:	19995, 20006 kHz, 143.625 MHz

VOSTOK NO. 4

Launch date:	8/12/1962
Apogee/Perigee:	254km/180km
Inclination:	65.0
Booster:	A-1
Code name:	Berkut (Golden Eagle)
Cosmonaut:	Popovich
Frequencies:	19990, 20006 kHz, 143.625 MHz

VOSTOK NO. 5

Launch date:	6/14/1963
Apogee/Perigee:	222km/175km
Inclination:	65.0
Booster:	A-1
Code Name:	Yastreb (Hawk)
Cosmonaut:	Bykovsky
Frequencies:	19998, 20006 kHz, 143.625 MHz

VOSTOK NO. 6

Launch date:	6/16/1963
Apogee/Perigee:	233km/183km
Inclination:	65.0
Booster:	A-1
Code Name:	Charyka (Sea Gull)
Cosmonaut:	Tereshkova
Frequencies:	19995, 20006 kHz, 143.625 MHz

Voskhod

Prior to the flights of Voskhod 1 and 2 there were two Cosmos precursor flights to test out the hardware the men would use.

On October 6, 1964, Cosmos 47 was put into an orbit 413 x 177 km with a 64.8 degree inclination. After just one day the Soviets initiated retrofire and the spacecraft returned although the carrier rocket remained aloft for another eight days. Spacecraft telemetry was noted on 19994 kHz.

Six days later a manned flight (Voskhod 1) was launched with orbital elements of 409 x 178 km; it also stayed up for one day.

On February 22, 1965, Cosmos 57 was put into orbit 512 x 177 km, but something went wrong, and the payload was exploded in orbit. Voskhod 2 did not follow as closely after this precursor; more time was required to assure that whatever went wrong with Cosmos 57 would be unlikely to occur in the manned flight to follow. Hence, the followup flight was delayed 24 hours and then entered a 495 x 173 km orbit. Cosmos 57 telemetry was monitored on 19997 kHz.

Voskhod 1, launched October 12, 1964, was the first multimanned flight. The crew was led by Col. Vladimir Komarov, accompanied by a military physiologist, Lt. Boris Yegorov, and a civilian technical scientist, Konstantin Feoktistov. The flight only lasted one day and was launched with the following orbital parameters: 409 x 178 km orbit, 65.1 degree inclination. This was the first manned use of the A-2 booster which is still used by the Soviets today.

Code name for the flight was Rubin (Ruby) and the following frequencies were monitored: 17365, 18035, 19994 kHz, and 143.625 MHz.

Voskhod 2, launched March 18, 1965, was commanded by Col. Pavel Belyayev, accompanied by Lt. Col. Alesksey Leonov. Leonov won a place in history by being the first man to walk in space. The crew orbited the earth for 16 orbits and was launched into a 495 x 173 km orbit. The A-2 booster was used and the mission had the code name of Almaz (Diamond). Telemetry noted during the flight included: 17365, 18035, 19996 kHz and 143.625 MHz.

After Voskhod 2 a number of statements suggested that further manned flights would occur, but we did not see any more manned Voskhod flights. It appears that at least one more Voskhod flew, but carried dogs instead of people. Designated Cosmos 110, the spacecraft was launched February 22, 1966, into a 904 x 187 km orbit inclined 51.9 degrees. The spacecraft carried the dogs Verterok and Ugolek. A television monitor sent back pictures of the dogs, while telemetry from biological and cabin environment sensors reported other information.

The flight set a duration record of 22 days and the dogs were successfully recovered. HF telemetry was monitored on 19894 kHz but no frequency is currently known to the author where TV transmissions occurred.

Soyuz

The Soyuz (Russian for "Union") program is a long term project aimed at the development of sustained flight, orbital maneuvering, docking, scientific, technical investigation, and the assembly of manned orbiting labs in earth orbit.

There have been 40 Soyuz flights and several precursors under the Cosmos label since the beginning of the series April 23, 1967. A detailed examination of the manned ferry system will be in the Soyuz-T section.

Cosmos 133/140 (Soyuz precursors)

After a 22-month lapse in the Soviet manned program, western analysts debated whether the Soviet program had paused or had been abandoned. The launch of Cosmos 133 on November 28, 1966, heralded the start of the Soyuz manned program. The unmanned spacecraft was put into a 232 x 181 km orbit and recovered after two days. Frequencies used for this mission included 10013 and 19995 kHz.

A second precursor was orbited on February 7, 1967, into a 241 x 170 km orbit inclined 51.7 degrees. Cosmos 140, like 133, was orbited for two days and the frequency observed in use was 20008 kHz.

Soyuz 1: The first Soviet space tragedy

On April 23, 1967, Russian cosmonaut Col. Vladimir Komarov blasted off from the A-2 launch pad at Tyuratam. Komarov was the first person ever to make a second trip into space. After the 18th orbit his retro rockets were fired commencing a return to earth. Most of the reentry proceeded normally, then his parachutes failed to deploy. The ship and the pilot were both lost on hard impact.

Most western analysts believe Soyuz 1 was going to rendezvous with another Soyuz crew. In fact James Oberg (author of *Red Star in Orbit*) went so far as to name the crew of what would have been Soyuz 2.

Komarov and Soyuz 1 was launched into a 224 x 201 km orbit inclined 51.7 degrees. His call sign for the mission was Rubin (Ruby) and transmissions were noted on 8035, 15008 and 20008 kHz.

Cosmos 186/188

Just in time to highlight the 50th anniversary of the Soviet State in November 1967, the Russians conducted a double launch with unmanned Soyuz prototypes. On October 27, Cosmos 186 was put into a 235 x 209 km orbit for a period of four days; while Cosmos 186 waited in orbit, Cosmos 188 was launched on October 30 into a 276 x 200 km orbit for a three day period.

The vehicles conducted docking operations in space and then returned to a safe landing in the Soviet Union. The only frequency noted for this mission was 20008 kHz for Cosmos 186.

Cosmos 212/213

On April 14 and 15, 1968, Cosmos 212 and 213 were launched as a test of Soyuz vehicles; they docked for three hours and 50 minutes. The only frequency noted during this mission was Cosmos 212 on 20008 kHz.

Cosmos 238

On August 28, 1968, Cosmos 238 was put into a 210 x 203 km orbit inclined 51.7 degrees; the craft was monitored on 15013 kHz. The flight was never commented on by the Russians after the initial launch announcement and after four days in orbit it was called down. Apparently, it represented a final checkout of onboard systems as a step in man-rating the Soyuz.

Soyuz 2

Soyuz 2 was launched on October 25, 1968 without any immediate announcement by the Russians and placed into a 224 x 185 km orbit. Telemetry was monitored on 15008 and 20008 kHz. The spacecraft remained in orbit for three days and was a target for Soyuz^O3.

Soyuz 3

On October 26, 1968, Col. Georgiy Beregovoy was launched aboard Soyuz 3 into a 225 x 205 km orbit. After achieving a co-orbit with Soyuz 2, the ship rendezvoused to within 200 meters. Beregovoy made several manual approaches toward Soyuz 2 but, for some unknown reason, never docked with the spacecraft.

Telemetry was monitored on 15008 and 20008 kHz and Soyuz returned to earth after four days. Call-sign for the mission was "Almaz" (Diamond).

Soyuz 6, 7 and 8

Launched on three successive days, Soyuz 6, 7, and 8 were to perform group flight with orbital assembly the prime mission. Soyuz 7 acted as a passive target for its active partner, Soyuz 8. The docking was never accomplished.

Soyuz 6 was launched on October 11, 1969, into a 223 x 186 km orbit. The flight was piloted by Georiy Shonin with flight engineer Valeriy Kubasov aboard. Call sign for the mission as "Artaeus."

The launch of Soyuz 7 occurred October 12 with a crew of Lt. Col. Anatoliy Filipchenko; Vladislav Volkov, flight engineer; and Viktor Gorbatko, research engineer aboard. Call sign for the mission was "Snowstorm."

Launched the day after Soyuz 7, Soyuz 8 commanded by Col. V. Shatalov and accompanied by flight engineer A. Yeliseyev--both veterans of Soyuz 4/5--was put into orbit. Call sign for the mission was "Granite."

Soyuz 9

On June 1, 1970, Soyuz 9 was launched on a solo mission from Tyuratam. On board the spacecraft was pilot Col. Andriyan Nikolayev and flight engineer Vitaliy Sevastyanov. The mission lasted 18 days and served to test both the hardware and human crew in prolonged weightlessness.

Experiments conducted during the flight included medical, earth resources, navigation, and astrophysical observations. Call sign for the flight was "Falcon."

Soyuz 10

On April 1, 1971, the Soviets launched Soyuz 10 towards a rendezvous with the Salyut 1 space station launched October 11, 1970. Details on Salyut 1 will appear in the section on Salyut space stations.

On board the Soyuz were Col. Vladimir Shatalov, Aleksey Yeliseyev, and Nikolay Rukavishnikov. Call sign for the flight was "Granite."

After 24 hours of free flight, the crew docked with Salyut 1 for 5-1/2 hours. The crew never entered the space station and there was never any explanation why they did not enter Salyut 1. A predawn recovery was made on the second day of the mission (a first for the Russians).

The Soviets launched the Soyuz 10 mission in the predawn hours at Tyuratam which gave western analysts the suggestion that the flight should have lasted at least 30 days. Telemetry was monitored on 20008 kHz.

Soyuz 11: The second Soviet space tragedy

The Soyuz 11 crew was launched on June 6, 1971, at 0455 GMT comprising Lt. Col. Georiy Dobrovolskiy, flight engineer Vladislav Volkov, and Salyut test engineer Viktor Patsayev. The crew successfully docked with Salyut 1 on June 7 at 0745 GMT. Patsayev entered the station first, followed by Volkov and Dobrovolskiy. The cosmonauts conducted experiments on board the space station for 24 days and on June 29 prepared to return to earth. The ships unlocked at 1828 GMT and retrofire occurred at 2234 GMT. Soyuz 11 telemetry was observed on 20008 kHz and the mission used the call sign "Amber."

Radio communications came to abrupt end at the moment of separating the work compartment, probably at 2247 GMT, even before the ionospheric blackout period. All parachute and systems functioned normally and the craft touched down at 2317 GMT.

Upon reaching the capsule, the recovery team was horrified to discover the three cosmonauts dead on their couches. The Soviets had done away with the practice of wearing spacesuits for launch and recovery by this time, primarily because the Soyuz could not accommodate three space-suited cosmonauts and suits were no longer considered necessary.

The 1973 the Soviets informed the U.S. negotiators for the Apollo-Soyuz project that a valve failed as the work module separated, venting the void of space into the ship's atmosphere. The crew tried manually to crank the valve shut but lost consciousness and subsequently died of pulmonary embolism. It was a year later before even an unmanned Soyuz would fly again.

Cosmos 496

After a long pause in man-related activities caused by the death of the Soyuz 11 crew, the Russians launched Cosmos 496 without announcing much more than routine parameters. The spacecraft, launched on June 26, 1972, from Tyuratam, has a 342 x 195 km orbital height. The craft was recovered after six days.

TASS in Moscow noted that it used the 20008 kHz frequency common to Soyuz spacecraft. Geoffrey Perry and Sven Grahn of the Kettering Group were not able to find signals on 20008 kHz, but they did discover that each time the ship reached the radio horizon of Yevpatoriya in the Crimeans, the ship sent signals on 922.750 MHz which had been used in the manned program previously. There were three carriers with high-speed commutated telemetry sidebands.

The strong inference was that the Russians were testing an improved Soyuz to correct the problems of Soyuz 11. Further manned flights were expected, although none came for over a year.

Cosmos 573

Almost a year after the launch of Cosmos 496, the Russians launched another unmanned Soyuz under the Cosmos label: Cosmos 573 was launched on June 15, 1973, flying a similar orbit to that of Cosmos 496. TASS announced it as having an apogee of 329 km, a perigee of 196 km, and that it used the 20008 Khz frequency common to man-related orbital flights. This time the ship only stayed up two days, a pattern that Soyuz 12 was to follow.

The Soyuz spaceship and carrier rocket being taken to the launching pad.

Soyuz 12: Ferry craft configuration

Soyuz 12, call sign "Ural," was the first manned spacecraft after the Soyuz 11 tragedy in 1971. Soyuz 12 was primarily a test of new designs, including introduction of a new launch escape rocket. This flight introduced the ferry craft version of Soyuz without solar panels which would be used to take crews back and forth to Salyut space stations beginning with Soyuz 14.

The spacecraft and crew were launched into a 249 x 194 km orbit at 1218 GMT on September 27, 1973; the ship was piloted by Lt. Col. Vasiliy Lazarev and flight engineer Oleg Makarov. It was inclined at 51.6 degrees with an orbital period of 88 minutes.

In a test of the control system the orbit was changed to 345 x 326 km, a 91 minute period on the second day of the mission. Sven Grahn suggested that this forecasted higher altitudes for Salyut space stations and his prediction was confirmed by the launch of Salyut 4.

The spacecraft transmitted telemetry on 20008 kHz and flew a two day mission, landing September 29, 1973, at 1134 GMT, 400 km southwest of Karaganda, Kazakhstan.

Soyuz 13

On December 18, 1973, at 1155 GMT, Soyuz 13 was launched from Tyuratam. The mission, code named "Kavkaz" (Caucasus), was primarily conceived as an orbiting astronomical observatory. Soyuz 13 cosmonauts Maj. Petr Klimuk and flight engineer Valentin Lebedev had undergone extensive training at the Byurakan Observatory in Armenia on the astronomical equipment on board called Orion 2.

On the fifth orbit of earth, Soyuz 13 was put into a 272 x 225 km orbit inclined 51.6 degrees with an orbital period of 88.2 minutes. It remained in space for eight days conducting astrophysical experiments with the

Orion 2 system. Soyuz telemetry was noted on the usual 20008 kHz.

Apollo-Soyuz Test Project

In 1972 the United States and Soviet Union agreed to conduct a joint manned mission in 1975 in which an American Apollo spacecraft would dock with a Soviet Soyuz. Four Soviet spacecraft made flights in connection with the project consisting of two unmanned test flights (Cosmos 638 and 672), a manned test flight (Soyuz 16), and the mission itself (Soyuz 19).

Cosmos 628 and 672

Cosmos 638 was launched on April 3, 1974, into a 325 x 195 km orbit, inclined 51.8 degrees, the same orbit intended for the ASTP mission. The unmanned spacecraft stayed up for ten days and was then recovered. The Kettering group noted telemetry signals on 20008 kHz.

Cosmos 672 was launched on August 12, 1974, into a 239 x 198 km orbit at 51.8 degrees. The orbit was adjusted to the ASTP altitude with the apogee was moved to 238 km and the perigee to 227 km. Later, as with Cosmos 638, the Soviets confirmed that this was an ASTP test flight.

Soyuz 16: The dress rehearsal

Soyuz 16, code named "Buran" (Snowstorm), announced as a precursor flight for the ASTP, was launched December 2, 1974, at 0940 GMT. The spacecraft was piloted by Col. Anatoly Filipchenko and Nikolay Rukavishnikov, the prime ASTP backup crew. This flight was a test of the new hardware systems installed for the ASTP mission as well as the docking procedures to be used.

This dress rehearsal by the Soviets included participation by NASA ground tracking facilities. NASA was told in advance that this was an ASTP test, but did not know the exact date or time of the launch. The Russians insisted on this secrecy because NASA refused to keep the information out of thre press. The mission lasted six days and telemetry was noted on 20008 kHz.

Soyuz 19: The first international flight

Soyuz 19, the Soviet half of the ASTP, was launched on July 15, 1975, at 1220 GMT, carrying Col. Aleksey Leonov and Valeriy Kubasov. The launch occurred nominally, placing the spacecraft in a 229 x 188 km orbit. On the 4th and 17th orbits, the crew raised the ship so it would be in a 225 km orbit.

At 1950 GMT, the U.S. Apollo spacecraft was launched with the crew of Gen. Thomas Stafford, Vance Brand, and Donald (Deke) Slayton. Apollo was placed in an initial orbit 173 x 155 km. Approximately one hour after attaining orbit, the American crew began transposition and docking with the module which had been specially designed to join the two dissimilar spacecraft. The Apollo orbit was then circularized to 172 km. Another burn was made to change the orbit to 234 x 173 km on July 16, and the next day it was circularized at 229 km.

The two ships rendezvoused and docked on July 17 at 1612 GMT where they remained for two days while the crews exchanged national mementos and conducted joint experiments. The ships completed an undocking and redocking maneuver and separated for the last time at 1526 GMT on July 19.

Soyuz 19 landed near Arkalyk at 1051 GMT on July 21. The Apollo remained in orbit to conduct additional experiments, splashing down in the Pacific at 2118 GMT on July 24.

The call sign for Soyuz 19 for this mission was appropriately "Union." Official frequencies listed for the Soyuz for the flight are as follows: 20008 kHz (CW); 18000 and 18008 kHz (AM); 121.750 and 142.400 MHz (FM); and 192.000 MHz (FM-TLM). Some other possible frequencies not confirmed by the author have been published, including: 15008 and 18060 kHz (AM) 142.6 MHz (Voice-FM), 463.00 MHz (FM-TV), 919.760 MHz (PM Command telemetry). The use of 121.750 was used for Soviet air-to-ground communica^^tions as well as ASTP ship-to-ship. The author would like to point out to readers that some, if not all, of these frequencies are still in use today.

Soyuz 14

Soyuz 14 was launched on July 3, 1974, at 1851 GMT; the mission was commanded by Col. Pavel Popovich while Lt. Col. Yuriy Artyukhin served as flight engineer. The crew docked with the Salyut 3 space station and entered the station on July 5.

Soyuz 14 validated the operational use of the reconfigured Soyuz as a ferry to Salyut space stations. It had a water capability and was touted as the first mission to be in continuous communications on all channels (voice and telemetry) with the flight control center near Moscow.

Western observers doubted the latter claim, however, since during several communications sessions the crew was heard to use the call sign for the Gagarin and Komarov tracking ships rather than that of the flight control center and, at one point, Popovich said he would relay greetings from the Komarov to Moscow.

The crew stayed aboard Salyut 3 for 16 days. Sovuz 14 undocked from Salyut 3 at 0903 GMT on July 19, and landed at 1221 GMT. Call sign for the flight was "Golden Eagle."

The Soyuz ferry vehicles carried the following frequencies aboard: 15008 kHz (AM-voice), 20008 kHz (telemetry), 121.750 MHz (FM-voice, 166.000 MHz (AM-telemetry), and 922.750 MHz (PM command telemetry.)

Soyuz 15

Launched on August 26, 1974, at 1958 GMT, Soyuz 15 was reportedly a continuation of the scientific research and experiments started by Soyuz 14.

The initial orbit of Soyuz 15 was 236 x 1173 km at 51.6 degrees.

Pilot Lt. Col. Gennadiy Sarafanov and engineer Col. Lev Demin ran into trouble on the second day of the mission when attempts to dock with Salyut 3 were unsuccessful.

First public reports of trouble with the mission came from the Kettering Group after visual observation showed that the two craft were still apart after the time they should have docked. The Kettering group also monitored telemetry from Soyuz which confirmed that no docking had occurred.

Since the ferry version of Soyuz did not have solar panels for energy but only chemical batteries, its life in space was limited to about 2-1/2 days. Thus, Soyuz 15 was forced to land at night on August 28.

The tracking ship Morzhovets, stationed in the Atlantic near St. Helena Island, reported the correct firing of the retrorockets and at 2010 GMT the cosmonauts landed 50 km southwest of Tselinograd in adverse weather conditions. Call sign for this mission was "Danube."

Soyuz 17

The launch of Soyuz 17 came on January 10, 1975, at 2153 GMT. The crew of Lt. Col. Aleksey Gubarev and flight engineer Georgiy Grechko docked with the Salyut 4 space station. During their 30 day mission the crew conducted a wide range of scientific experiments.

The crew returned to earth on February 9 and the crew landed in a blinding snowstorm. Despite the adverse weather conditions, rescue teams were on the scene immediately. Call sign for the mission was "Zenith."

The April 5 Anomaly

The next mission to dock with Salyut 4 resulted in the first abort of a manned mission during launch for the Soviet Union. On April 5, 1975, at about 1103 GMT, the Russians launched a spacecraft with the announced purpose of docking with the Salyut 4 and continuing scientific experiments.

A stage malfunction of the A-2 booster forced the mission to be aborted and the crew of Col. V. Lazarev and O. Makarov found themselves landing in cold, snowy Siberia, 1600 km away from their launch site. This location was only 320 km away from the Chinese border.

After the failure, the mission was renamed the "April 5 Anomaly" and the Soyuz 18 designation was given to the next craft in the series. Call sign for the mission was "Urals."

Soyuz 18

On May 24, 1975, six weeks after the failure of the "April 5 Anomaly," the Soviets launched Soyuz 18 at 1458 GMT. The mission was commanded by Col. Petr Klimuk accompanied by flight engineer Vitaliy Sevastyanov. The Soyuz docked with Salyut on May 25.

During the day the 63 day mission two communications sessions ensued between the Soyuz 18 crew and the orbiting ASTP Soyuz 19 crew. The Soyuz 18 crew returned to earth safely on July 26 after setting a new Soviet endurance record, although it did not surpass the American record of 84 days set by the Skylab 4 crew. Call sign for the flight was "Caucasus."

The Soyuz spaceship and carrier rocket on the launching pad.

Soyuz 20: Unmanned test flight

After the end of the Soyuz 18 mission, Salyut 4 remained in orbit, prompting speculation that a third crew might be sent to work on the space station.

On November 17, 1975, the Soviets did, indeed, launch another Soyuz toward the Salyut 4. Soyuz 20 tested rendezvous and docking equipment aboard the unmanned spacecraft and spend 90 days docked to the Salyut 4.

Telemetry for this mission was monitored on 20008 kHz.

Soyuz 21

Code named "Baykal" (Baikal), Soyuz 21 lifted off the A-2 pad at Tyuratam July 6, 1976, at 1309 GMT. Soyuz 21 carried the crew of Sol. Boriz Volynov and Lt. Col. Vitaliy Zholobov for a 1340 GMT docking with Salyut 5 on July 7.

The crew returned from 48 days in the military Salyut early according to some western experts. *Aviation Week and Space Technology* magazine concluded that the cosmonauts were forced back by an acrid odor from the Salyut 5 environmental control system.

Soyuz 22

The last of the free flying Soviet missions, Soyuz 22 was launched on September 15, 1976. The two man crew consisted of Col. Valeriy Bykovskiy and flight engineer Vladimir Aksenov.

The mission was primarily concerned with testing an East German multi-spectral camera, the MKF-6, for earth resources photography. It was the first foreign-made equipment to be flown on a Soviet spacecraft.

The Soviets wanted to fly over East Germany since the MFK-6 was an East German experiment, so the inclination of the orbit was changed to 64.7 degrees. This has been the only time such an inclination has been used by a Soyuz spacecraft.

The crew spent eight days in orbit and used the call sign "Hawk."

Soyuz 23

The Soviets attempted another mission to Salyut 5 on October 14, 1976, with the launch of the Soyuz 23 spacecraft. The crew comprised Lt. Col. Vyacheslav Zudov and Lt. Col. Valeriz Rozhdestvesnkiy.

According to the Soviets the Soyuz 23 was attempting an automatic docking with Salyut 5 but an operational malfunction caused the approach to the station to be cancelled.

The crew landed at 1746 GMT on October 16 on the surface of Lake Tengiz in a snowstorm with temperatures that fell below 0^To^T Fahrenheit. Call sign for the mission was "Radon."

Soyuz 24

The last mission to the Salyut 5 space station was launched February 7, 1977, carrying Col. Viktor Gorbatko and Lt. Col. Yuriy Glazkov.

It is believed that the crew performed some military recon activities and worked on shutting down the Salyut 5 station. After 16 days the crew returned to earth about 36 km northeast of Arkalyk. One day later the Soviets separated an unmanned reentry capsule from the Salyut 5, heightening western speculation that the crew returned important military information and explaining the short 16 day mission. Call sign for the mission was "Terek."

Soyuz 25-40

With the launch of the Soviet second generation space station, Salyut 6, on September 29, 1977, a new era in Soviet space operations commenced. Manned and unmanned spacecraft visited the Salyut on a regular basis until the station was finally deorbited July 29, 1982.

SOYUZ MISSIONS

Soyuz No.	Launch Date	Days in Orbit	Code Name	Cosmonauts	Mission Notes
25	10/9/1977	2	Photon	Kovalyonok/Ryumin	Linkup with Salyut 6 failed
26	12/10/1977	96	Taymir	Romanenko/Grechkp	First successful linkup with Salyut 6
27	1/10/1978	5	Pamir	Dzhanibekov/Makarov	First exchange of Soyuz after 2nd Soyuz docked.
28	3/2/1978	8	Zenith	Gubarov/Remek(Czech)	First visit on Soyuz to Salyut by a non-Soviet. (Interkosmonaut)
29	6/15/1978	140	Photon	Kovalyonok/ Ivanchenkov	Long duration crew aboard Salyut
30	6/27/1978	8	Caucasus	Klimuk/Hermaszewki (Polish)	Soviet Interkosmonaut mission to Salyut 6
31	8/26/1978	8	Hawk	Bykovski/S.Jahn (E. German)	Soviet Interkosmonaut mission to Salyut 6
32	2/25/1979	175	Proton	Lyakhov/Ryumin	Long duration crew aboard Salyut
33	4/10/1979	2	Saturn	Rukavishnikov/ Ivanov (Bulgarian)	Interkosmonaut mission that failed to rendezvous with Salyut 6
34	6/6/1979	74	-	Unmanned	Replacement & supply Soyuz for the long duration crew.
35	4/9/1980	185	Dneiper	Popov/Ryumin	Long duration crew aboard Salyut
36	5/26/1980	8	Orion	Kubasov/Farkas (Hungarian)	Interkosmonaut mission to Salyut
37	7/23/1980	8	Terek	Gorbatko/ P.Turn (Vietnamese)	Interkosmonaut mission to Salyut
38	9/18/1980	8	Taymyr	Romanenko/ Mendez (Cuban)	Interkosmonaut mission to Salyut
39	3/22/1981	6	-	Dzhanibekov/ Gurragcha(Mongolian)	Interkosmonaut mission to Salyut
40	5/14/1981	8	-	Popov/ Prunaria(Romanian)	Interkosmonaut mission to Salyut and the last Soyuz mission flown.

Soyuz 26-32, 34-40, Progress tankers 1-12, Cosmos 1267, and Soyuz-T1/T3/T4, all visited the Salyut 6. Soyuz 33 and Soyuz T2 were docking failures and never visited the space station.

A summary of the Soyuz missions closes out that era in the Soviet space program; the other vehicles mentioned above will be discussed in their respective sections.

Soyuz-T Spacecraft

Physical Details

In December 1979 the Soviets introduced a new, improved version of the Soyuz spacecraft, designated Soyuz-T. The T stands for "Troika" to indicate that it is a third generation spaceship.

The spacecraft is a manned ferry vehicle designed to carry up to three cosmonauts in space suits. The Soyuz-T, like its predecessor Soyuz, has four main sections: orbital module, descent module, instrument module, and the docking interface. But it differs in a number of details from the earlier Soyuz spacecraft.

Most of the onboard systems were designed anew, including a motion control system using an onboard digital computer. The crew can now control the Soyuz-T manually, not only in orbit, but also during descent into the atmosphere.

The radio system of the spacecraft--such as the command program downlink, radiometry and television systems--has been improved. Solar cells have been included in the power supply system.

Soyuz-T Frequencies

The Soyuz radio system appears to be a scaled down version of that used in the Salyut space station. The "Zarya" radiotelephone is intended to provide two-way earth station and space station communications in the HF and VHF bands. The system includes VHF and HF receivers, transmitters, and antenna feeder devices. The author believes that this system onboard the Soyuz-T is heard on 121.750 MHz (70 kHz wideband FM) and 18.060 MHz (AM voice).

The command radio link appears to use 922.750 MHz and consists of three carriers with high-speed commutated telemetry sidebands--apparently some form of pulse modulation.

The "Stroka" FSK teletype system on board Salyut space stations and Soyuz-T used to transmit alphanumeric text is probably 3rd shift cyrillic; monitors might wish to check appropriate Soviet manned frequencies for this type of activity.

The Soyuz also contains an orbital control radio system which performs trajectory measurements probably transmitting on 20.008 MHz. This frequency is normally quiet; reception of signal could indicate orbital maneuvers being made by Soyuz during independent flight and docking with Salyut.

Recent information reveals that the Soviet spacecraft are using 217-219 GHz for precise determination of orbital elements.

Telemetry has been successfully decoded by the Kettering Group and others on 20.008 MHz. A complete description of this method is beyond the scope of this book. Listeners interested in this exciting aspect of monitoring are referred to articles listed in Appendix A for further study.

Flight Profiles and Constraints

All launches to Salyut space stations as well as of the space stations themselves are made out of the Tyuratam launch facility into an orbit inclined 51.6 degrees. Routinely, the Soyuz spacecraft will make orbital adjustments on revolutions 4 and 17, with docking on revolutions 17-18. Docking with the Salyut occurs approximately 25 hours and 40 minutes ground elapsed time. Listeners should monitor the first five orbits for possible Russian AM voice activity on 18.060 MHz.

Using NASA prediction bulletins and home computers it is possible to predict possible launch windows for Soviet manned Soyuz flights and unmanned Progress cargo supply ships to Salyut space stations.

The rationale underlying the selection of opportunities for manned flights to Salyut space stations has been explained by British analysts Phillip S. Clark and Robert D. Christy.

Clark deduced empirically that landings are chosen to take place during the five hours before sunset; thus, for a given mission, there is a small range of launch times which can give landing under optimum conditions. This enabled Clark to make accurate predictions of launch and recovery times associated with Soyuz missions to Salyut.

The ascending node of the Salyut space stations orbit precesses with respect to the earth-sun line approximately once every 60 days, so there is a period of about ten days at two month intervals during which a Soyuz landing could be made under optimum conditions. If the revolution of the landing is known, the landing time can be estimated with reasonable accuracy.

The launch times and data for manned Soviet missions can also be calculated. An example of this is the visiting Interkosmonaut eight day missions; one could look for a flight about eight days before a recover "window" opened. Similar calculations chould be made for missions of longer duration.

Beginning with Soyuz 32, the long duration crews aboard Salyut 6 and 7 were launched during an optimum landing window in case Salyut proved unable to support the flight or the crew had a docking malfunction.

Christy's less empirical approach is based on two landing constraints which were revealed during ASTP mission preparations: First, the landing must take place at least one hour before local sunset and secondly, for the purpose of manual override of the automatic descent program, Soyuz must cross the terminator at least eight minutes before retrofire. These constraints must be fulfilled on the first pass over the landing site on the planned recovery day and the day after.

For solo missions like Soyuz 9, the choice of flight dates is completely open; with an orbiting space station, however, recovery dates are dictated by whether or not the Salyut orbit obeys the constraints.

A long time Soviet monitor and Kettering group member, Mark Severance, recently supplied the following equation to determine the exact launch time for a given Soyuz mission to a Salyut space station:

Soyuz launch time
= EQX* time + 10.8 + 4 (347.5 - EQX longitude)
* equator crossing of Salyut

Mark also provided possible EQX longitudes to watch for as possible Soyuz launch windows. Monitors should check NASA prediction bulletins for the orbiting Salyut in use and for possible windows for a Soyuz launch as follows:

340°-343° (EQX) for manned missions
334°-339° (EQX) for unmanned missions
These figures are ∓ 2°

As you can see, much of the veil of secrecy around the Soviet manned space program can be lifted with just a few elementary facts in hand.

Soyuz-T Missions

In December 1979 the Soviets introduced their new Soyuz variant. At least three unmanned precursors (Cosmos 869, 1001 and 1074) had already been flown, and the first Soyuz-T flight so identified was also unmanned. Two additional flights, Cosmos 670 and 772, may also have been related to the development of Soyuz-T.

Cosmos 670

Cosmos 670 was launched August 6, 1974, into a 307 x 217 km orbit and was recovered three days later. This flight was unique in that its inclination of 50.6 degrees had never before been used on a flight launched by an "A" class rocket.

Western experts had predicted that the 50.6 degree inclination would be used by the "G" class "supper booster"; there was some speculation that this could have been a developmental flight. External flight parameters seemed close to military recoverable observation flights.

There was also speculation that 670 might have been a test of a ferry vehicle to a new large space station which would be orbited by the "G" booster at some future time. With hindsight, it has now been classified as a Soyuz precursor.

Cosmos 772

September 29, 1975, the Soviets launched Cosmos 772 into a 299 x 195 km orbit inclined 51.8 degrees. Soyuz-type telemetry was monitored on 20008 kHz. Like the Soyuz ferry ships, Cosmos 772 had no solar panels but remained in orbit for three days like Cosmos 670. This suggested either a greater battery capacity or lessened electrical loads since it stayed one day later than most other flights of its kind. There was speculation that this might be a system test for returning to a three-man crew.

Cosmos 869, 1001, 1074

November 29, 1976, the Soviets launched Cosmos 869 into an orbit 307 x 202 km, inclined 51.8 degrees. 869 made several orbital maneuvers during its 18 days in orbit. Cosmos 1001 was launched April 4, 1978, into a 249 x 205 km orbit, later raised to 315 x 308 km. 1001 was deorbited by the Soviets April 15.

On January 31, 1979, Cosmos 1074 was launched into a 258 x 203 km inclined 51.6 degrees, later raised to 383 x 364 km. This Soyuz precursor remained in orbit for 60 days, reentering April 1. All three are now considered to have been related to development of Soyuz-T.

Progress Supply Tanker

On January 20, 1978, the Russians launched their first cargo ship into space, Progress 1. Since then the Progress cargo ships have been launched in a steady stream to the Salyut 6/7 space stations. An unmanned spacecraft based on the Soyuz design, Progress is a one-way vehicle--it will take material to the Salyuts, but it is not designed to survive reentry, so it cannot bring anything down.

Since 1978 twenty-three Progress supply ships, twelve to Salyut 6 and eleven to Salyut 7, have brought fuel, air, water, food, clothing, experiments, and replacement parts to the space station--all necessary items for maintaining long duration crews and long term space stations.

Progress vehicles also bring personal items to the crew, including mail and parcels from their friends and family, newspapers and other periodicals, and special items the cosmonauts request. For example, Soyuz 29 crew member Ivanchenkov requested that his guitar be sent up to Salyut 6. On August 7 Progress 3 was launched from Tyuratam with his guitar on board.

The Soviets have found that the crews use 45-65 pounds of expendable items each day, almost a ton per month!

Progress Description

Since the Progress is an unmanned, modified version of the Soyuz, all life support systems, heat shield, parachutes, and solar panels have been removed. This allows additional room and weight for cargo, 5,070 pounds of equipment, materials, fuel, water, and gas.

Progress has an independent flight capability of eight days and, when docked to a space station, Progress 11 stayed in space about 60 days.

Progress docks with Salyut after two days instead of one day in the case of a manned Soyuz craft; this requires less fuel to rendezvous and there is no hurry since no crew is aboard. The two day rendezvous also results in slower closing speeds which simplifies docking operations without a crew on board to make final adjustments.

These ships are not designed to dock at the forward end; thus, if a spacecraft already occupies the aft docking port (such as when a visiting crew exchanges ships with the long duration crew), that ship must be moved forward. This has required the Russians to do a number of "do-si-do" maneuvers to accommodate their Progress cargo ships.

A radar rendezvous system used for docking has warning indicators to ensure that plumbing and docking latches are connected properly.

There are three compartments in Progress: freight, refueling components and instruments. The hermetically sealed freight compartment contains all the dry cargo which is mounted on a special frame. The hatch can be opened automatically or manually.

The refueling compartment has four fuel tanks and spherical containers filled with compressed air and nitrogen, plus pneumatic and hydraulic systems for a total weight of fuel and gases around 2,204 pounds.

In the instrument compartment, there is almost twice as much instrumentation as with the Soyuz. Radio antennas, rendezvous apparatus, sensors and orientation engines are mounted externally. Three lights and two television cameras help with docking which is controlled by the Salyut crew, ground controllers, or both.

The Progress vehicle operates on two telemetry frequencies: 166 MHz and 922.750 MHz. There is probably no HF frequency for Progress.

After the Progress has been emptied of its cargo, the crew fills it with trash and old Equipment which burn up during reentry over the Pacific.

Cosmos Space Tug

One of the most interesting and easiest to hear elements of the Soviet manned space program is the Cosmos Space Tug. There have been three of these vehicles in the series so far: Cosmos 929, 1267 and Cosmos 1443.

The true mission for these satellites was revealed in the July 3, 1983, issue of *Pravda*. The article, written by D. Aleksyeyev, Candidate of Technical Science, described Cosmos 1443 as a freighter variant -- an "Interorbital tugboat." Aleksyeyev also revealed that 929 and 1267 were forerunners of Cosmos 1443.

SOYUZ-T MISSIONS

Soyuz-T No.	Launch Date	Days in Orbit	Cosmonauts	Mission Notes
1	12/16/1979	100	Unmanned	Docking test with Salyut 6
2	6/5/1980	4	Malyshev/Aksyonov	Visiting crew to Salyut 6
3	11/27/1980	13	Kizin/Makarov/ Strekalov	First three man crew since Soyuz 11. This was a Salyut 6 repair mission.
4	3/12/1981	75	Kovalenok/Savinyk	Long duration crew to Salyut 6. Final crew on Salyut 6.
5	5/13/1982	211	Berezovoy/Lebedev	First long duration crew to Salyut 7.
6	6/24/1982	8	Ivanchenkov/Djani-bikov/Chretien(Fr.)	Soviet Interkosmonaut mission to Salyut 7.
7	8/19/1982	8	Popov/Serebrov/ Savitskaya(woman)	Mission to Salyut 7 with second woman in space. Crew exchanged Soyuz with long duration crew (T-5)
8	4/20/1983	3	Titov/Strekalov/ Serebrov	Docking attempt with Salyut 7 failed because rendezvous radar did not deploy
9	6/27/1983	149	Lyakhov/Aleksandrov	Long duration crew aboard Salyut 7
10A	9/26/1983	0	Titov/Streakalov	Aborted launch to Salyut 7, rocket caught fire 90 min. before launch, cosmonauts used escape system, OK.
10	2/8/1984	237	Kizim/Solovev/Atkov	Long duration crew to Salyut 7
11	4/3/1984	8	Malyshev/Strekalov/ Sharma (Indian)	Soviet Interkosmonaut crew to Salyut 7. Returned to earth in Soyuz T-10
12	7/17/1984	12	Dzhanibekov/Savits-kaja(woman)/Volk	First woman to return to space a second time. First woman EVA.
13	6/6/1985	112	Dzhanibekov/ Savinykh	Rescue mission to save the failing Salyut 7 station. Dzhanibekov returned with Grechko after T-14 launch.
14	9/17/1985	65	Vasyutin/Grechko/ Vokov	Relief crew. Vasyutin/Volkov/Savinykh remained aboard Salyut 7 after T-13 departed. Vasyutin became ill and mission was terminated without notice 11/21/85.

Cosmos 929

The first step toward building modular space stations occurred July 17, 1977, with the launch of Cosmos 929. Once in orbit the spacecraft conducted extensive maneuvers and, on August 16, 1977, a capsule was returned to earth for recovery by Soviet space technicians. The main part of the spacecraft reentered February 2, 1978, after 201 days in orbit. Transmission of PDM-FSK telemetry was monitored on 19954 kHz; wideband telemetry was monitored on 166 MHz.

Cosmos 1267

June 19, 1981, the Soviets docked the second of the space tug series, 1267, with Salyut 6. The spacecraft, which weighed 15 metric tons, almost as much as the Salyut itself, had been launched on April 25, 1981, from Tyuratam. This space tug, as had 929, performed numerous orbital maneuvers. It separated a large modular section weighing 1322 pounds May 24 which reentered and was recovered. Cosmos 1267's orbital activity closely paralleled that of Cosmos 929.

Radio telemetry also paralleled the earlier mission with PDM-FSK telemetry monitored on 19954 kHz and wideband telemetry on 166 MHz.

The Soviets released a great deal of information 1267 with some unusual candor in July, September and December of 1981; they announced that 1267 was "designed to test systems and elements of design of future spacecraft and for training in the methods of assembly of orbital complexes of a big size and weight"--in other words, a modular space station.

During the time that Cosmos 1267 was docked to Salyut 6, I noted several instances from my records that the docked pair appeared to maneuver several times. I concluded that the 1267 was being used for orbital adjustments of the whole complex which meant that 1267 had on-board engines. This fact was later confirmed by the Soviets.

In October and November 1981 *Aviation Week and Space Technology* (sometimes nicknamed *Aviation Leak*) reported that the spacecraft was, in fact, an "anti-satellite battle station equipped with clusters of infrared-homing guided interceptors with firing ports to eject 1-meter-long miniature vehicles guided by infrared sensors."

I seriously doubt this claim in view of the fact that the Soviets do not normally release information on their military satellites at all and there was considerable coverage given in the Soviet press concerning 1267 and its mission to Salyut 6. Also, I know of no test firings that occurred with this weapon system.

Space surveillance data from NORAD never revealed any data of a launch of this kind, although I did have quite a few bulletins on trash bags the Soviets threw out of Salyut 6 after field days aboard the space station!

The Cosmos 1267/Salyut 6 combination was deorbited by the Soviets July 29, 1982.

Cosmos 1443

March 2, 1983, the Soviets launched Cosmos 1443; described as an operational version of Cosmos 1267, it docked with Salyut 7 on March 10. The fact that it retained a Cosmos designation instead of having a new name like Progress suggests it may still be in a preoperational phase.

The Russian press gave considerable press coverage to Cosmos 1443 and its mission which was described as serving several purposes: (1) a cargo craft both for bringing supplies and equipment to an orbiting space station (2.5 times the amount that Progress can carry) and returning up to 1100 pounds of cargo to earth in a descent module; (2) a space tug; (3) an extension to the space station to increase its habitable volume by 50 cubic meters; and (4) as an autonomous module for various tasks such as materials processing or astronomical observations.

Cosmos 1443 had two solar panels of its own with a total span of 52.5 feet and an area of 131 square feet and weighed in as much as Salyut 7 (20 tons). The spacecraft was 42.5 feet long and 13 feet in diameter at its widest point. Pictures published in the July 3 issue of *Pravda* show that the descent capsule resembles an American Gemini capsule from the 1960's.

In addition to 1143's cargo role, the spacecraft's engine was used to permit orbital corrections to the space station complex on several occasions. There is some circumstantial evidence that the Salyut 7 space station engines have either given poor performance or have a malfunction; most major maneuvers appear to occur when either a Progress vehicle or 1443 is docked with the station.

1443's engine was used to assure a correct docking flight profile for the Soyuz T9 mission. The Soyuz T9 crew load the main body of the spacecraft with trash and 1443 was deorbited September 19, 1983.

The descent module returned 775 pounds of material--over 45 experiments and pieces of equipment--August 23, 1983, a week after 1443 separated from Salyut 7.

As with its predecessors, 1443 transmitted telemetry continuously on 19.954 MHz including the "ON" data word; 19.953 MHz contained the "OFF" data word carrier. This is typical of the PDM-FSK system and future flights will give listeners a chance to recognize signal characteristics and telemetry charting techniques.

Cosmos 1686

Launched September 27, 1985, Cosmos 1686 continued to expand the roles of the Cosmos Space Tugs. Radio Moscow called the new Cosmos a "star module"; this could indicate the Soviets are developing the capability to add on modules to future generations of Salyut space stations. Radio Moscow reported that 1686 carried no reentry module to return material back to earth.

Initial orbit of the space tug was 319 by 177 km inclined 51.6^To^T. Transmissions were still noted as late as November 23, 1985, even after the Soyuz T-14 cosmonauts had returned to earth. Frequency of the transmissions was 19.954 MHz.

Future long duration missions to Soviet space stations should see continued use of the Cosmos space tug. Listeners should be alert to new signals on 19.954 MHz heralding this activity. NOTE: Do not confuse the FAX signals from the navy station in Washington, D.C. on the same frequency with those of the Cosmos Space tug; the two signals sound distinctly different.

Salyut Space Stations

The first Salyut space station was launched in April 1971 slightly more than ten years after Gagarin's flight and two years before the launch of the American Skylab space station. The vehicle was named in honor of the decennial "salute" to Gagarin.

The first generation of Soviet space stations spanned the years 1971 through 1976 and encompassed the successful launch of three stations, while two others were failures. The design of the station changed somewhat during this period, both externally in terms of the placement of the solar panels and internally to allow for different experiments. Each Salyut had a single docking port.

Salyut 1

On April 19, 1971, the Soviets launched the world's first space station, Salyut 1. The station initially was described by the Soviets as 65.5 feet long with a maximum diameter of 13 feet, and weighing 25 metric tons with a Soyuz attached.

Salyut 1 transmitted C/W PDM telemetry, similar to that of Soyuz on 15008 kHz. Toward the end of the manned phase of its operation it transmitted its own telemetry format on the Soyuz 11 frequency of 20008 kHz.

October 11, 1971, Salyut 1 was deorbited after a visit of only one crew, the ill-fated Soyuz 11 mission.

This space station had an initial orbit of 222 x 200 km inclined 51.6 degrees. Later analysis by monitors throughout the world confirmed the existence of a dual space station program: one military and one civilian. Based on the evidence, Salyut 1 was a civilian Salyut, first generation.

Salyut 2

April 3, 1973, Salyut 2 was launched into a 260 x 215 km orbit with a period of 89 minutes and inclined 51.6 degrees. Much debate ensued by western analysts over the lack of manned missions to Salyut 2; some experts claimed the Salyut was in trouble while others said that the Soyuz program had not fully recovered from the Soyuz 11 tragedy almost two years earlier. I lean toward the later hypothesis due to Soviet officials' boasting of new missions to Salyut that were in preparation.

The real trouble came on April 14 when Salyut was reported to have undergone a "catastrophic failure" which ripped off the solar panels, boom-mounted rendezvous radar and radio transponder, and left Salyut 2 tumbling in space without telemetry return.

The craft may have separated into many pieces, some large enough to be tracked. The NASA satellite situation report has identified 28 objects in conjunction with Salyut 2 (73-17A through B). Most of the objects were rather small and decayed quickly upon reentry.

While an explosion or misfiring thruster was blamed, the most widely held theory was that the upper stage D-1 booster had exploded and its debris had damaged the space station. The main body of the station decayed through airdrag May 28, 1973, and reentered near Australia.

Noting that the telemetry transmissions from Salyut 2 were similar to those used by Soviet reconnaissance satellites, *Aviation Week and Space Technology* concluded that the mission was not a Salyut at all, but that the Russians were simply trying to mislead the Soviet press and information agencies. The manned Salyut 3 a year later, however, used the same telemetry and experts now generally classify Salyut 2 as a military class, first generation Salyut. While my records show that 19.944 MHz was used for Salyut 2 telemetry, the Kettering Group never heard any transmissions on that frequency.

Cosmos 557

May 11, 1973, shortly after the failure of Salyut 2, the Russians launched Cosmos 557 into a 226 x 218 km orbit inclined 51.6 degrees. The Soviets offered no information on the flight, although Cosmos 557's telemetry resembled that of Salyut 1, typical of the manned programs rather than Salyut 2, typical of the unmanned military reconnaissance program.

Western experts thought there was a good chance that this was another Salyut, possibly of a different design, that failed so early in its mission that it was listed as a Cosmos. Tracking ships deployed for the expected manned flights to Salyut 2 were reported heading back for their home ports "before Cosmos 557 decayed" during the week of May 21.

Cosmos 557 transmitted telemetry on 15.008 and 922.750 MHz causing some analysts to believe that this was an unmanned Soyuz flight, but based on optical sightings the spacecraft was considerably larger than a Soyuz. Cosmos 557 appears to have been an unsuccessful civilian Salyut mission.

Salyut 3

Salyut 3 was launched June 24, 1974, and deorbited January 24, 1975. Two crews were launched to dock with the space station: The Soyuz 14 crew spent two weeks aboard, but the second crew, Soyuz 15, was unable to dock.

As noted previously, this is considered by most to be the first military operational Salyut. It spit out a recoverable capsule which was recovered September 23, 1974.

Salyut 3 telemetry on 19.944 MHz was FSK, like the early recoverable Cosmos satellites instead of the CW of the Soyuz and Salyut 1. This as well as the Salyut being in a lower earth orbit (better for recon work) and the all military crew has lead experts to conclude that Salyut 3 served as a military Salyut. Voice transmissions were monitored on 142.4 and 143.635 MHz.

Salyut 4

Salyut 4 was launched December 26, 1974, and was deorbited February 2, 1977. The space station complete 12,188 orbits of the earth. This space station completed the civilian studies begun by Salyut 1, and hosted two crews (Soyuz 17 and Soyuz 18). Telemetry was monitored on 20.008 MHz and the station definitely filled a civilian Salyut role. Other frequencies apparently used were 15008, 18060 kHz and 922.750 MHz.

Salyut 5

Salyut 5 was launched June 22, 1976, and deorbited August 8, 1977, after completing 6,630 orbits around the earth. This space station is considered to have been a military Salyut; it ejected a capsule February 26, 1977, the day after the last visiting crew left. There were three all-military crews sent to Salyut 5, only one of which could not dock (Soyuz 23).

Telemetry was again monitored on 19.944 MHz FSK, typical of the military Salyuts.

Salyut 6

Salyut 6, launched September 29, 1977, was the first second-generation space station launched by the Soviets, the most successful space station to date. Hosting both Soyuz, Soyuz-T Progress, and Cosmos space tug 1267, the Soviets came into their own in space research with Salyut 6. Telemetry was monitored on 15008 and 20008 kHz as well as 143.625 (Voice), 192.0 (AM-TM), 922.75, 925.24, and 926.06 MHz.

It would appear that this station was civilian in nature; if a military purpose was served by Salyut 6 it was very well hidden. One may assume that since no military Salyuts have been launched since 1976 and Salyut 6/7 appear to be operational versions of the Salyut program, that the military Salyut program is either on the back burner or has been scrapped altogether.

Salyut 7

During the final months of flight for Salyut 6/Cosmos 1267, the Soviets launched Salyut 7 aboard a Proton rocket from Tyuratam April 19, 1982. The Salyut 7's international designator is 82-33A and the NORAD number is 13138. The space station's initial orbital parameters were 302 x 281 km inclined 51.6 degrees.

The Salyut 7 is very similar in design to Salyut 6 with an improved navigation system, a strengthened forward docking port, and modifications to the solar panels to permit the attachment of additional panels.

Since this is the current operational Salyut in orbit a detailed analysis of the radio systems follows.

Salyut-7 orbital station in the assembly-and-testing block.

Command Radio Link

The primary command and control link between ground controllers and the Salyut space station passes trajectory measurements, two-way telephone conversations, television, and telemetry information, most likely on 922.750, 925.24 and 926.06 MHz.

The command radio link consists of receivers, transmitters, a programmed timer, and automated equipment.

Television Communications System

The Soviet space television system can transmit color and monochrome images from the Salyuts' on-board cameras, possibly using a 625 line, 8 MHz bandwidth (6 MHz wide video) FM audio signal much as their state television uses.

The television system includes:

* monochrome external stationary TV cameras
* color reporting TV camera
* one reserve monochrome camera
* several video monitors
* associated electronic equipment into the command radio link

Zarya Radiotelephone System

This radio system provides two-way earth-Salyut and Salyut-Soyuz communications in the HF and VHF spectrum. The Salyut primary VHF voice frequency is 142.4 MHz, the old ASTP voice channel. The Soyuz primary voice channel is 121.750 MHz. Both frequencies use 70 kHz wide bandwidth FM transmitters.

Possibly full duplex is used (Soyuz transmit 121.750, receive 142.4; Salyut transmit 142.4, receive 121.750). Listeners in Europe might wish to comment on this as they are in a better position to judge actual usage. Soyuz-Salyut docking operations are usually out of range here in the U.S. to confirm the VHF Zarya portion of the system.

The short-wave portion of the Zarya system is probably on 18.000 and/or 18.060 MHz AM, used during the first five orbits of the Soyuz mission for over-the-horizon backup communications with the on-board Salyut crew and ground stations.

The system also has "Stroka" RTTY capability, primarily for personal communications from family and friends to the crew, press reports, and basic information on orbital parameters.

The only descriptions of the system I have now indicates that this is a one-way circuit to Salyut (they only have a printer). Listeners might want to check out the above frequencies when Salyut is in range of ground stations and see if they are transmitting FSK signals.

Radiotelemetry Communications System (RTS)

RTS is designed to collect and transmit information from the spacecraft to earth. The Salyut 7 carries two RTS systems, one for spacecraft service information and the other to transmit scientific and experimental data to earth stations. A good guess is that the service system is the wideband telemetry signal heard on 166 MHz and the scientific data comprise the wideband telemetry on 192 MHz.

Orbital Control Radio System

The OCRS is utilized to obtain trajectory measurements of the Salyut 7's orbital parameters, likely transmitted on 15008 and 20008 kHz.

Signals from the Salyuts on HF would indicate that the Soviets only turn on the HF OCRS during Salyut complex maneuvers and Soyuz-T recoveries. The previously mentioned 217-219 GHz radio ranging system may also be used.

General Thoughts

The frequencies listed in this Soviet section are the author's conjecture based on monitoring and other information. I could be wrong and I invite anyone with thoughts, theories or hard information to contact the author.

There is one mystery frequency I have seen listed: 2802 MHz for Salyut 6. I would be very interested in its purpose and if Salyut 7 is currently using this frequency.

Missions are publicized in the U.S. press as well as Radio Moscow on shortwave. Those listeners with RTTY capability might want to monitor TASS HF press services for mission launch announcements and updates.

The Mir

A third generation space station, the Mir, was launched into orbit for the first of many unmanned tests February 20, 1986. Two cosmonauts were subsequently launched to the station March 13.

Equipped with six docking ports and separate crew compartments, the Mir is intended as a permanent manned space station.

Soviet Space Plane

Rumors have existed for years that the Soviets are developing a reusable space vehicle similar in concept to the American space shuttle.

In 1983 two tests of the prototype were conducted; launched under the Cosmos veil of secrecy these flights are most interesting. Cosmos 1445 was launched March 15 and landed in the Indian Ocean, as had an earlier test--Cosmos 1374. The Australian Navy was at the scene during the Soviet recovery operations and they released several pictures of the vehicle.

A third flight was made on December 27, 1983. The flight differed from the earlier two missions because it made a "controlled descent" into the Black Sea, indicating increased confidence in the system.

The Department of Defense released information early in 1984 that revealed the Soviets are developing two reusable vehicles: The first the same size as the U.S. STS; the second, a smaller space plane.

Western press reports based on intelligence photos indicate that the version equivalent to the U.S. STS had been seen mounted on the back of a Bison aircraft and is virtually identical to the U.S. shuttle.

Aviation Week, in their July 23, 1984, issue, said that U.S. analysts believe the smaller space plane could be used as a weapons platform to attack or inspect U.S. satellites in low earth orbit. It was also reported that the Soviet Saturn-5 class launcher at Tyuratam was nearing its first launch and that the final construction of three space shuttle launch facilities was nearing

Cosmos 1445 (Soviet spaceplane) being recovered March 1983 from Indian Ocean.

completion.

Listeners should be alert for the launch of all these vehicles and check past known Soviet frequencies for possible activity.

Soviet Energia Booster

After years of development, several false starts and three failures, the Soviet Union has successfully tested its "Energia" heavy-lift launch vehicle.

This Saturn-5 class launcher is a two-stage liquid hydrogen/oxygen system that weighs 4.4 million pounds at liftoff, develops 6.6 million pounds of thrust and is designed to loft 220,000 lbs into orbit in a piggyback cargo container. It also is the launch vehicle that is expected to boost Russia's reusable space shuttle on its flight.

Liftoff of the 198-ft.-tall vehicle was on May 15, 1987, less than 24 hours after the Russians broke years of secrecy on the project and announced that it was about to test "an new powerful rocket that could put in orbit reusable spacecraft and other heavy space vehicles." The announcement in advance of a high-risk Soviet space test was unprecedented.

After launch, the Soviets released television views of the Energia, its launch pad and liftoff of the giant rocket in darkness at 2130 Moscow time.

Television pictures released by the Soviets confirmed earlier U.S. recon satellite imagery that showed the new booster uses a large core element resembling the large external tank used by the U.S. Space Shuttle. Unlike the U.S. version, attached around the central core are four large liquid-fueled, strap-on boosters.

Mounted piggyback on the core section between two sets of boosters was a large cargo pod to be used in lifting unmanned payloads into orbit. U.S. intelligense officials indicate that the Energia could be used to lift payloads as heavy as 300,000 lbs in the future. For manned missions, the pod will be replaced by the Soviet space shuttle.

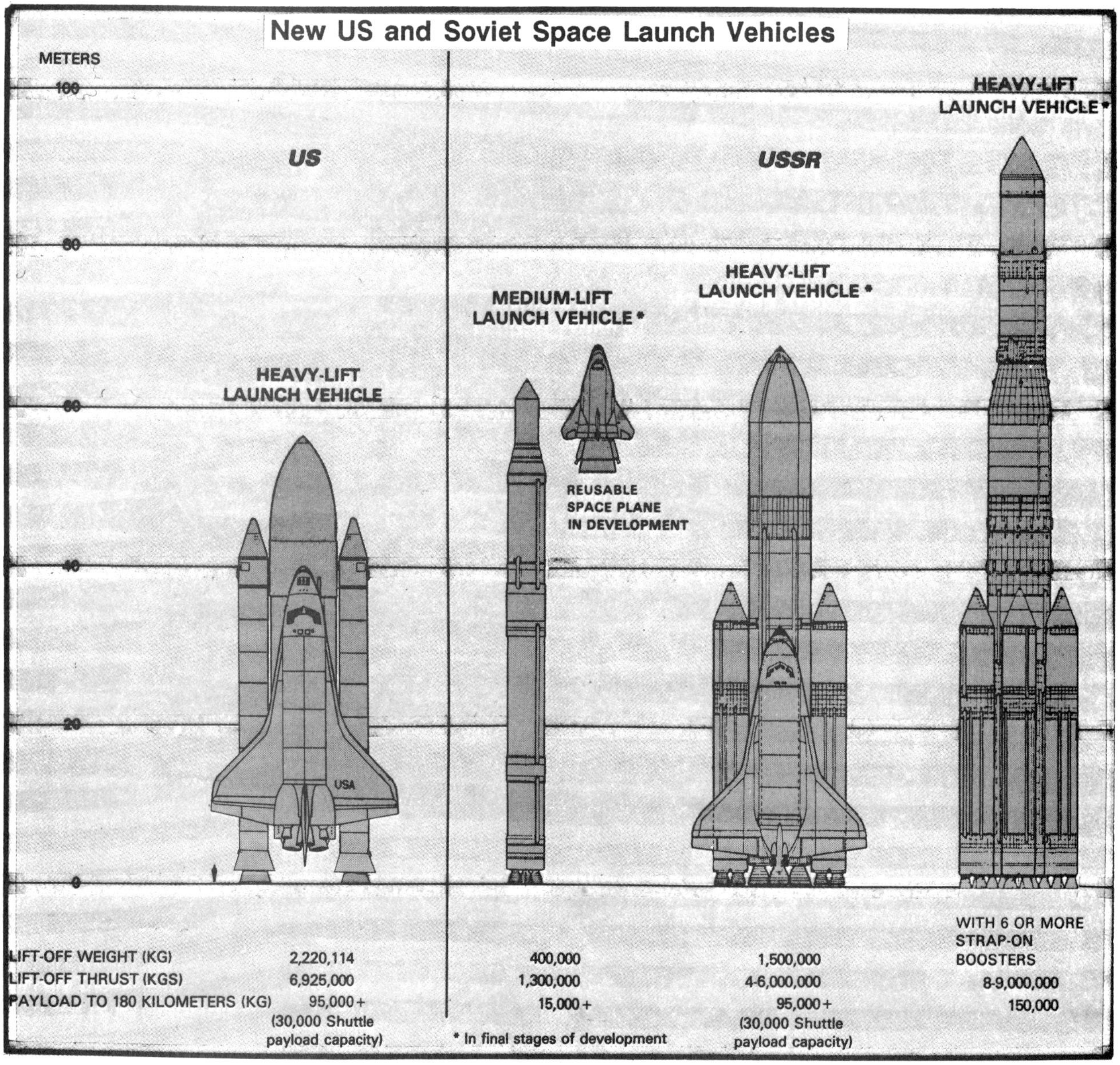

Unlike the U.S. shuttle, all of the boosters and main engines are mounted on the Energia, so the May 15 test proved the Soviet space shuttle propulsion system in flight.

U.S. recon satellites photographed a Soviet shuttle/booster mating test several months ago before Energia's reconfiguration for its first flight in the unmanned mode.

After liftoff from Tyuratam, Energia was powered through Mach 4-6 by the strap-on boosters and the core. The strap-on boosters then separated and the core vehicle with its cargo pod continued the ascent.

The core vehicle achieved its propulsion and guidance objectives, shut down four engines and then separated the payload complex. The payload was supposed to ignite its own rocket engines and continue into orbit while the core fell intothe Pacific Ocean. At this point a serious payload malfunction occurred, unrelated to the Energia booster.

The mission of the satellite payload is unknown but it was complex enough to have its own large propulsion system. The Soviet statement that the payload had "various problems" can be interpreted as another indication of satellite complexity.

Using TRW missile-early-warning satellites over the Pacific, the USAF Space Command was able to observe the reentry of the Energia payload. Data showed the payload as an "intense" hot object for an unusually long period of time as the satellite reentered the atmosphere. This could be an indication that the vehicle may have had a shallow reentry angle. In addition, it could also indicate a significant mass or a large propellant load that caused the unusually bright and lengthy fireball.

Chapter Eleven

SOVIET UNMANNED SPACE PROGRAM

Most of the Soviet spacecraft launched to date has been unmanned. These satellites have been orbited for a variety of purposes and under several program names. The most common name has been "Cosmos"; over 1700 have been launched since 1962.

The Cosmos program offers the hobby monitor more intrigue for the buck than any other space program I know. Just about the time you think you have them figured out, the Soviets manage a surprise or two! Appendix A contains further study material related to Soviet space activity to supplement information presented in this chapter.

Cosmos

In 1962 the Russians decided on a very simple cover plan which gave as complete privacy as technology would permit, as well as maintaining the "peaceful" image of the Russian's first five years of spaceflight.

Some 70 percent of all Russian flights were given the meaningless name of "Cosmos" and a serial number; at the same time over half the U.S. flights in that era were for the DOD.

The following text is extracted from the Soviet announcement for the launch of Cosmos 1 on March 16, 1962. References back to this original release have continued through all subsequent years for over 1600 Cosmos launches.

"A series of artificial Earth satellites will be launched from different cosmodromes of the Soviet Union in 1962. Another launching of an artificial Earth satellite was carried out in the Soviet Union on March 16, 1962...The launching of the artificial Earth satellite continues the current program of studying the upper layers of the atmosphere and outer space in the fulfillment of which a series of satellite launchings will be effected under this program from different cosmodromes of the Soviet Union in the course of 1962.

"The scientific program includes: The study of the concentration of charged particles in the ionosphere for investigating the propagation of radio waves; study of corpuscular flows and low energy particles; study of the energy composition of the radiation belts of the Earth for the purpose of further evaluating the radiation danger of prolonged space flights; study of the composition and intensity variations of cosmic rays; study of the magnetic field of the earth; study of the shortwave radiation of the Sun and other celestial bodies; study of the upper layers of the atmosphere; study of the effects of meteoric matter on construction elements of space vehicles; and the study of the distribution and formation of cloud patterns in the Earth's atmosphere. Moreover many elements of space vehicle construction will be checked and improved.

"The launching of sputniks of this series will be announced in separate reports. This program will give Soviet scientists new means for studying the physics of the upper atmospheric layers and outer space."

The first three Cosmos missions clearly came from a site different from the launches of the early Sputniks from Tyuratam: The new site was Kapustin Yar. Flight inclinations were 49 degrees, but Cosmos 4 was flown at the older Sputnik inclinations of 65 degrees. The vehicle was returned after three days of flight, heralding the advent of the photo recon missions and military use of the Cosmos cover name.

Broad Categories within Cosmos

With the advantage of hindsight, it is possible to sort out the Cosmos missions into broad categories. When operational systems for Meteor weather satellites or Molniya communications satellites appeared, it has been possible to find in the Cosmos series a family of precursor flights with almost the same characteristics. As discussed earlier in this chapter it has been possible to find manned precursor flights in the Cosmos series due to the distinctive nature of the program's operation.

Two categories which make up the bulk of Cosmos flights can give the monitor trouble identifying: unsuccessful deep space flights (earth orbiting Molniya 1 and 3, lunar probes Luna and Zond, and planetary probes Zond, Venera and Mars); and the larger group--the Soviet military space program.

Techniques for Defining Cosmos Missions

Soviet announcements coupled with later scientific publications permit positive identification of some of the Cosmos flights. The *Goddard Satellite Situation* report (published semi-annually) gives current orbital elements for all objects that are in orbit as well as decay dates on payloads and associated debris. If decay from orbit occurs before natural decay would dictate, it is reasonable to conclude that a method of retrofire was used to recover the satellite.

If piggyback payloads are separated later from the main payload, this fact can be generally noted in the situation report. Explosions in orbit can be detected in the NASA quarterly report by the large amount of debris, and the dispersion of new orbits from the original single orbit can tell something of the violence of the explosion.

The Royal Aircraft Establishment at Farnborough, England gives a much more detailed description of all world flights. They label which objects are payloads, spent rocket casings, or miscellaneous debris. The Establishment also gives the hour of launch (which helps to identify the launch site) and sometimes the purpose of the mission.

Because the Establishment also looks for repetitive patterns and adds data from optical observations, they are able to list the shape, weight, dimensions, apogee, perigee, inclination, period, semimajor axis, orbital eccentricity and the argument of perigee of each object.

The Royal Aircraft Establishment recently published a major work of their observations over the period 1957-1980. Called the *Table of Earth Satellites*, this is one of the finest reference books I use. While the book is somewhat expensive, a proper study of all space programs could not be started without one.

Still more information is available to private observers from the Kettering Group, headed by Geoffrey Perry in Northants, England. This group of dedicated Soviet space observers has established many of the Soviet Cosmos missions to a certainty.

Cosmos Missions

To aid the monitor hobbyist with identifying Cosmos missions the following is a summary of each category the author has identified. Particular attention should be focused on the orbital parameters for each type of mission as this is usually the best way to tell what the mission of the flight is and possible radio frequencies that will be in use.

Listeners are encouraged to share their loggings of new Soviet frequencies with the author; new information will be provided in future updates of this book.

Cosmos Photographic Recons

This mission is the single most important aspect of the Soviet space program; fully 50 percent of all Soviet Cosmos missions flown have been in the military photo recon category. Each new generation of satellite is an improvement over the earlier in terms of longer service, increase in maneuverability and technical capabilities. We would like to dedicate this section to Geoffrey Perry and the Kettering Group whose long and diligent efforts contributed extensively to this chapter.

First Generation, Low Resolution Recons

Launched by the Soviet A-1 booster, this series started in 1962 with the launch of Cosmos 4 and continued to be identified by the Kettering Group until 1967. Initial orbital durations started with three days, quickly stabilizing at eight days for the next seven years. The first generation apparently used the Vostok recovery shell and service vehicle modified for photo recon use. The Kettering group identified approximately 41 first generation recon missions.

Launched from Tyuratam were: Cosmos 4, 7 9, 10, 12, 13, 15, 16, 18, 20, 24, 28, 29, 32, 33, 35, 37, 46, 48, 50, 52, 64, 66, 68, 98, 99, 104, 105, 107, 115, 117, 132, 157, and 143.

First generation, low resolution recons launched from Plesetsk were: 112, 129, 136, 138, 147, and 153.

Second Generation, High and Low Resolution Recons

Appearing on the space scene about 1963, the second generation recons were presumably heavier, because they used the Soviet A-2 booster for a launch vehicle.

Two telemetry modes appeared among the second generation A-2 flights; since the flights were intermixed it was assumed that two different degrees of resolution were being used.

Second generation, high resolution flights were the first to be introduced. Over 71 missions were cataloged from 1963-1970. Launched from Tyuratam were:

22, 30, 34, 45, 67, 77, 109, 113, 126, 127, 128
130, 134, 155, 172, 182, 201, 224, 227, 234, 239, 267
274, 279, 284, 288, 296, 299, 310, 331, 345, 346, 352

From Plesetsk:

59, 65, 69, 79, 85, 91, 92, 94, 114, 121, 131
141, 150, 161, 175, 190, 194, 207, 214, 229, 232, 237
241, 246, 254, 270, 276, 297, 271, 282, 286, 289, 294
302, 322, 323, 349, 355.

Second generation, low resolution recons debuted in 1966 and flew until 1970 when the third generation recons were introduced. There were 33 recons identified in this category: launched from Tyuratam were Cosmos 120, 124, 162, 168, 177, 216, 231, 235, 240, 258, 287. Launched from Plesetsk were: Cosmos 164, 180, 181, 193, 195, 199, 205, 210 223, 247, 253, 255, 263, 266, 273, 278, 281, 290, 301, 325, 326, 344

Third Generation Recons

The third generation recons, maintaining flight durations of 12-13 days, have been divided into a number of subtypes.

The first to be flown were low resolution and

often carried supplemental payloads. These satellites were identifiable by their PDM (Pulse Duration Modulation) signals on 19.994 MHz and a recover beacon typically sending the Morse code characters "TG" on recovery. Kettering members found that if word "7" in the spacecraft HF telemetry frame was long at the onset of the satellite's mission, then the recon did not carry a supplemental payload. If word "7" was short at the beginning of the flight, then lengthened, the recon did carry a supplemental payload which was later separated during the mission.

These satellites were called (for identification purposes) third generation, low resolution, PDM science recons. The plain PDM recons made their appearance in 1969 and the PDM science type made their appearance in 1968. This type of Cosmos included:

(launched from Tyuratam)

208, 228, 243, 293, 306, 318, 350, 363, 366, 368, 377
392, 410, 428, 431, 473, 493, 517, 537, 547, 583, 599
658, 685, 702, 731, 780, 799, 809, 819, 856, 899, 904
914, 966, 973, 992;

(launched from Plesetsk)

309, 313, 329, 353, 394, 403, 439, 443, 477, 484, 490
512, 518, 525, 552, 555, 561, 575, 578, 596, 629, 635
640, 653, 669, 692, 696, 697, 721, 728, 741, 747, 751
769, 776, 784, 813, 834, 840, 848, 865, 879, 898, 935
995

Also in 1968, another third generation subtype made its debut: The high resolution Morse code maneuverable recons flew over 56 missions before being phased out in 1974. Telemetry consisted of Morse code in a series of three letters, actually quantitative data in binary form. Signals were typically broadcast on 19.150 MHz and at the end of the mission the Morse code ID beacon used the letters "TK" on either 19.995 or 20.005 MHz.

These flights maneuvered while in orbit and cast loose an engine late in the flight which usually lasted 13 days. They included:

(launched from Tyuratam)

251, 264, 280, 360, 370, 386, 399, 420, 429, 441, 432
452, 463, 466, 471, 474, 491, 492, 499, 513, 519, 543
551, 572, 581, 609, 632, 667;

(from Plesetsk)

328, 333, 361, 376, 396, 401, 406, 424, 430, 456, 454
464, 478, 483, 486, 495, 503, 522, 538, 548, 560, 563
577, 579, 584, 598, 600, 603.

A third generation, high resolution subtype introduced in 1969 is still in use today, emitting two-tone signals on about 19.989 MHz. They appear to have no detectable telemetry at this frequency.

Each tone transmitted on 19.989 MHz is of equal duration; one complete cycle of both tones is typically 1.5-1.75 seconds. Most likely the difference in tone cycle durations is to distinguish between different spacecraft in orbit simultaneously. Kettering group members in England report that amplitude modulated data is transmitted on command over Russian ground stations on 232 MHz in the UHF military aircraft spectrum.

When these recons are recovered, their beacon broadcasts the letters "TF". The flights maneuver and typically last 14 days. Recons of this type launched from Tyuratam are:

1259, 1264, 1272, 1277, 1303, 1319, 1352, 1373, 1381
1402, 1425, 1460, 1482, 1533, 1571, 1542, 1600, 1623
1644, 1760, 1775, 1781, 1804.

Launched from Plesetsk are:

867, 958, 1095, 1103, 1111, 1126, 1138, 1142, 1147
1149, 1155, 1166, 1178, 1183, 1216, 1221, 1224, 1237
1245, 1281, 1307, 1338, 1404, 1444, 1449, 1467, 1485
1488, 1493, 1497, 1499, 1505, 1512, 1530, 1545, 1549
1568, 1583, 1587, 1609, 1613, 1628, 1649, 1659, 1683
1685, 1702, 1705, 1740, 1742, 1765, 1772.

Other third generation, high resolution, maneuverable Cosmos satellites transmitting simple FSK on 19.989 MHz include:

(from Tyuratam)

390, 636, 652, 674, 691, 701, 710, 719, 722, 727, 740
754, 760, 774, 786, 802, 806, 817, 824, 835, 852, 859
866, 884, 888, 908, 920, 932, 957, 986, 989, 999, 1021
1113, 1120, 1170, 1173, 1268, 1279, 1313, 1316, 1329
1368, 1416, 1419, 1421, 1438, 1446, 1668, 1696, 1728
1747, 1787;

(from Plesetsk)

317, 364, 383, 427, 438, 442, 488, 515, 527, 550, 554
556, 559, 587, 597, 602, 607, 612, 625, 630, 639, 649
657, 659, 666, 671, 688, 694, 704, 709, 730, 742, 743
746, 748, 753, 757, 758, 771, 779, 788, 810, 815, 820
821, 833, 847, 854, 857, 863, 892, 896, 897, 902, 907
912, 915, 927, 934, 938, 948, 953, 964, 969, 974, 987
993, 1003, 1007, 1010, 1022, 1029, 1031, 1033, 1042
1047, 1049, 1050, 1050, 1059, 1068, 1071, 1073, 1078
1080, 1098, 1099, 1105, 1107, 1108, 1115, 1116, 1123
1127, 1128, 1165, 1187, 1189, 1200, 1202, 1205, 1213
1214, 1219, 1227, 1262, 1265, 1297, 1334, 1342, 1343
1396, 1411, 1422, 1469, 1509, 1551, 1573, 1592, 1632
1664, 1665, 1671, 1715, 1730, 1790.

Certain earth resource photographic recoverables form yet another subset of the third generation type recons; these can be distinguished by the "TK" recovery beacon. These missions originally flew at low altitude, inclined 81.3 degrees and were typically launched by the A-2 booster in time to observe the breakup of the Arctic ice pack along the northern Russian coast. These satellites have now been replaced by F-2 booster launched recons launched with an inclination of 82.3 degrees. High resolution recons launched by the F-2 booster from Plesetsk are:

1182, 1185, 1201, 1203, 1207, 1209, 1212, 1273, 1276
1280, 1301, 1314, 1353, 1369, 1376, 1387, 1401, 1406
1440, 1451, 1458, 1462, 1468, 1483, 1487, 1495, 1498
1537, 1557, 1572, 1575, 1582, 1584, 1590, 1591, 1597
1648, 1653, 1657, 1663, 1672, 1678, 1681, 1708, 1746
1757, 1762, 1768

Medium resolution recons launched are Cosmos 1210, 1283, 1284, 1385, 1472, 1789; medium resolution and geodesy satellites are Cosmos 1211, 1239, 1309,

1332, 1398.

First appearing in 1971, another special subset of the two-tone satellites transmits on 19.994 MHz and sends the Morse code letters "TL" at the end of their mission. These recons do not maneuver and are probably low resolution in nature, probably performing geodetic survey and mapping missions for the Russians. These missions probably enable correlation of international mapping grids for MIRVed warheads, allowing for precision targeting.

There are three remaining observations about these photo recons:

(1) The labeling of the supplemental or piggyback missions of these flights as "scientific" reflects that for the first few years after 1968 the Russians identified such extra experiments as geophysical or biological; however, since mid-1972, not a single additional identification has been supplied, probably because of the military nature of the flights.

(2) Flights identified as using 19.994 or 19.995 MHz may suddenly switch to 19.989 or 19.990 MHz to clear the first frequency for a second launch.

(3) From the earliest days of the military observation recoverable recons, a number of them carried extra experiments as mentioned before, but cast loose no piggyback capsules. Presumably these flights used Vostok/Voskhod hardware and the experiment was probably contained in the recoverable cabin. Most of the supplemental experiments today seem to be linked with a separate compartment which is cast loose, thus strengthening the notion that Soyuz hardware is being used by recons. What most often is cast off is a maneuvering engine, suggesting that the supplemental payloads are returned to earth in the main cabin, rather than cast off in orbit to decay through air drag. This, however, is only speculation.

Fourth Generation Recons

In 1975 a new generation of Soviet recons appeared which had a longer orbital life than earlier models. Kettering Group members have found no shortwave frequency in conjunction with these recons, although frequency modulated signals on 240 MHz have been observed at both high and low data rates transmitted from the long duration satellites.

These satellites do have a maneuvering capability and perigee heights are generally lower than previous generations requiring a frequent boost in orbit to overcome drag and maintain the satellite in orbit.

Imagery returned by this satellite is via the high data signals on 240 MHz and these are probably digital signals sent to ground stations over the Soviet Union. Film capsules are also returned on the ninth and eighteenth days of the mission.

Recovery beacons have been monitored from this recon sending a series of "TV"s at the end of the mission.

Fourth generation recons launched from Tyuratam are:

1218, 1240, 1246, 1270, 1282, 1298, 1330, 1336, 1347
1370, 1377, 1399, 1424, 1426, 1439, 1457, 1466, 1489
1504, 1516, 1552, 1585, 1608, 1611, 1616, 1630, 1643
1654, 1673, 1679, 1731, 1739, 1756, 1764, 1770, 1773
1784, 1792, 1810

Launched from Plesetsk are:

805, 844, 905, 949, 1028, 1079, 1097, 1121, 1144
1152, 1177, 1208, 1236, 1248, 1274, 1296, 1318, 1350
1384, 1407, 1442, 1454, 1471, 1496, 1511, 1532, 1539
1543, 1548, 1558, 1576, 1599, 1647, 1676, 1699, 1706
1713, 1724, 1734, 1807

Photo Recons as a Key to World Crises

If the Soviets are interested in monitoring a particular area of the world (i.e. invasion, overthrow, military buildup, etc.) what better way to find out what is going on that through the eyes of a high resolution photo recon satellite?

How can you use photo recons to determine the world's hot spots? It's really quite simple. First, acquire a copy of *The Satellite Experimenters Handbook* ($10 from the American Radio Relay League, 225 N. Main St., Newington, CT 06111). In the back of the book is an OSCARLOCATOR map projection. The clear plastic overlay will enable you to scribe the orbit of the common Soviet flight photo recon profiles so that you can see the ground track of each recon orbit.

Now, by using the NASA 30 day prediction service available from the Goddard Space Flight Service, you can determine the longitudinal crossing point for each satellite orbit over the northern hemisphere.

In the absence of maneuvers to produce a stabilized ground track, the only real clue to be found is in the location of the perigee point of the orbit. The best resolution for a given photo recon will occur at the lowest point in the orbit; hence, one should plot these perigee points for each day's orbits.

The perigee point will now define a band of latitude locations close to where the target of the satellite's interest lies. Greater confidence is obtained if it can be shown that the time of pass through the perigee takes place around their local noon. Use of the "DX Edge," (P.O. Box 834, Madison Square Stn, NY, NY 10159) can give you the sun/time picture for any location on earth. You can then tell with fair confidence the location the recon is most interested in.

If the ground track is subsequently stabilized, the analyst is then presented with sixteen fairly precise locations spaced at approximately 22.5 degrees around the latitude of perigee. A glance at your OSCARLOCATOR will show that some of these may be eliminated as they fall over the oceans or, in the northern hemisphere, within the territory of the Soviet Union.

Recons which have a maneuvering capability can also provide clues. The Soviets maneuver these satellites into areas of interest. Observers should be on the alert any time a major maneuver moves these satellites into a particular area and then back out again--this could be a hotspot!

Recent examples can be documented. Cosmos 1466, launched April 26, 1983, flew a fourth generation, long duration flight profile. After two days in space, the Soviets maneuvered the satellite into a low circular orbit where it stayed for two weeks. The ground track was analyzed, indicating that the Russians were watching the Iraq-Iran battlefield.

Another fourth generation, long duration recon, Cosmos 1489, launched August 10, 1983, appeared to be watching the Israeli departure from the mountains overlooking Beirut, Lebanon. Long-duration-recon Cosmos 1504 was also vectored in for a close look in this area.

Cosmos 1504 was launched September 14, 1983, from Tyuratam, and on the day of the Marine landing on Grenada, October 25, 1983, was some 1600 km east. The Soviets raised the orbit of 1504 the following day to accelerate its westward drift. After two days the orbit was again dropped to a nearly-ground-stabilized orbit to allow a noon day look at the situation in Grenada from October 28 until the first of November.

El Salvador has not escaped the attention of the Soviet eye in the skies. Cosmos 1471, launched June 28, 1983, for a long duration mission, maneuvered in for a four day look at that Central American country on July 3. Once its mission was completed, the Soviets raised the orbit of 1471 and the recon continued its global surveillance.

Clearly, the monitoring hobbyist equipped with basic equipment can watch international intrigue unfold right before his eyes (or ears).

Russian Military Comsats

The Russian military forces use a variety of communications satellites. Two of the programs are discussed in the International comsat section of this book; the other two fall under the Cosmos label.

At the bottom of the Soviet tier of communications satellites are two constellations whose frequencies remain a mystery to this author. The first constellation is probably the most exotic of the two systems.

The Soviets appear to have put up three large comsats in an 800 km circular orbit inclined 74 degrees, spaced 120 degrees apart, and launched on a C-1 booster out of Plesetsk. These satellites appear to perform a store-dump mission; as the satellite flies over a given area, transmissions from the ground are received by the satellite, stored on tape, and later replayed for ground stations over the Soviet Union.

Geoffrey Perry reported in the January 1982 *Military Electronics Countermeasures* magazine that in April 1977 an Iranian was arrested by the Shah's secret police. The arrest stemmed from the Iranian allegedly receiving coded communications from the KGB. The Iranian received these instructions via a satellite orbiting overhead which displayed the data on the LED output of a small pocket electronic calculator. These might be the kinds of satellites that fulfill this type of mission.

The second type of satellite in the Cosmos series is less mysterious. Eight 40kg satellites are launched on one C-1 booster out of Plesetsk into a 1700 x 1350 km orbit inclined 74 degrees. These satellites are reminiscent of the old DSCS Phase 1 comsats that the U.S. launched some years ago. It would not surprise me if they also utilize the 8/7 GHz military band for their communications links. The Soviets are very good at modeling their systems after U.S. satellites; this can sometimes offer clues as to missions and frequencies.

Soviet store-dump comsats launched by the C-1 rocket include:

158, 372, 407, 468, 494, 540, 614, 676, 773, 783, 836
841, 858, 923, 968, 990, 1023, 1048, 1110, 1125, 1140
1190, 1269, 1302, 1331, 1354, 1371, 1420, 1452, 1486
1503, 1538, 1570, 1624, 1680, 1741, 1763, 1777

Other tactical theater comsats launched by the C-1 rocket are Cosmos:

38-40, 41-43, 54-56, 61-63, 71-75, 80-84, 86-90
336-343, 411-418, 444-451, 504, 511, 528, 535, 564
571, 588, 595, 617-624, 641-648, 677-684, 711-718
732-739, 761-768, 791-798, 825-832, 871-878, 939-946
976-983, 1013-1020, 1034-1041, 1051-1058, 1081-1088
1130-1137, 1156-1163, 1192-1199, 1228-1235, 1250-1257, 1287-1294, 1320-1327, 1357-1364, 1388-1395
1429-1436, 1473-1480, 1522-1529, 1559-1566, 1635-1642, 1716-1723, 1748-1755, 1794-1801

Early Warning Satellites

These satellites are used by the Soviets to detect the infrared signature of the exhaust plume from the boost phase of a missile launch.

Launched into high, Molniya type orbits that can observe the northern hemisphere for long periods of time, these satellites fly 62.8 degree inclination orbits that range from 300 to 40,000 km in altitude.

A former source of confusion to me was being able to tell the difference between an Early Warning satellite and a failed Molniya communications satellite which also carries a Cosmos name. The answer became evident by looking at the argument of perigee.

Molniya satellites have arguments of perigee situated near 285 degrees; this gives them stable ground tracks over the central regions of Russia, and 12 hours later over the North American continent. Even if the Molniya's communications systems failed the satellite would still maintain this orbital factor (assuming, of course, successful final orbital insertion).

Early Warning satellites, on the other hand, exhibit different argument of perigee characteristics. Spaced every 40 degrees, nine satellites fill the orbital planes. The result is that the EW missions have broader northern hemisphere ground tracks than the Molniya satellites.

These satellites are capable of performing their mission over a considerable part of each orbit. Not only is a complete view of the U.S. possible during a satellite's Pacific loop, but a more evenly distributed coverage of the Atlantic Ocean and China is available on the adjacent loop.

Transmitting their signals from 2280-2304 MHz, in 1984 the Soviet EW satellites caused interference to amateur radio operators using 2304 MHz for EME work. This is not the first time that interference was caused to other services. NASA has worked out an agreement with the Soviets to shut off the EW satellite causing the racket whenever they are receiving signals from their deep space probes in the 2292-2297 MHz range.

Phase 1 testing launches include Cosmos 174, 260, 520, 606, 665, 706, and 775. Operational satellites (Phase 2) launched since 1976 are:

862, 903, 917, 931, 1024, 1030, 1109, 1124, 1164* 1172, 1188, 1191, 1217, 1223, 1247, 1261, 1278, 1285* 1317, 1341, 1348, 1367, 1382, 1409, 1456, 1481* 1518 1541, 1547, 1569, 1581, 1586, 1596, 1604, 1658, 1661 1675, 1684, 1687, 1698, 1701, 1729, 1761, 1774, 1783* 1785, 1793, 1806

(denotes launch failure)*

Ocean Surveillance Satellites

Up until 1978, the Soviet ocean surveillance program was unknown to the world except to intelligence agencies and amateur Soviet space observers. Then the flight of Cosmos 954 captured world headlines. As the nuclear powered satellite returned to earth, the world watched and waited to find out where its nuclear debris would be scattered. The satellite reactor ultimately spewed its radioactive material over northern Canada, causing the biggest space cleanup ever.

The Soviet ocean surveillance program went back to the drawing board, and there wasn't another launch of a nuclear-powered satellite until April, 1980. Since that time several missions have flown, all successfully injecting the reactor section into higher (965 km) orbit for safekeeping.

These 11,000 pound nuclear powered surveillance satellites are 46 feet long in orbit. Normally launched from Tyuratam on an F-1-m vehicle into a 265 x 250 km orbit inclined 65 degrees, they usually stay in orbit for several months providing the Soviets with "blob" type imagery 250 km orbit inclined 65 degrees, they usually stay in orbit for several months providing the Soviets with "blob" type imagery from the spacecraft's radar system. Their normal target is naval forces that could be a potential threat.

These satellites transmit their telemetry on 166 MHz and 19.954 MHz. At the end of the mission the Soviets shut down the nuclear reactor, then the satellite breaks into three pieces. It is at this point that the Soviets boost the orbit of the nuclear power supply into a higher orbit where it will remain safe from natural decay for about 600 years.

At the close of 1982, Cosmos 1402 brought world attention again to the Soviet ocean surveillance program. Launched on August 30, 1983, the satellite was immediately identified as a Soviet ocean surveillance radar type satellite by its characteristic orbital profile.

Cosmos 1402 appeared normal through December 27, 1982, and was in one piece. Kettering group analysis indicates that on the morning of December 28 the Soviets commanded the reactor separation and boost maneuver but a malfunction occurred in the procedure.

The satellite did separate into three pieces, but U.S. intelligence officials indicated that the propulsion system for the reactor section malfunctioned, preventing the boosting of the reactor system.

Transmissions grew weak and ceased altogether by January 1. With the satellite separated from the power supply, the batteries finally gave up.

Power to the satellite's radar system is provided by a nuclear reactor which uses about 100 pounds of uranium 235 fuel. Concern by worldwide experts over the 4,000-5,000 pound reactor section was that the uranium or its fission products (strontium, cesium and cerium) could pose a serious health hazard at reentry.

Soviet radar ocean surveillance satellites launched by the F-1-m rocket are:

198, 209, 367, 402, 469, 516, 626, 651, 654, 723, 724 785, 860, 861, 952, 954, 1176, 1249, 1266, 1299, 1365 1372, 1402, 1412, 1579, 1607, 1670, 1677, 1736, 1771

A second category of ocean surveillance satellite is more than likely a ferret, operating with a conventional solar power supply, and eavesdropping on enemy fleet transmissions. Satellites of this type include:

699, 777, 838, 868, 937, 1094, 1096, 1167, 1220 1260, 1286, 1306, 1337, 1355, 1405, 1461, 1507, 1567 1588, 1625, 1646, 1682, 1735, 1737, 1769

Nicholas Johnson of Teledyne Brown has noticed that the two ocean surveillance satellites appear to have an orbital phasing relationship 150 degrees apart; this could allow the Soviets to examine the shipping traffic under observation, then listen to the ships with the ferret on later orbits after the targets have been identified.

Ferrets

The Soviet Government has long had a reputation for giving special attention to the gathering of ELINT (electronic intelligence) data, also referred to as ferreting, or SIGINT (signal intelligence), COMINT (communication intelligence), and RADINT (radar intelligence).

By definition, all spacecraft which receive and report on electromagnetic radiation are performing the same basic task. This would include such satellites as solar studies, astronomy, weather reporting, earth resources, and communications.

The origin and the use of such signals, however, fall into two major categories: (a) those directed toward space deliberately to be picked up and relayed by communication satellites, and (b) those not intended to be picked up by the receiving satellites, such as private messages or inadvertent leakages of signals, and hence supporting the function of ELINT, SIGINT, RADINT, and COMINT.

Military interests extend to natural phenomena, partly to understand the difference between natural signals and those which are man-made, and partly because many natural emissions, such as reflected light or radiated heat, translate into pictures and data defining ground, air and space activity.

Those emissions which were generated by electronic devices such as radio stations, radar equipment, microwave towers, and other spacecraft provide a general category of signals whose frequencies, power levels, locations, directions, and times of emission may answer questions of military interest. Although detection of the emissions presents technical challenges, understanding those emissions may be an even bigger challenge.

Soviet interest in ELINT is evident in such activities as the maritime trawlers with their forests of antennas which follow NATO and U.S. naval maneuvers. These trawlers attend missile launchings and recovery areas, or cruise off the coasts of the United States.

Since Soviet trawlers, naval vessels, embassies, and air and space defense systems all engage in signal gathering, it can be assumed that they also gather, by spacecraft, signals which are then relayed, either in real time or after taped storage, to analytical centers in the Soviet Union.

Up until recently, no known signals were ever received from Soviet ELINT satellites; but a new generation of Soviet heavy-ELINT satellites has now been heard from. These satellites first made their appearance in 1981 launched from the F-2 rocket; they soon replaced the older heavy ELINTs launched by the A-1 rocket.

These new heavy-ELINT satellites are launched into 665 km by 630 km orbits inclined at 82.5 degrees. According to one report they "operate in a record-playback mode and can locate pulsed emitters to a best accuracy of about 10 kilometers," providing an electronic order-of-battle (EOB) capability.

These satellites have been heard transmitting a CW beacon on two frequencies: 51.140 and 51.160 MHz, as well as their third and fourth harmonics on 153,420/153.480 MHz and 204.560/204.640 MHz. The reason for two frequencies is to eliminate interference from two nearby satellites of the same class.

Soviet ELINT satellites launched by the C-1 rocket are:

103, 151, 189, 200, 236, 250, 269, 315, 330, 358, 387 395, 425, 436, 437, 460, 479, 500, 536, 544, 549, 582 610, 631, 655, 661, 698, 707, 749, 781, 787, 790, 812 845, 870, 899, 924, 960, 1008, 1062, 1114, 1345

Soviet ELINT heavy satellites launched by the A-1 rocket are:

389, 405, 476, 542, 604, 673, 744, 756, 808, 851, 895 925, 955, 975, 1005, 1043, 1063, 1077, 1093, 1116, 1143 1145, 1154, 1184, 1206, 1222, 1242, 1271, 1315, 1340 1346, 1356, 1400, 1437, 1441

All these satellites have been replaced by the 1603 type Cosmos.

Minor Soviet Military Craft

Based on orbital elements, there are five categories of minor military missions launched by the B-1 rocket, as follows: (a) 48.8 km low (b) 48.9 high (c) 71 low (d) 71 medium (e) 82 high. It is not known what missions each of these regimes represent but research continues.

Soviet minor military satellites launched by the B-1 rocket include:

31, 36, 70, 76, 93, 95, 101, 106, 116, 119, 123, 145 148, 152, 165, 173, 176, 191, 197, 202, 204, 211, 221 222, 233, 242, 245, 257, 265, 268, 275, 277, 283, 285 295, 303, 307, 308, 311, 314, 319, 324, 327, 334, 347 351, 357, 362, 369, 380, 388, 391, 393, 408, 421, 423 435, 440, 453, 455, 458, 467, 472, 485, 497, 498, 501 523, 524, 526, 545, 553, 558, 562, 580, 601, 608, 611 615, 633, 634, 662, 668, 686, 695, 703, 705, 725, 745 750, 801, 818, 849, 850, 901, 919

These minor military missions were phased out to those utilizing the C-1 rocket, namely:

660, 687, 752, 807, 816, 822, 885, 891, 913, 930, 933 965, 1006, 1065, 1075, 1112, 1146, 1169, 1179, 1186 1204, 1215, 1238, 1263, 1310, 1311, 1335, 1351, 1397 1418, 1427, 1450, 1453, 1463, 1465, 1494, 1501, 1502 1508, 1534, 1578, 1601, 1615, 1631, 1662, 1688, 1776 1788

There are seven orbital regimes within this classification of Cosmos satellites that I have identified:

a) 50.7 low -- Most missions range 548 x 347 km.

b) 50.7 high -- Only one mission in this orbital regime so far

c) 65-66 low -- Three classes as follows: 1) 515 x 470 km radar calibration satellites; 2) 495 x 445 km, mission unknown; 3) 410 x 374 km, could be linked with testing of a blue-green laser communications system, as each carries a reflecting mirror.

d) 74 low -- Radar calibration missions

e) 74 medium

f) 83 low

g) 83 high

Biosat/Scientific Cosmos

While the majority of the Soviet scientific satellites come under other programs, scientific and/or biosat flights are flown under the Cosmos label. During the early days of the Cosmos program, scientific results were published frequently that have been correlated to specific Cosmos missions.

It has been some time since this was the case. It is quite likely that the Russians have quit putting experiments on the recons.

Those satellites which could be identified as scientific are as follows:

1, 2, 3, 5, 6, 8, 11, 14, 17, 19, 23, 25, 26, 49, 51 53, 97, 108, 135, 137, 142, 149, 163, 166, 196, 215 219, 225, 230, 259, 261, 262, IK-1 (Interkosmos), IK-2 320, 321, 335, 348, IK-3, 356, IK-4, IK-5, 481, IK-9.

Geodesy and Mapping

Prior to 1957, geodesy, the study of the shape and size of the earth and the determination of geocentric coordinates of points on its surface, was dependent only on ground measurements over small areas limited by the borders of states and continents. The availability of artificial satellites has greatly increased the scope of geodesy, making satellite geodesy an important applied science.

The United States has flown many geodetic flights to tie together world maps and to better understand the true shape of the earth. If there have been strategic mapping flights as well, the results havenot been published.

From 1968 to 1978, the Soviets first used geodetic satellites launched by the C-1 rocket. It is believed that the Russian space geodesy program started with Cosmos 203 (launch inclination 74 degrees, orbital height 1180 x 1210 km). Satellites that have characteristics of a geodetic nature have been launched at the rate of one or two a year.

First generation geodetic satellites include Cosmos 203, 256, 272, 312, 409, 457, 480, 539, 585, 650, 675, 708, 770, 842?, 911? 963, 1045?, 1067.

The Soviets inaugurated their second genertion geodetic satellites in 1981 with the launch of Cosmos 1312. Follow-on Cosmos launches included: 1410 (1982), 1510 (1983), 1589(1984), and 1660 (1985). There have been two orbital inclinations noted for the second generation satellites over the last five years; the reason for this is not known, but they are summarized below. All are launched by the F-2 rocket from Plesetsk.

Cosmos	Apogee	Perigee	Incl	Intl Des
K1312	1499	1490	82.6	1981-098A
K1410	1502	1494	82.6	1982-096A
K1510	1525	1480	73.6	1983-115A
K1589	1502	1494	82.6	1984-084A
K1660	1526	1482	73.6	1985-047A
K1732	1526	1480	73.6	
K1803	1503	1498	82.6	

Cosmos Ocean Resources

Cosmos 1076, launched February 12, 1979, from Plesetsk using the F-2 vehicle, represented a newer class of satellites dedicated to observing and monitoring the world's oceans. Initial orbital parameters were 634 x 606 km, inclined 82.5 degrees.

A second launch in the series--Cosmos 1151--occurred on January 23, 1980, and on September 28, 1983, Cosmos 1500 was launched from Plesetsk using the same basic orbital parameter. The Soviets announced that the purpose of the latter flight was to "Continue testing new types of information-gathering equipment and methods of remote explorations of the world ocean and the surface of the earth in the interests of different branches of the national economy of the USSR and science." Cosmos 1500 was used to aid the Soviet merchant fleet caught in the ice on the Chunchi and East Siberian Seas.

According to the Soviets the satellite contains a 3cm side-looking radar system using vertical signal polarization with a resolution of 1.5-2 km. In addition to the side-looking radar, 1500 also carries a multichannel low-resolution scanner and a multichannel microwave radiometer.

The scanner can be operated simultaneously with the radar because of their uniform line scanning at the surface. The scanning frequency of the two instruments is 4 Hz.

Cosmos 1500 uses 137.4 and 466.5 MHz for downlink. Listeners equipped with weather satellite capability might wish to monitor these two frequencies as the signals can probably be decoded with relatively simple equipment. More information on these new Soviet ocean picture-takers is available in the October 15, 1984, issue of *Aviation Week and Space Technology*.

Anti-Satellites

In a recent edition of the Department of Defense's *Soviet Military Power*, DOD conceded that the Soviet Union now possess an operational, low-altitude, earth orbit, anti-satellite system. The Russians have launched over 20 missions since 1968 that could have been anti-satellite tests, nearly half of which have been rated a success by western experts.

One of the most unusual flights of 1984 seems related to the Soviet anti-satellite program. On September 28, 1984, Cosmos 1603 was launched from Tyuratam and exhibited a very unusual maneuvering capability. Launched initially on a 51.6 degree inclination, the vehicle stayed in an 88.4 minute orbit for about one day; it then fired its propulsion system for a long burn to change its orbital inclination and altitude significantly, resulting in a nearly circular orbit at 850 km, inclined 66.6 degrees.

The plane change from 51.6 to 66.6 degrees required a large propulsion system with a quantity of on-board propellant. The Cosmos 1603 payload was maneuvered again, from the 66.6 to 71 degree inclination.

Western experts wondered if this was the first flight test of the Soviet's long-awaited, medium lift booster. Further analysis ruled out this possibility.

A space vehicle exhibiting this kind of orbital maneuvering capability could have anti-satellite mission potential, capable of attacking U.S. spacecraft in geostationary orbit or other high altitude satellites now out of reach by the smaller anti-satellite described earlier.

Even more ominous was the report by *Aviation Week* in February 1983 that a U.S. recon satellite could have been damaged by a Soviet ground-based directed-energy weapon system while over the Soviet Union. Star Wars could be closer than we think.

Cosmos Navigation Satellites

Perhaps the easiest VHF Soviet satellites for newcomers to monitor are the Russian Navsats around 150 MHz, receivable on the simplest of setups. I have used a Bearcat scanner with a plug-in antenna indoors on the dining room table to get fair results.

These Navsats are launched from Plesetsk on the C-1 booster rocket into 1020 x 965 km orbits inclined 82.9 degrees. Orbital periods average 104 minutes.

Based on orbital data, Nicholas Johnson of Teledyn Brown has determined that there are two different Navsat systems. One system is comprised of four satellites spaced 45 degrees apart; the second system utilizes six satellites spaced 30 degrees apart. The latter is referred to as "constellation two" and the former as "constellation three."

It would appear that the second constellation is used by the Soviet military and the third by civilians. The more satellites in a particular constellation, the more accurate the navigation information provided.

In 1972, Geoffrey Perry and C.D. Wood conducted a systematic examination of Cosmos satellites having near-circular orbits and periods close to 105 minutes; these would have the proper characteristics for a Navsat. Their results are published in the "Philosophic Transactions of the Royal Society of London," A 294, 307-315 (1980). The article describes methods used to decode these Navsat signals...must reading for anyone wanting to tackle the Soviet Navsat telemetry. A copy of this paper should be available at any local university library.

Chris Wood's analysis of the Navsat telemetry indicated an FSK binary signal transmitted at a rate of 50 bits per second. The first 18 bits in any second consist of a pulse synchronized with Universal Time and a statement of Moscow Time in hours, minutes and seconds. In a one minute time frame there are eight blocks containing seven lines. The remaining 32 bits of information transmitted in each second of the first four blocks contain the geocentric coordinates and their rate of change, updated by the satellite every three minutes.

Each satellite also has a binary ID signal; Number 1 through 8 are used by the second constellation satellites, and 11 through 14 are used by the third constellation satellites.

Wood's and Perry's original article gave the following frequency breakdown for these satellites (in MHz):

149.9625	399.900	Cosmos
149.9750	399.933	Cosmos
149.9880	399.968	Transit
150.0000	400.000	Cosmos
150.0375	400.100	Cosmos

But something was amiss. In August 1984 I programmed the TIMEX-SINCLAIR 1000 with the W3IWI AMSAT satellite tracking program and the latest set of Soviet Navsat Kepplerian elements for AOS-LOS times for each of the satellites I had data for. After three days of on-the-air monitoring, I was convinced that either the computer had given up the ghost or my tracking element sets from Goddard were in error.

Over a period of three months I continued to monitor the Perry-listed satellite frequencies. I did note that my predictions for the 150.000 MHz channel satellites were quite accurate. This only muddied the waters that much more. I was not hearing any satellite telemetry from any other Kettering-listed frequency other than the U.S. Transit satellite on 149.988 MHz; in addition, the 150.000 MHz frequency was not supporting the proper number of satellites I expected to see, ten based upon the two constellations mentioned previously.

My thoughts at this time were that the Soviets had made a wholesale change in their Navsat frequencies. My only choice was to search 149.910-150.100 MHz for new frequencies. In December 1985 this effort was started and, much to my delight, the 15th was a red letter day:

Freq MHz	**Use**
149.910/399.760*	Military Soviet Navsats
149.940/399.840*	Military Soviet Navsats
149.970/399.920*	Military Soviet Navsats
149.978/399.9413*	U.S. Navsat maint chan
149.982/399.952*	U.S. Improved Transit Navsat channel
149.988/399.968*	U.S. Navsat channel
150.000/400.000*	Civilian Soviet Navsats
150.012/400.032*	U.S. Geosat channel plus unknown sats carrier only
150.030/400.080*	Military Soviet Navsats

* *The 400 MHz freqs are unmodulated beacons*

Cosmos navigation satellites from 1967 to the present include the following:

<u>Experimental</u>: 192, 220, 292, 304, 332, 371

<u>1st generation</u>: 385, 422, 465, 475, 489

<u>2nd generation</u>: 514, 574, 586, 627, 628, 663, 689, 729 800, 823, 846, 890, 962, 994, 1027

<u>3rd generation</u>: 700, 726, 755, 778, 864, 887, 894, 928 951, 971, 985, 991, 996, 1011, 1072, 1089, 1091, 1104 1141, 1150, 1153, 1181, 1225, 1244, 1275, 1295, 1308 1333, 1344, 1349, 1386, 1417, 1428, 1448, 1459, 1464 1513, 1531, 1535, 1550, 1577, 1598, 1605, 1610, 1627 1634, 1704, 1709, 1725, 1745, 1759, 1802, 1808

<u>4th generation</u>: 883, 926, 1000, 1092, 1168, 1179(fail) 1226, 1304, 1339, 1383, 1447, 1506, 1553, 1574, 1655 1727, 1791.

COSPAR SARSAT

On board Cosmos 1383, a standard Soviet Navsat launched June 29, 1982, was the first COSPAR emergency beacon location system called SARSAT (Search And Rescue Satellite-Aided Tracking). COSPAR 1 is part of an international effort to use satellites to locate aircraft and ships transmitting on international distress frequencies of 121.5/243.0 MHz. Downlink is 1544.5 MHz.

When fully operational, the system will consist of four satellites in low orbit launched jointly by the U.S. and Soviet Union. Host spacecraft for the U.S. satellites will be NOAA weather satellites; The Russians will fly their portion of the SARSAT system on Cosmos Navsats.

The SARSAT ground segment includes a ground station and mission control center to be operated by the U.S. Air Force at Scott AFB, Illinois. The U.S. Coast Guard will operate ground stations at Point Reyes, California, and Kodiak, Alaska.

The Air Force will be responsible for coordinating search and rescue operations within the continental U.S. and all adjacent waters; a station in Ottawa will extend coverage to Eastern North America; CNES is supplying a ground station in Toulouse, France.

Soviet stations at Moscow, Archangelsk, and

Cosmos Mission Profile Chart

To aid the monitor enthusiast in monitoring the Cosmos space program, the author has compiled this easy-to-use reference chart. All orbital heights are in kilometers; inclinations are in degrees. Launch sites are abbreviated as follows: TT = Tyuratam; PL = Plesetsk; KY = Kapustan Yar.

Incli-nation°	Altitude (km)	Launch Site	Launch Vehicle	Mission Profile
0.0	35800 x35800	TT	D-1-h	Geostationary Comsats
50.7	220 x 180	KY	C-1	Space plane
50.7	548 x 347	KY	C-1	Minor military; 50.7 low
50.7	1643 x 295	KY	C-1	Minor military; 50.7 high (K1578)
51.6	450 x 185	TT	A-2	Soyuz-T/Progress
51.6	450 x 185	TT	D-1-h	Salyut/MIR space stations
62.8	420 x 170	PL	A-2	Photo recon satellites
62.8	40180 x 580	PL	A-2-e	Early Warning (ArgPerigee-315°) Freq: 2298/2304 MHz CW beacon
62.8	40000 x 400	PL	A-2-e	Molniya 1/3 comsats (Arg-P=285°)
64.8	19100 x19000	TT	D-1-e	Glonass navigation sats
64.9	420 x 170	TT	A-2	Photo recon
65.0	265 x 250	TT	F-1-m	Radar ocean recon (nuc pwr)
65.0	445 x 425	TT	F-1-m	Elint ocean recon (conventional pwr)
65.8	515 x 470	PL	C-1	Minor military; 65-66 low/radar
65.8	495 x 445	PL	C-1	Minor military: 65-66 low/unk
65.8	410 x 375	PL	C-1	Minor military: 65-66 low/laser
65.8	1175 x 125	TT	F-1-m	ASAT interceptor 166.0 MHz (PPM-AM 300 kHz wide)
65.8	2106 x 144	PL	C-1	ASAT target; wideband "squiggers" every 4 kHz
67.1	420 x 170	PL	A-2	Photo recon
70.0	420 x 170	TT	A-2	Photo recon
70.4	420 x 170	TT	A-2	Photo recon
71.1	855 x 850	TT	D-1-e	Elint
72.9	420 x 170	PL	A-2	Photo recon
73.6	1525 x 1480 (1700 x 1500)	PL	F-2	Geodetic
74.0	553 x 469	PL	C-1	Minor Military: 74 low (83/K1453)
74.0**	570 x 501	PL	C-1	Elint: replace by F-2 launched s/c
74.0**	729 x 242	PL	C-1	Minor military: 74 med (76/K822)
74.0	823 x 764	PL	C-1	Military comsat: store-dump type
74.0	1550 x 1350	PL	C-1	Military comsat: 8 launch theater
81.2**	660 x 620	PL	A-1	Elint (phased out) (83/K1441)
81.2**	895 x 855	PL	A-1	Weather failure (phased out)
82.3	420 x 170	PL	F-2	Photo recon: Earth resources
82.5	668 x 630	PL	F-2	Elint
82.5	665 x 635	PL	F-2	Oceanographic
82.6	1420 x 1380	PL	F-2	Unknown
82.6	1525 x 1480	PL	F-2	Geodetic
82.9	1028 x 951	PL	C-1	Navigation satellite: civilian Freq: 150.00/400.00 MHz
82.9	1032 x 945	PL	C-1	Navigation sat: military Freq: 149.91/399.76; 149.94/399.84; 149.97/399.92; 150.030/400.080 MHz
83.0	513 x 444	PL	C-1	Minor military: 83 low
83.0	1995 x 300	PL	C-1	Minor military: 83 high
98.0	660 x 600	TT	A-1	Natural resources

* Indicates phased out mission profile

Vladivostok will track the NOAA and Cosmos satellites and will maintain communications with participating nations through the Mission Control Center at Scott AFB.

An improved emergency locating transmitter on 406.025 MHz is being developed for aircraft and ships. These beacons will not only give the exact location of the downed aircraft or sinking ship, but will also provide its national origin and aircraft or ship registration number.

Downlink for the SARSAT satellites is 1544.5 MHz.

Rescue of more than 250 persons in 100 distress situations have been credited to the SARSAT/ COSPAR program since the first rescue was performed by the system in September, 1982.

GLONASS

On October 12, 1982, the Soviets conducted the maiden launch of the GLONASS (Global Navigation Satellite System), designed to increase the navigational accuracy of their satellite systems, and apparently designed around the U.S. Navstar satellite system. In fact, Nicholas Johnson in his 1982 *Soviet Year in Space*, called the system "Navstarsky."

The first launch from Tyuratam carried Cosmos 1413, 1414 and 1415 into 20,000 km circular orbits, first inclined 51.6 degrees then maneuvered into an operational inclination of 63 degrees--identical to the U.S. Navstar system.

Even the operating frequencies are very close together:

NAVSTAR	1227.6/1575.42 MHz
GLONASS	1250.0/1603.50 MHz

These satellites will probably be phased in three planes with 120 degree spacing. As many as three satellites will operate in each plane.

There have been seven other launches in addition to the October 12, 1982, launch: Cosmos 1490-1492 (August 11, 1983); Cosmos 1519-1521 (December 29, 1983); Cosmos 1554-1556 (May 19, 1984); and Cosmos 1593-1595 (September 5, 1984); Cosmos 1650-1652 (May 17, 1985); Cosmos 1710-1712 (December 24, 1985); and Cosmos 1778-1780 (September 16, 1986).

Other Unmanned Programs

While the Cosmos satellite program is the largest unmanned program the Russians have, several other programs have been announced. Where these other programs have military applications, the civilian aspects of the program have been emphasized.

Some of these Russian programs will be discussed in their respective sections in other portions of this book for continuity; These include Iskra (amateur radio); RS (amateur radio); Meteor (weather); and all non-military communications satellites (communications satellite sections).

Elektron

This short-lived program saw launches for only one year: 1964. These launches were dual launches using the A-1 booster. The announced purpose of these satellites was to study the Van Allen radiation belt and earth's magnetic field.

Elektron 1 and 2 were launched from Tyuratam on January 30, 1964; Elektron 3 and 4 were launched July 11, 1964. No more flights were labeled by the Russians as a part of this series.

According to Bob Grove's *Shortwave Directory*, the following frequencies were noted for Elektron missions:

19430 kHz	Elektron 1
19540 kHz	Elektron 2
19943 kHz	Elektron 3 & 4
19954 kHz	Elektron 3 & 4
20005 kHz	Elektron 1, 3, & 4
30007 kHz	Elektron 1 & 4

Intercosmos

This series of international research satellites was adopted by nine communist nations meeting in Moscow in April 1967. Few details about these satellites have been released with the exception of Intercosmos 6 which was recovered. Electrical power is probably provided by batteries which accounts for the satellites' short lifetimes.

Monitors might want to check near 20, 30 and 40 MHz for possible signals when the listener knows of an Intercosmos launch.

Polyot

Only two satellites were ever announced in this series of satellites: Polyot 1 (launched November 1, 1963) and Polyot 2 (launched April 12, 1964). These were the world's first maneuverable satellites. There have been no more launches announced by the Soviets in this series. Frequencies used in this series are as follows:

19945 kHz	Polyot 1
19895 kHz	Polyot 2

Prognoz

Prognoz (Russian for "forecast") are designed to study solar radiation, the physical processes taking place on the sun, and the interference they can cause to space communications.

The first launch in the series occurred on April 14, 1973, using an A-2-e booster. The payload launched from Tyuratam had a highly eccentric orbit inclined 65 degrees (200,000 x 965 km orbit). There have been nine launches since, and the last was Prognoz 9 on July 1, 1983. This satellite produced a mild surprise when the spacecraft was put into a 720,000 x 380 km orbit inclined 65.5 degrees, allowing Prognoz 9 to orbit the earth every 27 days.

Telemetry signals from these spacecraft have been observed in the 928 MHz area.

Proton

Only four satellites have been identified by the Soviets with this program, beginning with the launch on July 16, 1965, of Protons 1-3. These satellites lasted only three months. Proton 4 was launched November 16, 1968, by a D-1 booster from Tyuratam and had an eight month lifetime.

The satellites were used to investigate and study the nature of cosmic rays of high and super-high energy. Intercosmos 6 appears to have been a carry-on mission for the Proton flights.

Frequencies used by these satellites were:

19545 kHz	Proton 2 & 3
19910 kHz	Proton 1 & 4

Astron

One satellite has been identified by the Soviets as an Astron spacecraft. Launched on March 23, 1984, the satellite was put into a 200,000 s 2000 km orbit, inclined 51 degrees, by the D-1-e launch vehicle from Tyuratam. The satellite makes one complete orbit around the earth every four days.

Astron appears to be an entirely new type spacecraft; its hardware, however, is based on that used in the Venera planetary program. The Astron satellite is being used to investigate X-ray and ultraviolet emissions in the universe.

Initial scientific studies by the Astron commenced on April 3, 1984. Telemetry is probably to be found in the 928 MHz area.

Sputnik

These satellites were the first used by the Soviets. Sputnik 1, launched October 4, 1947, shocked the free world--the first orbiting spacecraft around the earth was Soviet. The whole series of ten satellites appears to have been devoted to developing the ability to place men in earth orbit. Six dogs were placed into orbit by this series of satellites, all launched from Tyuratam over a 3.5 year period.

Soviet Space Surveillance

The Russians have a space defense system akin to NORAD. There have been frequent references in the western press to their ABM defense system which is not only a missile launching system, but is also an elaborate tracking system, built around large arrays of radar referred to as "Hen House." Any system which tracks long range strategic missiles also tracks space objects crossing Soviet territory or its approaches, regardless of nationality or absence of active signal emissions.

The perfect place to put early warning radar installations for the Russians would be in the far northern regions of the Soviet Union. Since these areas are in the polar regions, and the Molniya class 1 COMSATs have been turned over for military use, these birds are probably used by the Soviet network. In addition, the second tier of Soviet military comsats could also fill a communication need for a Soviet "NORAD."

Orbital Tracking in the USSR

During the early years, Soviet public statements made particular reference to optical tracking facilities in many parts of their own country; they also encouraged observers in Soviet bloc countries to send reports of observations to Moscow. Only a few telescope systems were pictured with no indication how many advanced systems existed.

Western reports were of a network consisting of a master station and 12 others equipped with receivers to measure Doppler shifts in radio signals, tracking radars, and phototheodolites, all transmitting data to a central computation center.

From the technical information released in connection with ASTP, it is known that an S-band radar transponder was installed on the Soyuz spacecraft. Satellites in the small Cosmos and Intercosmos series have standardized conical antennas of sizes suitable for these microwave frequencies, indicating that the 217-219 GHz range has been used for orbit determination from the outset.

The Soviets have, for whatever reason, decided not to use foreign tracking stations for their space program. What little information that has been released indicates that the Soviets have chosen only those countries with a political climate favorable to theirs. The following is a list of foreign tracking sites used by the Soviets in their space program.

Helwan, UAR
Mali (Africa)
Guinea (West Africa)
Havana, Cuba
Aswan, UAR
Fort Lamy, Chad
Khartoum, Sudan
Afgoi, Somali
Kerguelen (South Indian Ocean)
Mirnyy (Antartica)
India (Laser radar -Great Arc Project)
Bolivia(Laser radar-Great Arc Project)
Santiago de Cuba (Laser tracking station/satellite observation station)
Latvia (Laser tracking station)
Borowicz, Poland (Laser tracking)
Stara Zagora, Bulgaria (radiotelemetry station)
Panska Ves, Czechoslovakia (Ionospheric observatory)
Cracow, Poland (Meteorological satellite station)

Sea-Based Support

Because of Soviet reluctance to become too dependent upon foreign land-based stations, or perhaps because not all nations approached by the Soviets were willing to be host, the Soviet Union has put considerable emphasis upon developing a sea-based support system. Several classes of ships, one group in the mid-Pacific, are well equipped and are used frequently to record Soviet missile tests.

Other less equipped ships in comparison with the missile trackers have for some years operated in the tropical Atlantic and the Mediterranean along the path of orbital flights. Such ships would put into various ports in these parts of the world for supplies and crew rest, and when they left port it was usually an indication that new space launches were pending.

By noting which tracking ships are civilian and which are navy, it appears that the Pacific missile tracking vessels are under control of Soviet military authorities. By contrast, the ships operating in the Atlantic and Mediterranean operate for the Soviet Academy of Sciences.

The following ships have been identified as being in use by the Soviets in connection with the Soviet space program. Active maritime monitors are encouraged to contribute callsigns they might have for these ships so that fellow listeners can follow the Soviet sea-based space tracking network.

Akademik Sergey Korolov	Civil UISZ
Aksay	Civil UQRX
Apsheron	Military
Baskunchak	Military
Bezhitsa	Civil
Borovichi	Civil UVAU
Chazma	Military
Chukhotka	Military
Chumikan	Military
Cosmonaut Georiy Dobrovolskiy	Civil UZZV
Cosmonaut Pavel Belyayev	Civil UTDX
Cosmonaut Komarov	Civil UQBV
Cosmonaut Viktor Patsayev	Civil UZYY
Cosmonaut Vladmir Komarov	Civil UUVO
Cosmonaut Vladislav Volkov	Civil UIVZ
Cosmonaut Yuriy Gagarin	Civil UKFI
Dauriya	Military
Dikson	Military
Dolinsk	Civil
Donbass	Military
Ilichevsk	Civil
Kegostrov	Civil UKBH
Krasnodar	Civil
Morzhovets	Civil UUYG
Nevel	Civil UUYZ
Ristna	Civil UKHL
Sakhalin	Military
Sevan	Military
Sibir	Military
Sutchan	Military
Taman	Military
Tamal	Military

The foregoing paragraphs have referenced some of the places where Soviet tracking ships can be found during missions, but a comprehensive summary of these locations was prepared by James Oberg. He showed that during most Soyuz flights one of the high capacity civilian ships anchors off Sable Island, Nova Scotia (about 44.5° N, 59.5° W), where four successive orbits pass within easy direct communications range; later, this station appears to be abandoned in favor of a new location in the western Mediterranean close to Gibraltar.

Another location in connection with manned flights is in the Gulf of Guinea, West Africa. This site is probably used to monitor retrofire just before reentry and landing near Karaganda in the Soviet Union.

In this same Gulf of Guinea area, deep space flights frequently use a corridor over Africa-east across Asia and the Pacific, lose velocity, make a U-turn back over the South Atlantic, then head west over Central America. These flights usually cause tracking ships to be strung across the South Atlantic ground trace.

For flights returning from deep space to land in the Indian Ocean, ships are usually posted between South Africa and Australia. A large tracking ship, either in a Cuban port or in Trinidad, gives added coverage to deep space missions during the critical escape phase of the mission.

During the ASTP mission, a large tracking ship was located off the coast of Honduras (16° N, 87.5° W) to supplement the Sable Island tracking station. Tracking ships were observed off the upper Delaware coast during manned (Soyuz 35) missions.

Two locations of Soviet tracking ships pictured on Soviet postage stamps are east of the Phillipines and north of New Zealand. Monitors in these areas can expect to hear Soviet cosmonauts communicating with tracking ships and receive telemetry signals from Soviet spacecraft during overhead passes.

Soviet Ground Stations

The following list of known Soviet ground stations and their locations in the Soviet Union was plotted on an OSCARLOCATOR to track manned and unmanned missions for long distance reception of HF Soviet satellite signals.

Soviet Space Ground Stations

Location	Designator	Position
Yevpatoria	EVP	45.2° N 33.4° E
Tbilisi	TBL	41.7° N 45.0° E
Dzhusaly	DZhS	45.5° N 64.0° E
Kolpashevo	KLP	58.3° N 82.9° E
Ulan Ude	ULD	51.8° N 107.6° E
Ussuriysk	USK	43.8° N 131.9° E
Petropavlovsk	PPK	53.0° N 158.7° E

Soviet Launch Sites			
Tyuratam	TT	45.6° N	63.4° E
Plesetsk	PL	62.8° N	40.1° E
Kapustin Yar	KY	48.4° N	45.8° E

HF Soviet Tracking Station Frequencies

Common sense dictates that the Soviets should have an HF system similar to the U.S. communications system. As mentioned in an earlier section, the Zarya radiotelephone system uses both HF and VHF systems for communications between all units (Soyuz T, Salyut, Ground Stations) during a Soviet manned mission.

Possible frequencies to look for HF voice include 18000 and 18060 kHz. The frequencies 19990 and 19995 kHz are reportedly used by a Soviet ground station near Baikonour (47.3° N, 65.5° E), but this does not fit any of the Soviet sites listed above.

Some Cyrillic CW has been monitored on 20002 and 20004 kHz. In October 1982, during the Soyuz T7 mission to Salyut 7, several transmissions on 20008 kHz appeared to have Dopplershift, while a non-Doppler-shift station identifying as RLN4 was in contact on 20004 kHz. This could very well be a part of the Soviet space network.

As more missions are flown to Salyut 7 and future space stations, monitoring of the 20 MHz region should reveal much more.

Closing Remarks about the Russians

The Soviets probably have the most interesting space program from a monitor/hobbyist point of view. The challenges of following their space program can offer a lifetime of intrigue and enjoyment. About the time you think you have something figured out, the Russians are sure to rearrange the puzzle!

Those of you who get a thrill out of monitoring the numbers stations, single letter beacons, and other short-wave mysteries might want to look into monitoring the Soviet space program for the challenges it has to offer. Appendix A offers material for additional study.

Appendix A

List of Sources

While Communications Satellites *by Larry Van Horn ($12.95 from Grove Enterprises, 140 Dog Branch Rd., Brasstown, NC 28902) is the most comprehensive book on monitoring earth satellites ever written, other excellent sources of related information are recommended reading:*

GENERAL INFORMATION

1. The Satellite Experimenter's Handbook, Martin R. Davidoff, K2UBC. First Edition 1984. This is a must book for the newcomer and old pro alike. This should be the hobbyist's first satellite book. Available from ARRL, 225 Main Street, Newington, CT 06111.

2. The R.A.E. Table of Earth Satellites 1957-1980, D.G. King Hele, J.A. Pilkington, H. Hiller, & D.M.C. Walker. MacMillan Press, London, 1981. This is another must for the serious researcher of satellites. This work gives a detailed account of every object during the period covered. Expensive, but worth the cost.

3. The Illustrated Encyclopedia of Space Technology, Kenneth Gatland, chief author. Harmony Books, New York, 1981. The color pictures and diagrams are well worth the price of this book, a comprehensive history of space exploration.

4. The Observer's Spaceflight Directory, Reginald Turnill. Fredrick Warne, London, 1978. Information from the beginning of space exploration to 1977. Plenty of black and white pictures.

5. Observing Earth Satellites. D.G. King Hele. Van Nostrand Reinhold Co., New York, 1983. A practical guide for making visual observations of satellites. Excellent reference for the beginner.

6. The Dictionary of Space Technology, Joseph A. Angelo, Jr. Van Nostrand Reinhold Co., New York, 1983. Over 1,500 entries from aberration to Zulu time.

7. NASA Satellite Situation Report. This quarterly publication lists all objects currently in orbit and that decayed during the period. Annual compendium gives complete list of all objects that have been in orbit (date of launch, date of decay, NORAD and International designation). Copy may be obtained from NASA, Office of Public Affairs, Code 502, Goddard Space Flight Center, Greenbelt, MD 20771.

8. TRW Space Log. Published annually by the TRW Public AFfairs Office. Extensive data on satellite payloads back to 1957. Copies available to professional personnel, aerospace industry, military, and other government agencies; request on organizational letterhead from: Editor, Space Log, Public Relations Dept., TRW Systems Group, One Space Park, Redondo Beach, CA 90278.

9. Satellite News, Geoffrey Falworth. Five different newsletters that cover a lot of current ground on launches, orbital information, frequencies. For subscription information write to Geoffrey Falworth, 14 Wesley Drive, Banbury OX16 9LW, England.

10. Aviation Week and Space Technology. Published weekly by McGraw Hill; the ultimate in information sources for the active satellite sleuth. A subscription is expensive so check your local library.

11. Spaceflight and JBIS. Published monthly by the British Interplanetary Society. In-depth profiles on space programs and spacecraft. For subscription information contact: British Interplanetary Society, 27/29 South Lambeth Road, London SW8 1SZ, England.

12. Spacewarn Bulletin. Monthly bulletin gives the latest list of assigned international designators, text of

recent launch announcements, object decay information, and other miscellaneous information. Difficult to get on mailing list; for official use; write on letterhead to: IUWDS, World Warning Agency for Satellites, World Data Center A for Rockets and Satellites, Code 630.2, Goddard Space Flight Center, Greenbelt, MD 20771.

13. NASA Satellite Prediction Bulletins. Published in two forms: The NASA Prediction Bulletin issued daily gives two-line Kepplerian data and orbital prediction data for five days after issue; the other bulletin gives two-line orbital elements only. Requesters must provide NASA with international designations for the objects they wish bulletins on. From NASA Goddard Space Flight Center, Code 5413, Greenbelt, MD 20771.

ORBITAL MECHANICS

14. Fundamentals of Astrodynamics, by R. Bate, D. Mueller, and J. White. Dover Press, New York, 1971. If you are really interested in Astrodynamics this is the book to buy--and it is cheaper than most others.

15. Methods of Orbital Determination, P. Escobal. John Wiley & Sons, New York, 1976. Excellent, but considerably more expensive than number 14. After you have mastered number 14, you might consider the purchase of this book or borrow through interlibrary loan.

16. AMSAT Software Exchange, P.O. Box 27, Washington, D.C. 20044. This is the place to go for computer software and satellite tracking. Send them a number 10 SASE and they will send you their latest catalog.

17. Using Microcomputer Programs for Radio Amateur Satellite Prediction, Bob Diersing, N5AHD. An excellent reference for those who wish to used micros in satellite tracking. Address same as No. 16.

AMATEUR RADIO SATELLITES

18. AMSAT, P.O. Box 27, Washington, D.C. 20044. This is an excellent organization; membership is open to amateurs and non-amateurs alike. If you want to keep current on amateur satellites this is the place to be. Their publication--*AMSAT Satellite Journal*--is included with membership.

19. Amateur Satellite Report, 221 Long Swamp Road, Wolcott, CT 06716. Excellent bi-weekly publication on amateur radio satellites, up-to-date. Occasional information on other satellites.

20. AMSAT Computer Bulletin Board Service: 512-852-8194.

21. Satellite Info Service, by Heinz-Juergen Wehle, DB5ER, Postf. 170313, 4000 Duesseldorf, 1 Federal Republic of Germany. This is an excellent German language publication. Write for details.

22. QST Satellite Biography, ARRL, 225 Main Street, Newington, CT 06111, Attn: *QST* Satellite Biography. Send an SASE business sized envelope to receive this 20 page biography of relevant *QST* articles since 1961.

23. Amateur Satellite Keplerian Data, available from AMSAT Headquarters. Send an SASE for the latest Keplerian data. Address is the same as No. 18.

24. K2ZRO Satellipse, Kaz Deskur, ZRO Technical Devices, P.O. Box 11, Endicott, NY 13760. This $10 manual tracker for OSCAR 10 can be used with the Project OSCAR calendar listed below.

25. Project OSCAR Calendar, Project OSCAR, P.O. Box 1136, Los Altos, CA 94022. Computer-generated information on currently-active amateur radio satellites. Information useful for those with manual tracking devices such as No. 24 listed above. Write for price and availability.

26. AMSAT-UK, Ronald J.C. Broadbent, 94 Herongate Road, Wanstead Park, London E12 5EQ England. The British version of AMSAT-US. Their quarterly magazine is called OSCAR News.

27. QST, ARRL. This monthly informative amateur radio journal often contains amateur satellite information. Membership is open to all amateurs and non-amateurs. Write for current information; address same as No. 22.

28. 73 Magazine, Subscription Dept, P.O. Box 931, Farmingdale, NY 11737. This excellent amateur radio publication has occasional articles relevant to satellites and construction projects.

WEATHER SATELLITES

29. APT Notes, Coordinator, Direct Readout Services, OA/S131, NOAA/NESS, Washington, D.C. 20233. Excellent information on current weather satellites. Also available is a current list of NOAA Technical Memorandums and Reports (TM/TR).

30. NTIS, Department of Commerce, 5285 Port Royal Rd., Soringfield, VA 22141. Technical information on a variety of satellite subjects, including the above-mentioned TM/TR's. Write for information on how to purchase.

31. Weather Satellite Handbook, R.E. Taggart, 1981; now out of print. This is an excellent book for the weather satellite monitor. No self-respecting weather buff should be without one--if he can find it!

32. Worldview, Raul J. Alvarez, 2512 Arch Street, Tampa, Florida 33607. Quarterly newsletter for the environmental satellite user. Excellent for the beginner and old pro as well.

DOMESTIC/INTERNATIONAL/DBS SATELLITES

33. Coop's Satellite Digest, Satellite Television Technology, P.O. Box G, Arcadis, OK 73007. Bob Cooper's publication is the leader in the pack. If you aren't a subscriber, you should be. Bob's been at it since the beginning and so has his magazine. It's expensive, so you might want to share the price with a friend.

34. The World of Satellite Television, by Mark Long and Jeffrey Keating. The Book Publishing Company, Summertown, TN, 1983. For the beginner in the TVRO field, I highly recommend this one. Down to earth explanations of how this TVRO business works.

35. The Home Satellite TV Book, by Anthony T. Easton, Wideview Books, 1982. Another excellent book by an expert in the TVRO field.

36. The Hidden Signals on Satellite TV, by Thomas P. Harrington and Bob Cooper, Jr. Universal Electronics, 1984. Tom and Bob's book is a must for those of you tired of watching TVRO and want more from your dish and receiver.

37. STV-Satellite Television Magazine, Triple D, Inc., 501 N. Washington St., Shelby, NC 28150. An excellent monthly magazine at a down-to-earth price.

38. Satellite Orbit, P.O. Box 1700, Hailey, Idaho 83333. This monthly is the domestic satellite TV guide. Updated cable satellite listings and feature articles. Their toll-free subscription number is 1-800-321-2100.

39. Satellite Orbit International. From the people who brought you *Satellite Orbit*, now comes the international version. If you are tired of U.S. programming, this is the magazine for you. Same toll-free number as No. 38.

40. Birkill's International Satellite TV Reception Guidebook, STTI, P.O. Box G, Arcadia, OK 73007 ($40). Recommended for those who want to monitor international satellites.

41. STTI Satellite Study Package, same address as No. 40. Excellent reference material including a Home Satellite TV Handbook and Worldwide Satellite wall chart.

SOVIET SPACE PROGRAM

42. Soviet Space Programs: 1976-1980, Library of Congress, Vol.1-December 1982, Vol.2-October 1984, Vol.3-Not yet released. Available through the Superintendent of Documents, Washington, D.C. The ultimate in Soviet information.

43. The Soviet Year in Space, by Nicholas Johnson, Teledyne Brown Engineering; annual since 1981. Nicholas is probably the top analyist of the Soviet Space program in this country.

44. The Russian Satellite Navigation System, by C.D. Woods and G.E. Perry, Phil. Trans. Royal Society London, A294,307-315(1980). Excellent source information on decoding Russian 150 MHz NAVSATs.

45. Telemetry from Russian Spacecraft, G.E. Perry and R.S. Flagg, JBIS Vol.23, pp.4510464, 1970. This excellent article on decoding Soviet spacecraft appears in the Hournal of the British Interplanetary Society. See No. 11 for address. Should be available on interlibrary loan.

46. Red Star in Orbit, James E. Oberg. Random House, New York, 1981. An excellent look at the Soviet manned space program peering through the veil of secrecy.

47. The New Race for Space, James E. Oberg. Stackpole Books, Harrisburg, PA, 1984. Takes up where Red Star left off, comparing the Russian and U.S. shuttle program.

48. Foreign Broadcast Information Service: Soviet Union, National Technical Information Service (NTIS), see address No. 30. Translations of Soviet news broadcasts, several pages per week on space activities. Issued daily.

The author of Communications Satellites also writes two monthly columns on space communications:

Space Communications, *RCMA Newsletter*, P.O. Box 542, Silverado, CA 92676. This column appears on an as-information-is-received basis. RCMA is an excellent scanner club newsletter, founded in 1975. Scanner buffs will find the information in its pages valuable.

Signals from Space, *Monitoring Times*, Grove Enterprises, Inc., 140 Dog Branch Road, Brasstown, NC 28902. *Monitoring Times* is the leader in SWL/monitor hobby publications. Due to its tabloid style publication, information is made available faster than from slick cover magazines. A must for the monitor/hobbyist.

Appendix B

SATELLITE COMPENDIUM

The following is a list of satellites, launching agencies and mission acronyms -- past, present and future. Not all satellites listed here have been discussed in the book and additional information would be appreciated by the author. Sponsor abbreviations are identified in the Glossary section.

Satellite Name	Launching Sponsor	Acronym/Mission
A ROCKET	URS	DOD rocket designation SL 1-2. Core vehicle rocket that launched Sputniks 1-3.
A-1	F	See ASTERIX.
A-1	URS	DOD rocket designation SL-3, launched Cosmos sats
A-2	URS	DOD rocket designation SL-4, launched Cosmos, Soyuz T/TM and Progress spacecraft
A-2-e	URS	DOD rocket designation SL-6, launched Cosmos, Prognoz and Molniya Spacraft
A-2-m	URS	DOD rocket designation SL-5, launched early Venera and maneuverable Cosmos 379, 398, 434
ABS	US	Advanced Business Systems (Ku-band)
ACTS	US	Advanced Communications Technology Satellite; RCA/NASA (Ka-band) launch in 1989
AD EXPLORER	US	Air density Explorers; Explorers 9, 19, 24, 39
ADVANCED SATCOM	US	Advanced Geostationary domsat series launched for RCA Americom (C-band). See SATCOM
AE	US	Atmosphere Explorer, Explorer series; Explorers 17, 32, 51, 54, 55
AEM	US	Applications Explorer Mission (See HCMM/SAGE)
AEROS	D	AEROnomy Satellite, 2 scientific satellite series
AEROS	US	Earth resources satellite
AEROSAT	F/ESA	Experimental aircraft comsat, cancelled due to budget cuts.
AFP-888	USAF	See Teal Ruby.
AFSATCOM	USAF	Air Force comsat package that is on a variety of platforms
AGENA	US	Upper stage vehicle for various programs.
ALOUETTE	CAN	First Canadian satellite, scientific sat series (2 sats in series)
ALSEP	US	Apollo Lunar Surface Experiments Package
AMPTE	F/US	Active Magnetospheric Particle Tracer Explorer, 3 scientific sats launched by the same rocket.
AMS	ISR	Geostationary comsat (Ku/C-band sat), has yet to be launched.
AMS	USAF	Advanced Metsat (DMSP Block 5D), see DMSP.
ANIK	CAN	Geostationary domsat series; means "brother." See TELESAT.
ANNA	USN/NASA	Army, Navy, NASA, Air Force joint geodetic satellite
ANS	HOL	Astronomical Netherland Satellite, astronomical scientific research
APPLE	IND/ESA	Ariane Passenger Payload Experiment, geostationary experimental communications package (C-band)
APOLLO	US	U.S. manned lunar series (3 men)

Satellite Name	Launching Sponsor	Acronym/Mission
APOLLO CM	US	Apollo Command Module
APOLLO LEM	US	Apollo Lunar Excursion Module
AQUACADE	USAF/CIA	Second Byeman code name for the Rhyolite series, changed after Boyce and Lee disclosures.
ARABSAT	ARS	Geostationary comsat (C-band), operated by an Arab consortium.
APEX	F	Geostationary comsat (3, 6/4, 7 GHz transponders)
ARGENTINE SAT	ARG	Geostationary comsat; managed by Argentina
ARGUS	USAF/CIA	Byeman code name for Advanced Rhyolite satellites
ARIANE	ESA	Primary launch vehicle for ESA
ARIEL	G	Ionospheric scientific research satellite, 6 sats in series. See UK.
ARYUBHATA	IND/URS	Atmospheric research satellite, named for Indian astronomer.
ASAT	USAF	Anti-satellite rocket
ASC	US	Geostationary domsat, American Satellite Company (C and Ku-band) series satellites
ASTERIX	F	First French satellite, technical research sat
ASTEX	USAF	Advanced Space Technology Experiments, DOD payload
ASTP	US/URS	Apollo-Soyuz Test Project, US-USSR manned cooperation flight.
ASTRA-1	L	Geostationary Ku-band comsat
ASTRO-A	J	7th Japanese scientific sat, in-orbit name HINOTORI
ASTRO-B	J	Japanese scientific satellite, studied solar flares in X-ray region, in-orbit name TENMA.
ASTRON	URS	Scientific sat to study ultra-violet and X-ray regions of space.
ATCOS	US-DOD	Atmospheric Composition Satellite; DOD payload.
ATDA	US	Augmented Target Docking Adapter, part of Gemini 9 mission.
ATLANTIS	US	5th named Space Shuttle, 4th to fly
ATLAS AGENA-D	US	U.S. launch vehicle
ATLAS CENTAUR	US	U.S. launch vehicle
ATLAS E	US	U.S. launch vehicle
ATLAS F	US	U.S. launch vehicle
ATS	US	Application Technology Satellites; geostationary communications experimental series.
AURA	F	See D2-B.
AUREOLE	F/URS	Scientific satellite to study the northern lights. Three in series; also called OREOL.
AURORA	USA/USN	Army/Navy scientific satellite
AURORA	US	RCA Americom Satcom 5 renamed and now owned by Alascom, Inc; geostationary domsat
AURORAE	ESRO	Polar ionosphere measuring scientific satellites. Also ESRO-1A.
AUSSAT	AUS	Geostationary domsats (Ku-band), three in series.
AVL802	USAF	Aerodynamic data research; DOD payload.
AXAF	US	Advanced X-ray Astrophysics Facility; scientific X-ray telescope to be launched in 1990's.
AYAME	J	In-orbit name for ECS satellites; means "sweet flower."
AZUR	D	Scientific research satellite; studied Earth-Sun relationships.
B-1	URS	DOD rocket designation SL-7; launched Cosmos and early Intercosmos satellites.
BE	US	Beacon Explorer; part of Explorer series, BE-A/S-66, BE-B/Explorer 22, BE-C/Explorer 27.
BEANSTALK	USAF	UHF emergency communciation repeater for strategic forces carried by Minuteman missiles.
BHASKARA	IND/URS	Earth Resources sat; named for early Indian astronomer
BIOS	US	See Biosatellite.
BIOSATELLITE	US	Three scientific research satellites that carried animals into orbit.
BIG BIRD	USAF	Fourth generation multi-function recon; nickname for Project Code 467 (LASP), Byeman code Hexagon (KH-9).
BLACK ARROW	G	First UK booster rocket, also see PROSPERO.
BLOCK-5D	USAF	See DMSP and AMS.
BLOSSOM	USAF	Suborbital program in late 1940's launched monkeys.
BLUEBELL	USAF	DOD sat, mission unknown, inflated spheres/cylinders.

Satellite Name	Launching Sponsor	Acronym/Mission
BMEWS	USAF	Ballistic Missile Early Warning Satellite. Also DSP Code 949.
BOLD ORION	USAF	First U.S. ASAT program, also called USAF 7795.
BOREAS	ESRO	Polar research; also known as ESRO-1B.
BRASILSAT	B	Geostationary comsat, two C-band satellites.
BRIDGETTE	US-DOD	Early U.S. military ferret program name.
BS	J	Geostationary comsat, two Ku-band sats; also known as YURI.
BSE	J	Experimental geostationary comsat (Ku-band); also YURI.
BULGARIA 1300	BUL/URS	Part of Intercosmos scientific series (22); Communist bloc
BYEMAN	US-DOD	One of more than 30 SCI (Sensitive Compartmented Information) that deal with recon aircraft/satellites systems.
C-1	URS	DOD rocket desig SL-8, launched Cosmos & Intercosmos 10-20.
CALSPHERE/THORBURNER	USN	Early anti-satellite association program, radar calibration spheres.
CAMEO	US	DOD; Chemically Active Materials Ejected in Oribt (barium/lithium)
CANNONBALL	USAF	DOD payload, atmospheric density
CASTOR	F	Technology sat released with Pollux; also known as D-5B
CAT	F/ESA	Capsule Ariane Technologique; Ariane test sat
CBS	US	DBS sats that will be launched in future by CBS, Inc.
CCC	US	Columbia Communications Corp; dual band comsats
CCE	US	Charge Composition Explorer; U.S. part of AMPTE.
CEP	US	Cylindrical Electrostatic Probe; scientific sat
CHALET	USAF/CIA	Class of geostationary SIGINT sats; Byeman code name changed to VORTEX following *New York Times* article.
CHALLENGER	US	Third named space shuttle, second to fly. Destroyed Jan. 28, 1986, during 51L disaster.
CHINA #	CHN	Chinese satellite program name.
CHINASAT	CHN	See STW.
CLASSIC WIZARD	USN	See White Cloud/NOSS
CLIPPER BOW	USN	Possible future Navy ocean surveillance series, current status unknown.
COBE	US	Cosmic Background Explorer; launch in 1989.
COLUMBIA	US	Second named space shuttle; first to fly.
COMPOSITE	USN	U.S. Navy sat; mission unknown
COMSAT	--	General acronym for COMmunication SATellite
COMSTAR-D	US	Geostationary domsat; domestic satellite program of Comsat, Inc.; four C-band satellites.
COPERNICUS	US	OAO-3 renamed after launch.
CORONA	USAF/CIA	Byeman code name for Discoverer/KH-4 series.
CORSA	J	COsmic Ray SAtellite; scientific; also HAKUCHO
COS-B	F/ESA	Celestial Observation Satellite; astronomy.
COSMOS	URS	A multi-mission (over 1800) satellite cover name; also written KOSMOS.
COSPAS	URS	COsmos Satellite for Program of Air and Sea rescue; SAR locator packages carried aboard Cosmos navigation satellites.
COURIER	US	First active comsat repeater; DOD payload.
CRRES	US	Combined Radiation and Release Satellite; NASA/USAF research probe, launch in 1989-1992.
CS	J	Communication Satellite; Geostationary inter-regional comsat; also SAKURA. Two C/Ku-band satellites.
CSE	J	Communication Satellite Experimental; CS precursor; also SAKURA-2. C-band and 20/30 GHz.
CSL-1	CHN	Chinese rocket derived from CSS-3. Launched China 1 and 2. Also called Long March 1.
CSL-2	CHN	Chinese rocket derived from CSS-X4. Launched China 3-8. Also called FB-1.
CSL-3	CHN	Chinese rocket derived from FB-1. Launched STW-1/2. Also called Long March 3.

Satellite Name	Launching Sponsor	Acronym/Mission
CTS-HERMES	CAN/USA/ ESA	Communication Technology Satellite; geostationary comsat; Ku-band transponder.
CYGNUS	US	Proposed geostationary comsat op by Cygnus Satellite Corp.
CZ-3	CHN	See CSL-3.
D-1	URS	DOD rocket designation SL-9/10; launched Proton, Salyut 1-8, Cosmos spacecraft.
D-1A	F	Geodetic research; also called DIAPASON.
D-1C	F	Geodetic research; also called DIADEME 1.
D-1D	F	Geodetic research; also called DIADEME 2.
D-1-e	URS	DOD rocket designation SL-12; launched Cosmos Luna, Zond, Venera, and Mars.
D-1-h	URS	DOD rocket designation SL-13; launched Raduga, Gorizont, Ekran and Cosmos comsat spacecraft.
D-2A	F	Solar UV scientific sat; also called TOURNESOL.
D-2B	F	Same as D-2A; also called AURA.
D-5A/B	F	See POLLUX and CASTOR, respectively.
DAPP	US-DOD	Defense Acquisition and Processing Program
DASH	US-DOD	Density and Scale Height; DOD payload
DATS	USAF	Despun Antenna Test Satellite
DBS	--	Generic acronym for Direct Broadcast Satellite
DBSC	US	Direct Broadcast Satellite Corp; geostationary DBS
DE	US	Dynamic Explorer; two satellites in series
DELTA	US	U.S. satellite launching system
DENPA	J	Japan's second scientific sat; also called REXS/SS-2
DFS	D	Deutscher Fernmeld Satellit; Proposed geostationary comsat; two in series
DIADEME	F	See D-1C/D.
DIAL/WIKA	G/F	Scientific satellite
DIAMANT	F/G	French rocket, launched DIAL/WIKA
DIAPASON	F	See D-1A.
DIGISAT	US	Proposed geostationary satellite operated by Digital Telesat
DISCOVERER	CIA/USAF	First generation film return recon; also called KH-4, Byeman code named Corona.
DISCOVERY	US	Fourth named space shuttle, third to fly.
DME	US	Direct Measurement Explorer, Explorer 31
DMSP	USAF	Defense Meteorological Satellite Program, DOD metsat series
DODECAPOLE	USN	Calibration satellite; DOD payload
DODGE	USN	DOD gravity experiment; DOD payload
DOMINIC	USAF	High altitude nuclear test program
DOMINION	US	Proposed DBS comsat; managed by Dominion Video Services.
DOMSAT	--	Generic acronym for DOMestic SATellite
DRIMS	--	Delta Redundant Inertial Measurement Unit
DSCS	USAF	Defense Satellite Communication System; geostationary milcomsat constellation
DSP Code 467	USAF	Defense Support Program; See Big Bird
DSP Code 647	USAF	See IMEWS
DSP Code 949	USAF	See BMEWS
DSP Code 1010	CIA/USAF	See KH-11
EARLY BIRD	US	World's first active commercial comsat/Intelsat 1
ECHO	US	Experiment communications experiment, large metallic balloon used to reflect signals, two in series.
ECS	J	Experimental Communications Satellite; also called AYAME. Geostationary comsat, two in series.
ECS	ESA	European Communication Satellite; geostationary comsat (Ku-band) several in series planned.
EGP	J	Experimental Geodetic Payload; one in series.
EGRS	US-DOD	Electronic Geodetic Ranging System; See SECOR.
EGS	J	Experimental Geodetic Satellite; geosurvey.
EKRAN	URS	Geostationary UHF DBS comsat; means "Screen."

Satellite Name	Launching Sponsor	Acronym/Mission
ELEKTRON	URS	Scientific satellite series to study the Van Allen radiation belts; four in series.
ENTERPRISE	US	First named space shuttle, used for atmospheric test, never has flown in earth orbit.
EOLE	F	French Cooperation Applications Satellites; meteorological experiments.
EQUASTAR	US	Proposed geostationary (C-band) comsat managed by Equatorial
ERBS	US	Earth Radiation Budget Satellite; launched by Space Shuttle part of the ERBE program.
ERS	ESA	European Remote Sensing satellite; follow-on earth resources program after the French Spot.
ERS	USAF	DOD Environmental Research Satellite
ERS-1	J	Proposed remote sensing spacecraft
ERS-1	ESA	Proposed remote sensing of ocean and ice zones
ERTS 1	US	Earth REsource Technology Satellite; renamed LANDSAT-1
ESA-GEOS	ESA	European Space Agency GEOStationary sat; scientific
ESRO-1A	ESRO	See AURORAE.
ESRO-1B	ESRO	See BOREAS.
ESRO-2	ESRO	See IRIS.
ESRO 4	ESRO	Earth Scientific Research Satellite
ESSA	US	Environmental Survey SAtellite; early series weather satellite
ETS	ESA	Experimental Technology Satellite
ETS	J	Engineering Test Satellite; also called KIKU.
EURECA	ESA	Automatic reusable multi-purpose platform to be launched by Space Shuttle
EUTELSAT	ESA	Geostationary comsats; ECS frame
EUVE	US	Extreme UltraViolet Explorer; proposed astronomical satellite to be launched by Space Shuttle.
EXOS-A	J	See KYOKKO.
EXOSAT	ESA	Scientific satellite; astronomical studies
EXOS-B	J	See JIKKEN.
EXOS-C	J	See OHZORA.
EXPLORER	US	Extensive series of scientific satellites.
F-1-r	URS	DOD rocket designation SL-11A; launched Cosmos FOBS
F-1-m	URS	DOD rocket designation SL-11B; Cosmos sats
F-2	URS	DOD rocket designation SL-14; Meteor & Cosmos
FARRAH	USAF	Late 70's code name for electronic ferret series
FARSIDE	USAF	Early 50's suborbital program that launched balloons
FB-1	CHN	See CSL-2.
FEDEX	US	Proposed geostationary comsat managed by Federal Express
FENGYUN 1	CHN	Polar orbiting Metsat
FINANSAT	US	Proposed geostationary comsat; financial network
FLEETSATCOM	USN	U.S. Navy geostationary milcomsat program; carries AFSATCOM transponders, also newer models carry EHF experimental packages (MILSTAR precursors)
FOBS	URS	Program under Cosmos label to put nuclear weapons into space on satellites.
FORDSAT	US	Geostationary comsat; dual band sats by Ford Aerospace Corp.
FRANCE 1	F	See FR-1.
FR-1	F	Studied VLF wave propagation experiments; See FR-1, French Research-1
F-SAT	ESA	Proposed multi-band geostationary comsat for France
G-1	URS	Giant Soviet super booster that was to carry cosmonauts to the moon. Three attempts to launch, all failures; appears abandoned.
GALAXY	US	Geostationary domsat; C-band sats for Hughes Aerospace Corp, three satellites in series.
GALAXY DBS	US	Proposed geostationary DBS; managed by Hughes.
GALAXY K	US	Geostationary comsat; Ku-band managed by Hughes.
GALAXY KA	US	Proposed geostationary comsta; Ka-band managed by Hughes.
GALILEO	US	Jupiter orbiter and entry probe.

Satellite Name	Launching Sponsor	Acronym/Mission
GALS	URS	Russian maritime and milcomsat package to ride on host satellites much like the AFSATCOM system.
GAMBIT	CIA/USAF	Byeman code name for KH-8, 3rd generation close look sat.
GAMMA-1	URS	Soviet gamma ray research satellite.
GDL	L	Proposed geostationary comsat.
GEMINI	US	U.S. two-man series of earth orbit missions.
GEOS	US	Geodetic Earth Orbiting Satellite; two satellites in series: GEOS-1 (Explorer 29)/ GEOS-2 (Explorer 36)
GEOS-1/2	ESA	Earth space physics research satellites
GEOS-3	US	Geodynamic Experimental Ocean Satellite
GEOSAT	USN	Geodesy satellite
GEOSTAR	US	Proposed civilian satellite location system
GEOTAIL	J	Proposed geophysic satellite; launch in 1991
GGSE	USN	Gravity Gradient Stabilization Experiment; DOD payload
GGTS	USAF	Gravity Gradient Test Satellite; DOD payload
GIOTTO	ESA	Halley's Comet research satellite
GLOMR	US-DOD	Proposed low earth orbit; low-cost milsatcom sats
GLONASS	URS	Advanced Soviet navsat constellation; often referred to as "NAVSTARsky" due to similarities to the U.S. system.
GMS	J	Geostationary Metsats; similar to U.S. GOES Metsat series, also called HIMAWARI.
GOES	US	Geostationary Operational Environmental Satellite; geostationary metsat series.
GOMS	URS	Proposed Russian geostationary Metsat series.
GORIZONT	URS	Geostationary Soviet TV/telephone relay satellites
GPS	US-DOD	Global Positioning Satellites; also called NAVSTAR.
GREB	US-DOD	DOD payload; also called SR-6B.
GRID SPHERE	US-DOD	Calibration satellite; DOD payload
GRID SPHERE DRAG	USAF	Drag experiment satellite; DOD payload
GRO	US	Gamma Ray Observatory; to be launched by Shuttle
GRS	USAF	Geophysical Research Satellite; DOD payload
GRS-A	G	German Research Satellite. See AZUR.
GSC	US	Geostationary comsat; proposed constellation by Graphic Scanning
GSTAR	US	Geostationary comsat; managed by GTE Corp.
H-1A	J	Latest generation of Japanese launch vehicle based on N-2 first stage.
H-1B	J	1990's proposed Japanese launch vehicle.
HAKUCHO	J	See CORSA-B; means "Swan."
HAWKEYE	US	Part of the Explorer series (Explorer 52)
HCMM	US	Heat Capacity Mapping Mission. See AEM.
HEAD	US	High Energy Astronomy Observatory; three in series.
HELIOS	D	Solar scientific probe
HEOS	ESRO	Highly Eccentric Orbit Satellite; two in series.
HERMES	CAN	See CTS.
HERMES	ESA	Proposed manned space plane.
HESP-1	J	Solar physics spacecraft; launch in 1992.
HEXAGON	USAF	Byeman code name for KH-9 Big Bird recon; See Big Bird
HILAT	USAF	Ionospheric research satellite
HI-LO	USN	U.S. Navy early experimental ASAT program
HINOTORI	J	See ASTRO-A.
HIMAWARI	J	See GMS.
HIPPARCOS	ESA	High Precision Parallax Collecting Satellite; proposed ESA astronomy satellite.
HITCHIKER	USAF	Radiation research sat; DOD payload
HS-376	US	Popular spacecraft mainframe developed by Hughes Aerospace, used by many orbiting geostationary comsats today.
HUBBLE TELESCOPE	US	See Space Telescope.
IDCSP	USAF	Initial Defense Communication Satellite Program; DSCS Phase 1 constallation. See DSCS.
IDEALIST	US-DOD	Byeman code word for the U-2 recon aircraft

Satellite Name	Launching Sponsor	Acronym/Mission
IMEWS	USAF	Integrated Missile Early Warning Satellite; also called DSP Code 647, geostationary early warning satellite series.
IMP	US	Interplanetary Monitoring Program; part of Explorer program, Explorers 18, 21, 28, 33, 34, 35, 41, 43, 47, 50
INDIGO	US-DOD	Byeman code name for an unknown satellite program
INJUN	USN	Radiation research; DOD payload
INJUN-4/5	US	Science sats part of Explorer program; Expl 25/40
INMARISAT 2	ESA	Next generation of maritime geostationary comsats
INSAT	IND	Geostationary comsat/Meteosat 3 in series.
INTASAT	E/USA	First Spanish satellite; technology & science
INTELSAT	USA/IT	Geostationary comsat; constellation used to relay TV/telephone traffic worldwide.
INTERCOSMOS	URS	Eastern bloc countries scientific sat program (22 satellites launched in program)
IONDS	USAF	Nuclear detection satellite system; to be carried on GPS/NAVSTAR satellites
IRAS	HOL/USA	Infra-Red Astronomy Satellite
IRIS	ESRO	Infra-Red Interferometer Spectrometer. See ESRO-2.
IRM	G	Scientific sat; part of AMPTE package
IRS	IND	Proposed Indian remote sensing satellite.
IRT	US	Integrated Rendezvous Target
ISEE	US	International Sun-Earth Explorer; three in series. Proposed geostationary comsat constellation by International Satellite, Inc.
ISIS	CAN	International Satellite for Scientific Studies; two in series
ISKRA	URS	Experimental comsats; aeronautical & amateur radio
ISO	ESA	Infrared Space Observatory; proposed science astronomy mission.
ISPM	ESA	International Solar Polar Mission
ISS-1	J	Ionospheric Sounding Satellite; See UME.
ITALSAT	I	Proposed geostationary comsat
ITOS	US	Improved Tiros Operational System; NOAA metsat pre-launch name
IUE	US	International Ultraviolet Explorer
IUS	US	Inertial Upper Stage; used by payloads launched from the Space Shuttle
JAS-1	J	Japanese amateur radio satellite
JERS-1	J	Proposed Japanese Earth Resources Satellite
JIKKEN	J	See EXOS-B.
JUMPSEAT	USAF	Byeman code name for a very secret high elliptic orbit ABM radar SIGINT satellite. Characteristics of this orbit have been confused with SDS program satellite series.
JUNO II	US	U.S. launch vehicle
JUPITER C	US	U.S. launch vehicle
KEYHOLE	US	General code name for U.S. recon satellite programs KH-1 SAMOS - 1st gen radio/area (USAF) KH-4 DISCOVERER/CORONA - 1st gen close look (CIA) KH-5 SUPER SAMOS - 2nd gen radio/area (USAF) KH-6 2nd gen close look (CIA) KH-7 3rd gen radio/area (USAF) KH-8 GAMBIT - 3rd gen close look (CIA) KH-9 BIG BIRD/HEXAGON/DSP-467 - 4th gen KH-10 MOL(Manned Orbiting Laboratory)-scrapped KH-11 KENNAN/DSP-1010 - 5th gen radio transmission KH-12 To be launched by Space Shuttle - 6th gen
KIKU	J	See ETS.
KYOKKO	J	See EXOS-A.
LAGEOS	US	Laser Ranging Geodetic Satellite; two in series.
LAMBDA 4S	J	Japanese booster rocket
LAMPO	US-DOD	DOD payload
LANDSAT	US	Earth resources satellite series.
LASERCOM	US-DOD	Proposed laser communications payload under development

Satellite Name	Launching Sponsor	Acronym/Mission
LASP	USAF	See Big Bird.
LCS	USAF	Lincoln Calibration Satellite; DOD payload.
LDEF-1	US	Long Duration Exposure Facility; reusable passive space platform designed for experiments requiring long periods of space exposure, shuttle launched.
LEASAT	USN	Follow-on military comsat to Fleetsatcom; also called Syncom IV.
LEASECRAFT	US	Reusable automatic platform designed by Fairchild Research and Technology, Shuttle launched.
LES	USAF	Lincoln Experimental Satellites; DOD experimental comsats
LIDOS	USAF	Large Inclination Doppler Only Satellite; DOD payload
LM	US	Lunar Module; Apollo program spacecraft
LOFTI	USN	Low Frequency Trans-Ionospheric satellite; DOD payload
LOGACS	US-DOD	Low Gravity Accelerometer Calibration Satellite; DOD payload.
LONG MARCH 1	CHN	See CSL-1.
LONG MARCH 3	CHN	See CSL-3/CZ-3
LRV	US	Lunar Roving Vehicle, Apollo program
L-SAT	ESA	Proposed DBS comsat to be launched by ESA. Now Olympus.
LTA	US	Lunar Module Test Article
LUCH (LOUTCH)	URS	Geostationary comsat; 14/11 GHz
LUNA	URS	Unmanned lunar exploration probes
LUNAR ORBITER	US	Unmanned lunar exploration series; five probes
LUNOKHOD	URS	Soviet unmanned moon-roving probes
LUX-SAT	LUX	Proposed DBS comsat for Luxembourg
MAGIC	URS	MAGnetospheric InterCosmos
MAGION	URS/TCH	MAGnetospheric and IONospheric satellite; Russian/Czech scientific satellite
MAGNUM	US-DOD	Byeman code name for follow-on program for Rhyolite/AR/ Aquacade geostationary Sigints
MAGSAT	US	MAGnetic Field SATellite, scientific
MARECS	ESA	Geostationary Maritime comsat
MARILYN	US-DOD	Code name for early DOD ferret program
MARINER	US	Scientific planetary satellite series
MARISAT	US	Geostationary maritime comsat series managed by COMSAT
MARECS	ECS	Geostationary maritime comsat series managed by ESA.
MARS	URS	Soviet unmanned scientific Martian probes.
MERCURY	US	First U.S. manned spacecraft program.
MERINO	US-DOD	Byeman code name for the DOD satellite facility at Pine Gap, Australia.
METEOR	URS	Soviet orbiting metsat series.
METEOR-PRIRODA	URS	First earth resources sat; means "nature"
METEORSAT	ESA	Geostationary Metsat series operated by ESA.
METSAT	--	Generic acronym for METeorological SATellite
MIDAS	USAF	MIssile Defense Alarm Satellite; first DOD early warning series.
MIKU	F	Rocket test capsule
MILSATCOM	--	Generic acronym for MILitary SATellite COMmunications
MILSTAR	US-DOD	MILitary Strategic TActical Relay; future DOD comsat series
MIR	URS	Modular space station in Salyut series, Salyut 8; means "Peace"
MIRANDA	G	Called the X-series satellites; they tested three-axis control.
MMC	US	Proposed geostationary comsats; managed by Martin Marietta.
MOBILESAT	US	Geostationary comsat; Intended to provide low-cost cellular-type communication to rural areas.
MOLNIYA	URS	High eccentric orbit type comsats; two generations in service now.
MOP	ESA	Op Meteosats; next generation of advanced geostationary metsats
MORELOS	MEX	Geostationary comsat (C-band), two in series
MOS-1	J	Marine Observation Satellite; designed for ocean research.
MPF	US	Materials Processing Facility, commercial material processing lab launched from the Shuttle
MS-F2	J	See SHINSEI.
MS-T1/2/3/4	J	See TANSEI.
M-SAT	CAN	Proposed geostationary multi-band mobile comsats

Satellite Name	Launching Sponsor	Acronym/Mission
MTS	US	Meteoroid Technology Satellite
MU 4	J	See TANSEI, SHINSEI, DENPA.
MUDFLAP	USA	Early cancelled U.S. Army ASAT program; also called Program 505.
MUSES-1	J	Proposed lunar survey spacecraft; launch in 1993.
MUSKETBALL	US-DOD	Density measurement satellite; DOD payload.
NATO	US/NATO	Geostationary milsatcom series launched by the U.S. for NATO.
NAVSTAR	US-DOD	Navigation System Using Timing and Ranging; Advanced follow-on navigation satellite for U.S. Transit program. Also called GPS.
NAVSAT	--	Generic acronym for NAVigation SATellite
NDS	US-DOD	Navigational Development Satellite; also called NAVSTAR 1.
NIMBUS	US	Orbiting metsat series
NNSS	USN	Navy Navigation Satellite System; general term describing the Navy Transit satellite program.
NOAA	US	Current orbiting metsat series.
NOSS	USN	Navy Ocean Surveillance Satellites; also called White Cloud/Classic Wizard.
NOVA	USN	Follow-on to Navy Transit navsat program
N-ROSS	USN	Oceanographic research information; no launch date
NTS	USN	Navigation Technology Satellite; another early Navstar precursor. NTS-1 also called TIMATION-3.
NUDETS	USAF	Early developmental suborbital nuclear detection program.
NUSAT	US	Northern Utah Satellite; experimental orbital sat
OAO	US	Orbiting Astronomical Observatory; scientific
OFO	US-DOD	Orbiting Frog Otolith; DOD payload
OGO	US	Orbiting Geophysical Observatory; scientific
OHSUMI	J	First Japanese payload; scientific research
OHZORA	J	Scientific sat studied earth's middle atmosphere; See EXOS-C.
OLYMPUS	ESA	See L-SAT.
OMNISTAR	US	Proposed multi-sensor earth resources satellite to be launched by the Shuttle in the late 1990's for EOSAT.
OMV	US	Orbital Maneuvering Vehicle; proposed satellite retrieval and repair vehicle; launch in 1991.
OPERATION #	US-DOD	Classified DOD payload designation, has been replaced by USA designation.
ORBIS	USAF	Orbiting Radio Beacon Ionospheric Sat; DOD payload
ORBIS-CAL	USAF	Calibration satellite for ORBIS; DOD payload
OREOL	URS/F	Magnetospheric and Ionospheric Explorer; see AUREOLE.
ORION	US	Proposed geostationary comsats; managed by Orion.
ORS	US-DOD	Octahedron Research Satellite; DOD payload.
OSCAR	US/Others	Orbiting Satellite Carrying Amateur Radio; series built by amateur radio community.
OSCAR	USN	TRANSIT satellite name once in orbit.
OSO	US	Orbiting Solar Observatory; scientific
OTS	ESA	Orbiting Test Satellite; 1st satellite in ECS geostationary comsat prog
OV	US-DOD	Orbiting Vehicle; series of DOD satellites.
OXCART	US-DOD	Byeman code word for the SR-71 spy recon aircraft.
P-76-5	US-DOD	DOD communications research payload; modified Transit navsat.
P-78-1	US-DOD	See SOLWIND.
P-78-2	US-DOD	DOD research stellite; also called SCATHA.
PAC-1	US-DOD	Semi-active gravity gradient stabilization (SAGS) tests; DOD payload, Package Attitude Control.
PACSAT	US	Geostationary comsat; Pacific area satellite.
PAGEOS	US	Passive Geodetic Orbiting Earth Satellite, balloon
PALAPA	INS	Geostationary comsat; six in series for Indonesia.
PAM-D/D2	US	Payload Assist Module; upper stage vehicle designed for shuttle satellite launches.
PANAMSAT	US	Proposed geosttionary comsat; managed by PanAm SC
PAPUA NEW GUINEA SAT		Proposed geostationary comsat

Satellite Name	Launching Sponsor	Acronym/Mission
PEGASUS	US	Three winged meteoroid research satellite series.
PEOLE	F	Preliminaire a EOLE; scientific satellite.
PIONEER	US	Solar system scientific research satellite series
PIONEER-VENUS	US	Venus orbiting research probe
PLANET A	J	See SAKIGAKE.
POGO	US	Polar Orbiting Geophysical Observatory
POLLUX	F	Technology satellite; See Castor/D-5A-B
POLYOT	URS	Early series of Russian maneuverable satellites
PORCUPINE	USN	Calibration sphere; DOD payload
POSEIDON	USA/F	Ocean Topography Experiment satellite; also called TOPEX, to be launched 1991.
POTOK	URS	Geostationary communication platform (C-band)
PROGNOZ	URS	Solar radiation research series; means "forecast"; ten so far in series
PROGRAM 437	USAF	Early USAF ASAT research program
PROGRAM 505	USA	See MUDFLAP.
PROGRESS	URS	Resupply and tanker spacecraft for Soviet manned program.
PROSPERO	G	Known as X3 before launch; UK technology satellite
PROTON	URS	High energy cosmic ray research stellite series.
PROTON (D)	URS	Commercial name given to D class Soviet rocket.
RADARSAT	CAN	Proposed remote sensing satellite; launch 1991.
RADCAT	USAF	RADiation CAlibraTion satellite; DOD payload.
RADOSE	USN	RADiation DOSimeter Experiment; DOD payload.
RADSAT	USAF	RADiation research SATellite; DOD payload.
RADUGA	URS	Geostationary comsat; means for "Rainbow."
RAE	US	Radio Astronomy Explorer; Explorer 38/49.
RANGER	US	Series of lunar probes.
RACHEL	US-DOD	Late 70's code name for series of orbiting ferrets
RCA-BSS	US	Proposed series of geostationary DBS comsats for RCA Americom.
RCA-SATCOM	US	Geostationary comsat (C-band), for RCA Americom.
RELAY	US	Comsat experimental satellites in the 1960's
REP	US	Radar Evaluation Pod; ejected from Gemini 5
REXS	J	See DENPA.
RIGID SPHERE	USAF	Radar calibration satellite; DOD payload.
RM-18	USAF	Radiation Meteoroid; research payload used to measure Earth's UV background; DOD payload
ROHINI	IND	Small scientific research satellite
RONTGENSATELLIT	G	See ROSAT.
ROSAT	G	Proposed X-ray observtory; also called RONTGENSATELLIT
RHYOLITE	USAF	Byeman code name for geostationary sigint; See AQUACADE.
RMS	US	Radiation Meteoroid Satellite
RI	URS	Radio Sputnik (Sport) amateur radio satellites
RSI	US	Proposed geostationary comsat; to be managed by Rainbow Satellite, Inc.
SABS	ARS	Geostationary comsat (Ku-band), Arab states series
SAGE	USAF	Stratospheric Aerosol and Gas Experiment (AEM); scientific satellite.
SAINT	US-DOD	Early Satellite Interceptor; satellite surveillance
SAKIGAKE	J	Interplanetary probe.
SAKURA	J	See CS.
SALYUT	URS	Russian manned space station program.
SAMOS	USAF	Satellite And Missile Observation Satellite; 1st generation radio recon, KH-1
SAMRO	F/ESA	Proposed French photo recon
SAN MARCO	I	Scientific satellite series, five total
SARIT	I	Proposed geostationary comsat; Ku-band
SARSAT	--	Search And Rescue Satellite-Aided Tracking; piggyback package that hears emergency ELTs.
SAS	US	Small Astronomical Satellite; part of Explorer series.

Satellite Name	Launching Sponsor	Acronym/Mission
SATCOL	CLM	Geostationary comsat; managed by Columbia
SATCOM	US	Geostationary comsat managed by RCA Americom
SAX	I	Satellite for X-Ray Astronomy; proposed X-ray research satellite.
SBS	US	Geostationary comsat; managed by Satellite Business Systems.
SBTS	B	Geostationary comsat managed by Brazil
SCATHA	USAF	Spacecraft Charging AT High Altitudes; See P-78-2.
SCORE	US	System of Communicating by Orbiting Relay Experiment; world's first comsat
SCOUT	US	U.S. rocket used to life small payloads into orbit.
SDI	US-DOD	Strategic Defense Initiative; DOD's space based nuclear defense system commonly called Star Wars
SDRN	URS	Geostationary comm package; Russian version of our TDRSS system.
SDS	USAF	Satellite Data System; high eccentric orbit comm relay satellite; AFSATCOM transponder, and KH-11 imagery.
SEASAT	US	SEA survey SATellite; experimental ocean survey satellite
SECOR	USA	SEquential Collation Of Range satellites; DOD payload
SENTRY	USAF	Byeman code word for SAMOS program
SEO	IND	Satellite for Earth Observation. See Bhaskara.
SERT	USAF	Space Electric Rocket test; orbital test of an ion engine.
SESP	US-DOD	Space Experiments Support Program; celestial mapping
SESP 73-5	US-DOD	See SESP.
SESP 72-2	US-DOD	See SESP.
SEWS	US-DOD	DOD follow-on for early warning satellites
SGC	US	Proposed advanced WESTAR geostationary comsat
SHINSEI	J	SS-1 scientific satellite; means "New Star" (MS-F2)
SHOTPUT	NASA	Suborbital Echo program test
SIGNE 3	F/URS	Solar-Interplanetary Gamma Neutron Experiment; French satellite launched by Russians; scientific
SIRIO	I	Satellite Italiane Ricerca Industiale Orientale; experimental
SKYLAB	US	U.S. 1970's Space Station
SKYLINK	US	Geostationary comsat managed by Skynet, Inc.
SKYNET	G	Geostationary milsatcom
SM	US	Service Module
SME	US	Solar Mesosphere Explorer; scientific satellite
SMM	US	Solar Maximum Mission; scientific satellite
SMS	US	Synchronous Meteorological Satellite; 1st gen geostationary metsat
SN	US-DOD	Radiation satellite; DOD payload
SNAP	US-DOD	Supplementary Nuclear Power
SNAPSHOT	USAF	Experimental nuclear power supply; DOD payload
SOICAL	USN	Space Object Identification CALibration; DOD payload
SOLRAD	USN	SOLar RADiation research satellite; DOD payload
SOLWIND	US	Scientific research satellite
SOOS	USN	Stacked Oscars On Scout; See Transit/Oscar
SOYUZ	URS	3rd generation manned spacecraft; Russian for "Union"
SOYUZ-T	URS	4th generation manned spacecraft; T for "Troika"
SOYUZ-TM	URS	5th generation manned spacecraft; M for "Modify"
SPACELAB	US/ESA	Orbiting laboratory launched aboard space shuttle
SPACENET	US	Geostationary comsat; South Pacific Corp
SPACE TELESCOPE	US	Orbiting Astronomical Observatory to be launched by the space shuttle; Also called Hubble telescope
SPADES	USAF	Solar Perturbation of Atmospheric Density Experiments Satellite; DOD payload
SPARTA	AUS	See WRESAT
SPAS	US	Shuttle PAllet Satellite
SPASUR RUD	USN	Calibration sphere, 60 meters long; DOD payload
SPM	ESA	Solar Polar Mission; scientific satellite
SPOT	ESA	Satellite Probatoire de l'Observation de la Terre; Geostationary comsat; ESA earth resources satellite
SPOTNET	US	Geostationary comsat; managed by National Exchange

Satellite Name	Launching Sponsor	Acronym/Mission
SPUTNIK	URS	Sputnik 1 earth's first artificial satellite; scientific series
SR	US-DOD	Solar Radiation; DOD payload
SRATS	J	Solar Radiation And Thermospheric Satellite; see Taiyo
SRET	F/URS	Satellite for Research on Environment and Technology
SSS	US	Small Scientific Satellite; EXPAM
SSU	USN	See Classic Wizard/White Cloud
STARFLASH	US-DOD	DOD payload
STARLETTE	F	Satellite de Taille Adaptee avec Reflecteurs Laser paur les Etudes de la TErre; geodetic satellite
STARAD	USAF	STarfish ARtificial RAdiation sat; DOD payload
STATSIONAR	URS	Russian orbital position designations for their geostationary comsats.
STC	US	Geostationary DBS system; Satellite TV Corp.
STP	US	Space Test Program; formerly SESP
STS	US	Space Transportation System (Space Shuttle)
STW	CHN	China's first gostationary comsat series; experimental
SUNRAY	US-DOD	See SR.
SURCAL	USN	SURveillance CALibration system; DOD payload
SURVEYOR	US	Soft landing moon probes
SVS	US	Staged Vehicle System
SYLDA	ESA	Container used to hold sat on multi-payload ESA Ariane launches.
SYMPHONIE	ESA	Geostationary experimental comsats
SYNCOM	US	SYNchronous COMmunication; early '60s NASA project
SYNCOM	US-DOD	In orbit name for LEASAT series satellites
TACSAT	US-DOD	Geostationary experimental milsatcom; formerly TACMOSAT.
TAD	US	Thrust Augmented Delta; upgraded Delta rocket
TAIYO	J	3rd Japanese scientific sat; also called SRATS
TANSEI	J	Test satellites; also called MS-T1/2/3/4
TANSEI 2	J	See MU-3C.
TARGET AGENA	US	Agena upperstage vehicle used as target for Gemini 10
TAT-DELTA	US	Thrust Augmented Thor-Delta
TAVE	US	Thor Agena Vibration Experiment. Launched with Alouette 1.
TD-1A	ESRO	Thor Delta; scientific satellite
TDRS	US	Tracking & Data Relay Sat; geostationary comsat shuttle link
TDF-1	ESA	French/German geostationary DBS project
TEAL RUBY	USAF	Special experimental infrared recon platform to test for observation of aircraft/cruise missile in flight.
TELECOM	ESA	Geostationary advanced comsat; includes French military comm transponders
TELESAT	CAN	See ANIK. Telesat 1/Anik-A1 Telesat 2/Anik-A2 Telesat 3/Anik-A3 Telesat 4/Anik-B Telesat 5/Anik-D1 Telesat 6/Anik-C3 Telesat 7/Anik-C2 Telesat 8/Anik-D2 Telesat 9/Anik-C1
TELE-X	ESA	Geostationary DBS project of Scandinavian countries: Sweden, Norway, and Finland
TELSTAR	US	Active comsat of the 1960's
TELSTAR	US	Geostationary comsat managed by AT&T
TEMSAT	USN	Black Calibration Sphere; DOD payload
TENMA	J	See Astro-B.
TETR	US	Apollo tracking net and training satellite
THOR	US	1st stage rocket, base system appeared in the following combinations: Thor-Able, Thor-Able Star, Thor-AD/Agena-AD, Thor-Agena, Thor-Agena B, Thor-Agena D, Thor-Delta.

Satellite Name	Launching Sponsor	Acronym/Mission
TIMATION	USN	TIMe navigATION; navigation satellite forerunner to NAVSTAR
TIPS	USN	Transit Improvement Program; DOD NAVSAT payload
TIROS	US	Television & Infrared Observation Sat; orbiting metsat series.
TIROS-N	US	Improved version of the TIROS metsats
TITAN-IIIC	US	U.S. rocket used to place military payloads in orbit
TOPEX	USA/F	See Poseidon.
TOPO	USA	TOPOgraphic satellite; Army SECOR type tactical survey satellite
TOS	US	Tiros Operational Satellite; name of ESSA prior to launch.
TOURNESOL	F	See D2-2A; scientific satellite
TRAAC	USN	Transit Research And Attitude Control satellite; DOD payload
TRANSAT	USN	Navy navigational satellite; DOD payload
TRANSIT	USN	Navy navigational satellite system; DOD payload
TRIAD	US-DOD	DOD payload
TRS	US-DOD	Tetrahedral Research Satellite; DOD payload
TSIKADA	URS	Name given in conjunction with Cosmos 1000 (Soviet civilian navsat) 1st announced navsat mission
TTS	US-DOD	Test and Training Satellite; DOD payload
TV	US	Target Vehicle
TV-SAT	ESA	French geostationary DBS comsat
TV-SAT A3 & A5	D	Proposed geostationary DBS
UARS	US	Upper Air Research Satellite; proposed scientific sat
UHURU	US	See SAS (Explorer 42).
UKS	G	United Kingdom Subsatellite; scientific; part of AMPTE project.
UK-X4	G	See MIRANDA.
ULYSSES	ESA	Scientific probe to measure interplanetary medium out of the ecliptic plane. Launch in 1989.
UME/ISS-1	J	Scientific satellite
UNISAT	G	Proposed geostationary DBS for Britain
UOSAT	G/US	University Of Surrey Satellite; scientific amateur radio series.
USA	US-DOD	Latest public designation for DOD payloads, replaces the Operation # method of labeling.
U.S.A. SAT	US	See ISI.
USAT	US	Geostationary comsat (Ku-band) managed by United States Satellite Systems
USGCSS	US-DOD	ITU designation in official list for DSCS satellite positions
USSB	US	Proposed geostationary DBS; to be managed by Hubbard BCing.
USSS	US	See USAT.
UV RADIOMETER	US-DOD	Experimental ultra-violet radiation payload; DOD
VANGUARD	US	First series of U.S. scientific satellites; plagued by problems.
VEGA	URS	Dual planetary probes to Venus and Halley's comet
VELA HOTEL	US-DOD	High orbit nuclear detection satellites; DOD payload
VENERA	URS	Russian Venus planetary probes
VIDEOSAT	F	Geostationary comsat
VIKING	US	Mars planetary orbiter/lander program
VIKING	S	Swedish scientific research sat; electrical magnetic auroral studies
VOLNA	URS	Maritime comm package that will ride on different host satellites; some will carry mil transponders
VOSKHOD	URS	Second generation Soviet manned spacecraft; means "Sunrise."
VOSTOK	URS	1st generation Soviet manned spacecraft; means "East"
VOYAGER	US	Two solar system planetary explorers; now heading for Neptune.
VSS	US	Proposed geostationary DBS; managed by Video Satellite System.
WESTAR	US	Geostationary comsat managed by Western Union
WESTFORD NEEDLES	US-DOD	Experiment to test a possible communication medium using metallic needles as a passive reflector; DOD payload
WHITE CLOUD	USN	See Classic Wizard/SSU/NOSS.
WIKA/DIAL	G/F	See DIAL/WIKA.
WRESAT	AUS	Weapons Research Establishment SATellite; 1st Australian sat.
X3	G	See Prospero and Black Arrow.
X4	G	See Miranda.
YURI	J	See BS/BSE comsats.

Satellite Name	Launching Sponsor	Acronym/Mission
ZOND	URS	Unmanned lunar spacecraft; prob up-rated Soyuz intended to send men to the moon during US-USSR space race in 60's.
ZOHREH	IRN	Geostationary comsat proposed for Iran.

Appendix C

SATELLITE LOCATION GUIDE

International Satellite Positions

The following list represents the current and future state of the geostationary ring at this writing. Current launch developments can be followed through the pages of *Aviation Week and Space Technology*, *Satellite News* by Geoffrey Falworth and the author's monthly column in *Monitoring Times*.

Frequency Band Key
U 0.3 - 1 GHz
L 1 - 1.7 GHz
S 1.7 - 3 GHz
C 3 - 6.5 GHz
X 6.5 - 10.7 GHz
K 10.7 - 20 GHz
E 20 - 60 GHz

Service Key
F Fixed Satellite
M Maritime Mobile Satellite
B Broadcast Satellite
R Space Research
X Experimental

Data Key
u Data undetermined or unknown

Longitude	Name	Country/Sponsor	Service	Band	Orbit Status, Due Date, Int'l Designator
		EASTERN ARC			
1.0E	GDL-5	Luxembourg	F	U,X,K	u
4.0E	Telecom-1C	France	F	S,C,X,K	86
5.0E	OTS-2	ESA	F	U,K	in orbit/78-44A
5.0E	Tele-X	Nor,Swe,Fin	B,F	K,E	86
7.0E	F-SAT 1	France	F	S,C,E	87
7.0E	Eutelsat 1-3	ESA (ECS 3)	B,F	K	Lost on launch
10.0E	Apex	ESA	B,F	S,C,E	u
10.0E	Eutelsat	ESA	B,F	K	u
10.0E	Eutelsat 1-2	ESA (ECS-2)	B,F	K	in orbit/83-58A currently at 6E
12.0E	Prognoz-2	USSR	X	S,C	u
13.0E	Italsat	Italy	F	E	86
13.0E	Eutelsat 1-1	ESA (ECS-1)	B,F	K	in orbit/84-81A
14.0E	Nat Sat-1	Nigeria	F	C	u
15.0E	AMS-1,2	Israel	F	C,K,E	86
16.0E	Nat Sat-2	Nigeria	F	C	u
16.0E	Sicral-1A	Italy	M,B,F	U,X,K,E	87
17.0E	SABS	Saudi Arabia	B,F	K	u
19.0E	GDL-6	Luxembourg	F	X,K	86
19.0E	Arabsat-1	Arab Lg.	F,B	S,C	in orbit/85-15A
20.0E	Nat Sat-1	Nigeria	F	C	u
22.0E	Sicral-1B	Italy	M,F,B	U,X,K,E	87

Longitude	Name	Country/Sponsor	Service	Band	Orbit Status, Due Date, Int'l Designator
23.5E	DFS-1	Germany	F,B	S,K,E	87
26.0E	Arabsat-2	Arab Lg.	F,B	S,C	in orbit/85-48C
26.0E	Zohreh-2	Iran	F,B	K	u
28.5E	DFS-2	Germany	B,F	S,K,E	87
32.0E	Videosat-1	France	B	S,K	87
34.0E	Zohreh-1	Iran	B,F	K	u
34.3E	Raduga 9	USSR	F	C	in orbit/81-69A
35.0E	Prognoz-3	USSR	X	S,C	u
35.0E	Statsionar 2	USSR	--	--	--
35.2E	Raduga 11	USSR	F	C	in orbit/82-113A
35.3E	Cosmos 1629	USSR	--	--	in orbit/85-16A
38.0E	Paksat-1	Pakistan	F,B	K	86
40.0E	Statsionar 12	USSR	--	--	--
40.0E	Gorizont 3	USSR	F	C,X	in orbit/79-105A
41.0E	Zohreh-4	Iran	F,B	K	u
41.0E	Paksat-2	Pakistan	F,B	K	86
44.6E	Raduga-13	USSR	F	C	in orbit/83-88A
45.0E	Statsionar 9	USSR	--	--	--
45.0E	Loutch P2	USSR	F	K	u
45.0E	Volna-3	USSR	M	U,L	u
45.0E	Gals-2	USSR	F	X	u
47.0E	Zohreh-3	Iran	F,B	K	u
52.3E	Gorizont 9	USSR	F	C,X	in orbit/84-41A
52.8E	Gorizont 5	USSR	F	C,X	in orbit/82-20A
53.0E	Statsionar 5	USSR	--	--	--
53.0E	Loutch 2	USSR	F	K	u
53.0E	Volna 4	USSR	M	L	u
57.0E	IS V MCS-C	Intelsat	M	L,C	86
57.0E	IS V F-1	Intelsat	F	C,K	in orbit/81-50A
57.0E	IS V-A Ind 2	Intelsat	F	C,K	86
60.0E	IS V MCS-B	Intelsat	M	L,C	u
60.0E	IS V F-7	Intelsat	F,M	L,C,K	in orbit/83-105A
60.0E	IS V-A Ind 1	Intelsat	F	C,K	86
60.0E	IS VI Ind	Intelsat	F	C,K	88
62.9E	IS V F-5 MCS	Intelsat	F,M	S,C,K	in orbit/82-97A
63.0E	IS V-A Ind 3	Intelsat	F	C,K	u
64.5E	Marecs Ind 1	Inmarisat	M	U,L,C	86
66.0E	IS IV-A Ind 4	Intelsat	F	C	u
66.0E	IS V Ind 4	Intelsat	F	C,K	u
66.0E	IS MCS Ind D	Intelsat	M	L,C	u
69.0E	Ekran 9	USSR	B	U,C	in orbit/82-93A
70.0E	STW-2	PR China	F	C	u
72.5E	Marisat F-3	US-ComGen	M	U,L,C	in orbit/76-101A
73.0E	Marecs Ind 2	Inmarisat	M	L,C	u
73.6E	Insat-1B	India	F	U,S,C	in orbit/83-89B
77.0E	Palapa A-2	Indonesia	F	C	in orbit/77-18A(drifting)
80.0E	Statsionar 1	USSR	--	--	--
80.0E	Statsionar 13	USSR	--	--	--
80.0E	Potok	USSR	F	C	u
80.0E	Prognoz	USSR	X	S	u
80.0E	Prognoz-4	USSR	R	S,C	u
80.1E	Cosmos 1540	USSR	--	--	in orbit/84-22A

Longitude	Name	Country/Sponsor	Service	Band	Orbit Status, Due Date, Int'l Designator
80.2E	Cosmos 1366	USSR	--	--	in orbit/82-44A
80.5E	Gorizont 10	USSR	F	C,X	in orbit/84-78A
83.0E	Palapa A-1	Indonesia	F	C	in orbit/76-66A(drifting)
84.2E	Raduga 14	USSR	F	C	in orbit/84-16A
85.0E	Statsionar 3	USSR	--	--	--
85.0E	Loutch P-3	USSR	F	K	u
85.0E	Volna-5	USSR	M	U,L	u
85.0E	Gals-3	USSR	F	X	u
85.2E	Raduga 10	USSR	F	C	in orbit/81-102A
87.1E	Sirio	Italy CNR	B,F	U,K	in orbit/77-80A
90.0E	Statsionar 6	USSR	--	--	--
90.0E	Loutch-3	USSR	F	K	u
90.0E	Volna-8	USSR	M	L	u
90.1E	Gorizont 8	USSR	F	C,X	in orbit/83-118A
94.0E	Insat-1C	India	F	U,S,C	86
95.0E	CSDRN	USSR	R	K	u
98.3E	Ekran 13	USSR	B	U,C	in orbit/84-90A
99.0E	Statsionar T	USSR	--	--	--
99.0E	Statsionar T2	USSR	--	--	--
99.4E	Edran 14	USSR	B	U,C	in orbit/85-24A
99.6E	Ekran 11	USSR	B	U,C	in orbit/83-100A
108.0E	Palapa B-1	Indonesia	F	C	in orbit/83-59C
108.0E	BS-3A	Japan	B	K	89
110.0E	BS-2A	Japan	B	U,K	in orbit/84-05A
110.0E	BS-2B	Japan	B	U,K	u
110.0E	BS-3B	Japan	B	K	90
113.0E	Palapa B-2	Indonesia	F	C	launch failure on ground
118.0E	Palapa B-3	Indonesia	F	C	u
125.0E	STW-1/China15	PR China	F	C	in orbit/84-35A
129.1E	Raduga 15	USSR	F	C	in orbit/84-63A
130.0E	Statsionar	USSR	--	--	--
130.0E	ETS-2	Japan	R	U,L,S,E	in orbit, out of service
132.0E	CS-3	Japan	F	U,S,C	88
132.2E	CS-2A	Japan	B	U,C,E	in orbit/83-06A
135.9E	CS-2B	Japan	B	U,C,E	in orbit/83-81A
140.0E	Statsionar 7	USSR	--	--	--
140.0E	Loutch 4	USSR	F	U,K	u
140.0E	Volna-6	USSR	M	L	u
140.0E	Gorizont 6	USSR	F	C,X	in orbit/82-103A
140.3E	Gorizont 11	USSR	F	C,X	in orbit/85-07A
145.0E	Statsionar 16	USSR	--	--	--
150.0E	CSE	Japan	F	C	in orbit/77-118A
156.0E	Aussat-1	Australia-OTC	B,F	K	in orbit/85-76B
160.0E	Aussat-2	Australia-OTC	B,F	K	in orbit/85-109C
164.0E	Aussat-3	Australia-OTC	B,F	K	90
168.0E	Potok	USSR	F	C	u
170.0E	Statsionar 10	USSR	--	--	--
170.0E	Loutch P-4	USSR	F	K	u
170.0E	Volna 7	USSR	M	U,L	u
170.0E	Gals 1	USSR	F	X	u
173.0E	IS V-A Pac 1	Intelsat	F	C,K	87
174.0E	IS IV-A F-6	Intelsat	F	C	in orbit/75-42A
174.0E	IS V-A Pac 1	Intelsat	F	C,K	88
176.0E	IS V-A Pac 2	Intelsat	F	C,K	87
176.0E	IS V Pac 2	Intelsat	F	C,K	86
176.1E	Marisat 2 Pac	US-ComGen	M	U,L,C	in orbit/76-53A
177.5E	Marecs Pac 1	Inmarisat	M	U,L,C	in orbit/84-114B
179.0E	IS IV-A F-3	Intelsat	F	C	in orbit/74-93A
179.0E	IS V Pac	Intelsat	F	C,K	u
179.0E	IS-V-A Pac 3	Intelsat	F	C,K	u
179.0E	IS V, V-A	Intelsat	F	C,K	87

Longitude	Name	Country/Sponsor	Service	Band	Orbit Status, Due Date, Int'l Designator

WESTERN ARC

Longitude	Name	Country/Sponsor	Service	Band	Orbit Status, Due Date, Int'l Designator
1.0W	IS IV-A F-2	Intelsat	F	C,K	in orbit/76-10A
1.0W	IS V	Intelsat	F	C,K	86
1.0W	IS V Cont 4	Intelsat	F	C,K	u
1.0W	IS V-A Cont 4	Intelsat	F	C,K	u
4.0W	IS IV-A Atl 1	Intelsat	F	C	86?
4.0W	IS V Cont 3	Intelsat	F	C,K	87
4.0W	IS V-A Cont 3	Intelsat	F	C,K	87
5.0W	Telecom 1B	France	F	L,C,X,K	
8.0W	Telecom 1A	France	F	L,C,X,K	in orbit/84-81B
8.5W	Statsionar 11	USSR	--	--	--
10.0W	GDL-4	Luxembourg	F	X,K	86
11.0W	F-Sat-2	France	F	S,K,E	86
11.5W	Symphonie 2	Fr,W.Germ	R	C	in orbit/75-77A
13.9W	Gorizont 7	USSR	F	C,X	in orbit/83-66A
14.0W	Loutch 1	USSR	F	K	u
14.0W	Statsionar 4	USSR	--	--	--
14.0W	Volna-2	USSR	M	L	u
14.4W	Potok-1	USSR	F	C	u
14.8W	Marisat-1	US-ComGen	M	U,L,C	in orbit/76-17A
16.0W	WSDRN	USSR	F	C	u
16.5W	IS IV-A	Intelsat	F	C	86
16.5W	IS V-A or V-B	Intelsat	F	C,K	87
18.5W	IS V-A Atl 4	Intelsat	F	C,K	87
18.5W	IS V F-6 MCS	Intelsat	F	C,K	in orbit/83-47A
18.5W	IS V-B	Intelsat	F	C,K	86
19.0W	L-Sat	ESA	B	K,E	86
19.0W	TDF-1,2.	France	B	S,K	86
19.0W	LuxSat	Luxembourg	B	K	86
19.0W	Helvasat	Switzerland	B	K	86
19.0W	TV-Sat A3,A5	W. Germany	B	S,K	86
19.0W	Sarit	Italy	B	K	86
20.0W	GDL-4	Luxembourg	B,F	X,K	u
21.5W	IS V Atl 5	Intelsat	F	C,K	86
21.5W	IS IV-A F-4	Intelsat	F	C	in orbit/77-41A
21.0W	IS MCS	Intelsat	M	L,C	u
23.0W	Marecs Atl 2	Inmarisat	M	L,C	u
24.0W	Prognoz 1	USSR	R	C	u
24.0W	Cosmos 1546	USSR	--	--	in orbit/84-31A
24.4W	Raduga 7	USSR	F	C	in orbit/80-81A
24.5W	IS VI Atl 1	Intelsat	F	C,K	86
24.5W	IS V-A Atl 1	Intelsat	F	C,K	u
24.5W	IS V MCS Atl-D	Intelsat	M	L,C	u
24.5W	IS V Atl 1 F-3	Intelsat	F	C,K	in orbit/81-119A
25.0W	Statsionar 8	USSR	--	--	--
24.0W	Loutch P-1	USSR	F	K,E	u
25.0W	Volna 1	USSR	M	U,L	u
25.8W	Marecs Atl 1	Inmarisat	M	L,C	in orbit/81-122A
26.8W	IS V-A F-10	Intelsat	F	C,K	in orbit/85-25A
27.4W	IS V F-2	Intelsat	F	C,K	in orbit/80-98A
27.5W	IS VI Atl 2	Intelsat	F	C,K	87
27.5W	IS V-A Atl 2	Intelsat (F-11)	F	C,K	in orbit/85-55A
27.5W	IS MCS Atl-B	Intelsat	M	L,C	u
31.0W	Unisat 1	UK BTI	F,B	K	86
31.0W	IS V Atl-6	Intelsat	F	C,K	u
31.0W	IS IV-A F-1	Intelsat	F	C	in orbit/75-91A
34.5W	IS V-A Atl-3	Intelsat	F	C,K	86
34.5W	IS MCS Atl-E	Intelsat	F	C,K	in orbit/82-17A
34.5W	IS MCS Atl-E	Intelsat	M	L,C	u
34.5W	IS VI	Intelsat	F	C,K	88
37.5W	Videosat-2	France	B	K	87
37.5W	Orion	US-Orion	F	K	86

Longitude	Name	Country/Sponsor	Service	Band	Orbit Status, Due Date, Int'l Designator
40.5W	IS V-B	Intelsat	F	C,K	86
40.5W	IS V-A Atl	Intelsat	F C,K	86	
41.0W	TDRS-East	US-Spacecom	R	L,K	in orbit/83-26B
41.0W	TDRS-East	US-SysGen	F	C	in orbit/83-26B
43.0W	Cygnus II	US-Cygnus	F	K	88
45.0W	Cygnus I	US-Cygnus	F	K	88
47.0W	Orion	US-Orion	F	K	87
49.1W	IS IV Atl 1 F-1	Intelsat	F	C	in orbit/75-42A
50.0W	IS V-A Cont 2	Intelsat	F	C,K	86
50.0W	IS V Cont 2	Intelsat	F	C,K	u
50.0W	IS V-b	Intelsat	F	C,K	87
50.0W	Orion	US-Orion	F	K	87
53.0W	IS IV-A Atl 3	Intelsat	F	C	u
53.0W	IS V-B Cont 1	Intelsat	F	C,K	86
53.0W	IS V-A Cont 1	Intelsat	F	C,K	u
53.0W	IS V F-8 MCS	Intelsat	F,M	L,C,K	in orbit/84-23A
56.0W	USASat 13E	US-ISI	F	K	88
56.0W	IS IV-A	Intelsat	F	C	87
45.0W	IS V,V-A,V-B	Intelsat	F	C,K	87
57.0W	Digisat 1A	US-Digi TI	F	C	87
57.0W	Digisat 1A,1B	US-Digi TI	F	C,K	88
57.0W	Pan Am Sat	US-Pan AM SC	F	C,X,K	87
58.0W	USASat 8-C	USA	F	C	87
58.0W	USASat 13-D	US-ISI	F	K	88
60.0W	IS IV-A	Intelsat	F	C	86
60.0W	IS V,V-A,V-B	Intelsat	F	C,K	87
61.0W	RCA Amer 9	US-RCA	F	C	89
61.0W	Satcom K-4	US RCA AM	F	K	89
61.0W	TDRS-Cent	US-Spacecom	R	L,K	u
61.0W	VSS PH II	US-Video Sat	B	K	u
61.5W	SDT-1	US-Sat DT	B	K	87
62.5W	Colsat(2 sats)	US-Col.C	F	C,K	86
63.0W	RCA Amer 8	US-RCA	F	C,K	89
65.0W	SBTS A-2	Brazil	F	C	in orbit/85-15B
65.0W	RCA Amer 7	US-RCA	F	C,K	89
66.0W	USASat 8A	USA	F	C	u
67.0W	Satcom 6	US-RCA	F	C	86
69.0W	Spacenet 2	US-GTE	F	C,K	in orbit/84-114A
70.0W	SBTS A-1	Brazil	F	C	85
70.0W	USASat 7C	USA-CGC	F	C	u
72.0W	Satcom 2R	US-RCA	F	C	in orbit/83-94A
73.0W	Galaxy H5	US-Hughes	F	K	87
73.0W	MMC-1	US-Martin Marietta	F	K	88
73.0W	Ford Aero 3	US-Ford	F	C,K	88
74.0W	USASat 7A	USA	F	C	u
75.0W	Rainbow 3	US-Rainbow	F	K	87
75.0W	Satcol 2	Colombia	F	C	u
75.0W	MMC-2	US-Martin M	F	K	88
75.0W	USAT III	US-USSI	F	K	89
75.0W	USSSI	US-USBI	F	K	u
75.0W	Galaxy H6	US-Hughes	F	K	89
75.0W	Spotnet C2	US-NEX	F	C	87
75.0W	Spotnet K2,4	US-NEX	F	K	87
75.4W	Satcol 1A/1B	Colombia	F	C	u
76.0W	Comstar 1	US-ComGen	F	C	in orbit/76-42A(drifting)
	Comstar 2	US-ComGen	F	C	in orbit/76-73A
	Comstar 3	US-ComGen	F	C	in orbit/78-68A
	Comstar 4	US-ComGen	F	C	in orbit/81-18A
76.0W	Telstar 3B(304)	US-AT&T	F	C	88
79.0W	Westar 2	US-WUTC	F	C	in orbit/74-75A
7O.0W	USASat 7D	USA	F	C,K	u
79.OW	Westar 9	US-WUTC	F	C	89
79.0W	RSI-1	US-Rainbow	F	K	86
81.0W	Amer Sat-1	US-ASC	F	C,K	in orbit/85-76C

Longitude	Name	Country/Sponsor	Service	Band	Orbit Status, Due Date, Int'l Designator
81.0W	Satcom K-2	US-RCA	F	K	in orbit/85-109D
81.7W	USASat 5C	USA	F	C	u
83.0W	Usat 4	US-USSSI	F	K	88
83.0W	ABCI	US-ABCI	F	K	86
83.0W	STSC-1	Cuba	F	C	88
83.0W	Satcom IV	US-RCA	F		in orbit/82-04A
85.0W	Usat 1	US-USSSI	F	K	86
85.0W	Mobilesat 1	US-Mobilesat	M	K	88
86.0W	Westar 6	US-WUTC	F	C	u
87.0W	Telstar 3C(302)	US AT&T	F	C	in orbit/84-93D
87.0W	Westar 7	US-WUTC	F	C	u
87.0W	Westar E	US-WUTC	F	C	88
87.0W	Satcom K-1	US-RCA	F	K	Launch STS-61C
87.0W	Westar-B	US-WUTC	F	C	88
89.0W	SBS-4	US-SBS	F,B	K	in orbit/84-93B
91.0W	Spacenet 3	US-GTE	F	C,K	u
91.0W	Galaxy Ka-2	US-Hughes	F	E	88
91.0W	Westar A	US-WUTC	F	K	88
91.0W	Westar 3	US-WUTC	F	C	in orbit/79-72A
91.0W	Westar 9	US-WUTC	F	C	89
93.0W	Comstar K-1,3	US-ComGen	F	K	87
93.0W	RSI-4	US-Rainbow	F	K	u
93.0W	ASC-3	US AmerSat	F	C,K	89
93.0W	Ford Aer 2	US Ford	F	C,K	87
93.0W	Equastar 2	US-Equit	F	C	88
93.0W	Galaxy H7	US-Hughes	F	C	88
93.0W	SBS-7	US-SBS	F	K	88
93.0W	Westar C	US-WUTC	F	K	89
93.0W	Westar 8	US-WUTC	F	C	88
93.0W	Galaxy Ka1	US-Hughes	F	C	89
93.0W	Galaxy Ka2	US-Hughes	F	E	89
93.5W	Galaxy 3	US-Hughes	F	C	in orbit/84-101A
95.0W	SBS-9	US-SBS	F	K	90
95.0W	SBS-3	US-Hughes	F	K	in orbit/82-110B
95.0W	RCA-DBS	US-RCA	B	K	87
95.0W	USASat 6C	USA	F	K	u
96.0W	Telstar 3a(301)	US-AT&T	F	C	in orbit/83-77A
97.0W	SBS-8	US-SBS	F	K	88
97.0W	SBS-2	US-SBS	F	K	in orbit/81-96A
99.0W	Westar IV	US-WUTC	F	C	in orbit/82-14A
99.0W	Mobilesat	US-Mobilesat	M	K	u
99.0W	SBS-6	US-SBS	F	K	87
100.0W	SBS-1	US-SBS	F	K	in orbit/81-91A
101.0W	Spacenet 3V	US-GTE	F	C,K	u
101.0W	G-Star 3	US-GTE	F	K	u
101.0W	Comstar K-2	US-ComGen	F	K	88
101.0W	Fednet	US-Fed Ex	F	K	88
101.0W	USSB(1 of 2)	US-USSBCI	B	K	87
101.0W	Ford Aero 1	US-Ford	F	C,K	87
101.0W	Spotnet K1	US-NEX	F	K	87
101.0W	Spotnet K3	US-NEX	F	K	88
101.0W	Spotnet C1	US-NEX	F	C	88
101.0W	STC (1,2)	US-STC	B	K	87
101.0W	RCA Eastern	US-RCA	B	K	87
101.0W	ACC	US-Adv. CC	B	K	87
101.0W	DBS C-1	US-DBS	B	K	87
101.0W	USSB-1	US-USSBCI	B	K	87
101.0W	NEX DBS-1	US-NEX	B	K	88
101.0W	NEX DBS-2	US-NEX	B	K	90
101.0W	NCN	US-N.C.N.	B	K	87
101.0W	SCS(1)	US-Space C.S.	B	K	87
103.0W	G-Star 1	US-GTE	F	K	in orbit/85-35A Malfunction on launch (Ariane)
104.5W	Anik D-1	Canada-Telesat	F	C	in orbit/82-82A Telesat 5
105.0W	G-Star 2	US-GTE SN	F	K	u

Longitude	Name	Country/Sponsor	Service	Band	Orbit Status, Due Date, Int'l Designator
106.0W	USASat	USA	F	K	u
106.5W	Msat	Canada	M,F	U,E	u
107.2W	Anik C-2	Canada-Telesat	F	K	in orbit/83-59B Telesat 7
107.4W	Anik C-1	Canada-Telesat	F	K	in orbit/85-28B Telesat 9
108.0W	Anik B	Canada-Telesat	F	C,K	in orbit/78-116A Telesat 4
110.0W	WU-DBS	US-WUTC	B	K	87
110.0W	SSS	US-S.S.S.	B	K	87
111.5W	Anik E	Canada-Telesat	F	C	90
111.5W	Anik D-2	Canada-Telesat	F	C	in orbit/84-113B Telesat 8
113.5W	Morelos 1	Mexico	F	C,K	in orbit/85-48B
116.5W	Morelos 2	Mexico	F	C,K	in orbit/85-109B drifting next two years to orbital slot
117.5W	Anik C-3	Canada-Telesat	F	K	in orbit/82-110C Telesat 6
119.0W	DBS-HS 1,2	US-Hughes	B	K	89
119.0W	RCA-Central	US-RCA	B	K	87
119.0W	VSS-1	US-VSSI	B	K	87
119.0W	NCN	US-NCNI	B	K	87
119.0W	GraphSat	US-Graphic	B	K	87
119.0W	W.U. DBS	US-WUTC	B	K	88
120.0W	Usat 2	US-USSSI	F	K	87
120.0W	Spacenet 1	US-GTE SN	F	C,K	in orbit/84-49A
122.0W	Equastar 1	US-Equit	F	C	87
122.5W	Westar 5	US-WUTC	F	C	in orbit/82-58A
124.0W	SBS-5	US-SBS	F	K	86
125.0W	Telstar 3d(303)	US AT&T	F	C	in orbit/85-48D
125.0W	Mobilesat 1	US-Mobile	M	K	88
126.0W	Satcom K-3	US-RCA	F	K	87
127.0W	Comstar D-4	US-ComGen	F	C	in orbit/81-18A moving to 76W
128.0W	ASC-2	US-AmerSat.	F	C,K	86
130.0W	USA Sat 10D	USA	F	K	86
131.0W	Satcom-3R	US-RCA	F	C	in orbit/81-114A
132.0W	RSI-2	US-Rainbow	F	K	86
132.0W	Satcom 3	US-RCA	F	C	in a useless orbit
134.0W	Galaxy 1	US-Hughes	F	C	in orbit/83-65A
134.0W	FedNet 2	US-Federal Exp.	F	K	88
134.0W	Digisat 2	US-Dig C	F	K	89
136.0W	Amigo 1	Mexico	B	K	u
137.0W	Aurora 3	US-Alascom	F	C	86
137.0W	Galaxy 4	US-Hughes	F	C	91
138.3W	Satcom 2	US-RCA	F	C	in orbit/76-29A
139.0W	Satcom 1R	US-RCA	F	C	in orbit/83-30A
139.0W	Satcom 6	US-RCA	F	C	u
141.0W	Aurora 2	US-Alascom	F	C	89
141.0W	Ilhicahua 3	Mexico	F	C,K	u
143.0W	Aurora 1	US-Alascom	F	C	in orbit/82-105A
145.0W	Ilhuicahua 4	Mexico	F	C,K	u
146.0W	Amigo 2	Mexico	B	K	u
147.5W	Colsat(CCC)2	US-Col C.	F	C,K	87
148.0W	STC-2	US-STC	B	K	87
148.0W	CBS-2	US-CBS	B	K	87
148.0W	RCA-Mtn	US-RCA	B	K	87
148.0W	USSB-West	US-USSBCI	B	K	87
148.0W	VSS-DBS	US-DVSI	B	K	87
148.0W	ACCODBS	US-Adv. C.C.	B	K	87
148.0W	DBS-C2	US-DBSC	B	K	87
148.0W	GraphSat	US-Graphic	B	K	87
148.0W	NCN-DBS	US-Nat.C.N.	B	K	87
148.0W	WU-DBS	US-WUTC	B	K	88
157.0W	SSS	US-Synd.	B	K	87
147.0W	WU-DBS	US-WUTC	B	K	88
157.0W	RCA-Pacific	US-RCA	B	K	u
160.0W	ESDRN	USSR	R	K	u
166.0W	SCS-2	US-Space C	B	K	87

Longitude	Name	Country/Sponsor	Service	Band	Orbit Status, Due Date, Int'l Designator
167.0W	Ekran 8	USSR	B	U,C	in orbit/82-09A
168.0W	Potok 3	USSR	F	C	u
170.0W	Volna 7	USSR	M	U,L	u
170.0W	Statsionar 10	USSR	--	--	--
170.0W	Loutch-P4	USSR	F	K	u
171.0W	TDRS-West	US-Spacecom	R	S,K	86
171.0W	TDRS-West	US-SysGen	F	C	86
175.0W	VSS PH 2	US-Dominion VS	B	K	u

Domestic Comsat Geostationary Positions

Present and Future

(A) Orbital Position (deg. west longitude)
(B) Satellite band authorized at position
(C) Present satellite in position/ (transponders)
(D) Future satellites in position/ (transponders)

A	B	C	D
57 W	Ku		Digisat-1A(24)
	C		Digisat-1B(24)
61	Hybrid		SATCOM 9/K6(40)
62.5	Hybrid		CCC-1 (48)
63	Hybrid		SATCOM 8/K5(40)
65	Hybrid		SATCOM 7/K4(40)
67	C		SATCOM 6 (24)
69	Hybrid	Spacenet 2(18)	
70	UHF		Geostar 1 (2)
71	Ku	Unassigned	
72	C	SATCOM 2R(24)	
73	Ku		Galaxy H5 (16) Ford 3(Hybrid-48)
74	C	Galaxy 2 (24)	
75	Ku		RSI-3(20) Spotnet K2(24) Spotnet K4(24) Spotnet C2 (C band-24) Usat-3(20) Galaxy H6(16)
76	C	COMSTAR-D1/D2/D3/D4	Telstar 304(24)
78.5	C	WESTAR 2(12)	WESTAR 6(24) RSI-2(16) WESTAR 12 (C band-24)
81	Hybrid	ASC-1 (24)	WESTAR 9 (C band-24)
83	Ku		Usat-4 (20) ABC-1 (20)
83.5	C	SATCOM 4 (24)	
85	Ku		Usat-1 (20) Mobilesat-1(UHF /L-band/4) Telstar 302(24)
87	Ku	COMSTAR D3 (C-band/24) SATCOM K1(16)	WESTAR C/F(16) WESTAR 18 (C-band/24)
88.5	C		
89	Ku	SBS-4 (10)	
91	Hybrid	WESTAR 3 (C-band/12)	SPACENET 3 (18) WESTAR 11 (C-band/24) WESTAR 17 (C-band/24)
93	Ku		RSI-4 (20) SBS-7 (19) Galaxy H7 (16) COMSTAR K1/K3 (16) WESTAR A/D (16) WESTAR 10 (C-band/24) WESTAR 16 (C-band/24) Equastar 2 (C-band/24) Ford-2 (Hybrid/48)
93.5	C	Galaxy 3 (24)	
95	Ku	SBS-3 (10)	SBS-9 (19)
96	C	Telstar 301(24)	WESTAR 7 (24)
97	Ku	SBS-2 (10)	SBS-8 (19)
98.5	C	WESTAR 4 (24)	WESTAR 8 (24)
99	Ku	SBS-1 (10)	SBS-6 (14)
100	UHF		Geostar-2 (2)

A	B	C	D
101	Ku/DBS		CBS-1 (DBS-3) DBSC-1(DBS-18) RCA-BSS 1 (DBS-6) STC-1(DBS-3) USSB-1(DBS-6) FORD-1 (Hybrid-48) ASC-3 (Hybrid-24) Spotnet C1 (C-band/24) Spotnet K1/K3 (Ku/24) Gstar A3(Ku/44) COMSTAR K2 (Ku/16)
103	Ku		Gstar A1 (44) (Lost due to malfunction)
104.5	C	Anik D-1 (24)	
105	Ku		Gstar A2 (44) Skylink-1 (UHF/2)
106.5	UHF		M-SAT (5)
107.5	Ku	Anik C-1(16)	Anik C2 (16)
108	C	Anik B(18)	Canada
109	Hybrid		
110	Ku		Anik D-2 (C-band/24) WU BSS (DBS/4)
111.5	C	Anik D-2	Canada Anik E
112.5	Ku		Canada
113.5		Hybrid Morelos-1 (22)	Mexico
114	C		Canada
116.5	Hybrid	Morelos-2 (22)	Mexico
117.5		Anik C-3(16)	Canada
119	C/DBS	SATCOM-2 (24)	RCA BSS-2 (DBS/6) VSS-1 (DBS/2) WU BSS-2 (DBS/4) WESTAR 5 (24) will move here from 123

A	B	C	D
120	Ku	Spacenet-1 (18)	Usat-2 (20)
122	Hybrid		Equastar-2 (C-band/24)
123	C	WESTAR 5 (24) will move to 119.5	WESTAR 14 (24)
124	Ku		SBS-5 (14)
125	C	Telstar 303(24)	Mobilesat-1(UHF, L-band/4)
126	Ku		RCA K3 (16)
128	Hybrid		ASC-2 (24)
130	Ku		ABC-2 (20) WESTAR 15 (C-band/24) Geostar-3 (UHF/2)
131	C	SATCOM 3R (24)	
132	Ku		Rainbow-2 (16)
134	C	Galaxy-1 (24)	Digisat-2 (Ku/24)
135	UHF		Skylink-2 (2)
136	C	SATCOM-1(no longer	in use)
137	C		Galaxy 4 (24) Aurora 3 (24)
139	C	SATCOM 1R (24)	
141	C		Aurora 2 (24)
143	C	SATCOM 5/ Aurora 1 (24)	
147.5	Hybrid		CCC-2 (48)
148	DBS		CBS-2 (3) DBSC-2 (18) RCA BSS-3 (6) STC-2 (3) USSB-2 (6) VSS-2 (2) WU BSS-3 (4)
157	DBS		RCA BSS-4 (6) WU BSS-4 (4)
166			SSS(?)
175	DBS		SCS-2(?) VSSPH2(?)

NOTES

DBS:12.2-12.7 GHz downlink; 17.3-17.8 GHz uplink
Ku-band: 11.7-12.2 GHz downlink; 14.0-14.5 GHz uplink
C-band: 3.7-4.2 GHz downlink; 5.925-6.425 GHz uplink

WEATHER SATELLITES

U.S. Environmental Satellites 1960-1980

Satellite	Purpose[1]	Launch	Average[2] Altitude	Ceased Operation	Remarks[3]
TIROS 1	R	04/01/60	720	06/19/60	First weather satellite providing cloud cover photography
TIROS 2	R	11/23/60	672	02/01/61	
TIROS 3	R	07/12/61	760	10/30/61	
TIROS 4	R	02/08/62	773	06/12/62	
TIROS 5	R	06/19/62	694	10/11/63	
TIROS 6	R	09/18/62	694	10/11/63	
TIROS 7	R	06/19/63	645	02/03/66	
TIROS 8	R	12/21/63	749	01/22/66	First APT satellite
Nimbus 1	R	08/28/64	S/677	09/23/64	Carried AVCS, APT, and High Resolution Infrared Radiometer for night pictures.
TIROS 9	R	01/22/65	S/1630	02/15/67	First TIROS satellite in Sun-synchronous orbit
TIROS 10	O	07/01/65	S/792	07/03/66	
ESSA 1	O	02/03/66	S/765	05/08/67	First satellite in the operational system; carried 2 wide-angle TV cameras
ESSA 2	O	02/28/66	S/1376	10/16/70	Carried APT cameras. APT carried on all even-numbered ESSA satellites.
Nimbus 2	R	05/15/66	S/1136	01/18/69	
ESSA 3	O	10/02/66	S/1427	10/09/68	Carried first AVCS cameras. ABCS carried on all odd-numbered ESSA satellites.
ATS 1	R	12/06/66	G/35,765	10/16/72 (pictures)	WEFAX discontinued December 31, 1978
ESSA 4	O	01/26/67	S/1373	10/06/67	
ATS 2	R	04/05/67	-	-	Unstable attitude-data not useful
ESSA 5	O	11/05/67	S/1379	02/20/70	
ATS 3	R	11/05/67	G/35,815	10/30/75 (pictures)	WEFAX discontinued December 31, 1978
ESSA 6	O	11/10/67	S/1437	11/04/69	
ESSA 7	O	08/16/68	S/1440	07/19/69	
ESSA 8	O	12/15/68	S/1429	03/12/76	
ESSA 9	O	02/26/69	S/1456	11/15/73	
Nimbus 3	R	04/14/69	S/1100	01/22/72	Provided first vertical temperature profile data of the atmosphere on a global basis.
ITOS 1	R/O	01/23/70	S/1456	06/17/71	Second generation prototype
Nimbus 4	R	04/08/70	S/1108	09/30/80	
NOAA 1	O	12/11/70	S/1438	08/19/71	First NOAA funded second generation satellite.
ITOS B	O	10/21/71	-	-	Failed to orbit.
Landsat 1	R	07/23/72	S/918	01/16/78	
NOAA 2	O	10/15/72	S/1460	01/30/75	First operational satellite to carry all scanning radiometer.
Nimbus 5	R	12/12/72	S/1110	03/29/83	
ITOS E	O	07/16/73	-	-	Failed orbit
NOAA 3	O	11/06/73	S/1510	08/31/76	First operational satellite to permit direct broadcast of VTPR data.

Satellite	Purpose[1]	Launch	Average[2] Altitude	Ceased Operation	Remarks[3]
SMS 1	R/O	05/17/74	G/35,788	01/29/81	Deactivated. Boosted out of geosynchronous orbit
NOAA 4	O	11/15/74	S/1460	11/18/78	Deactivated
Landsat 2	R	01/22/75	S/918	03/31/83	On standby
SMS 2	R/O	02/06/75	G/35,800	08/05/82	Deactivated. Boosted out of geosynchronous orbit.
Nimbus 6	R	06/12/75	S/1110	03/29/83	
GOES 1	O	10/16/75	G/35,796	-	First NOAA operational geostationary satellite; 130°W on standby
NOAA 5	O	07/29/76	S/1511	07/16/79	Deactivated
GOES 2	O	06/16/77	G/35,787	-	Second NOAA operational geostationary satellite; 113°W supporting Central WEFAX
Landsat 3	R	03/05/78	S/918	03/31/83	First Landsat with infrared capability. Now on standby.
GOES 3	O	06/16/78	G/35,784	-	On standby. Moving to 135°W.
Seasat 1	R	06/26/78	850	10/10/78	Electrical failure
TIROS-N	R/O	10/13/78	S/850	02/27/81	Deactivated
Nimbus 7	R	10/24/78	S/954	-	Carrying Coastal Zone Color Scanner
NOAA 6	O	06/27/79	S/807	-	First NOAA funded TIROS-N system satellite.
NOAA B	O	09/09/80	G/35,782	-	First geostationary satellite to carry the VISSR Atmospheric Sounder (VAS) which has now failed. At 139°W. Provides west WEFAX and DCS.
GOES 5	O	05/22/81	G/35,785	-	At 75°W; also carried VAS. Now failed. Provides east DCS, WEFAX, and relay of GOES 6 imagery.
NOAA 7	O	06/23/81	S/847	-	Second NOAA funded TIROS-N system satellite.
Landsat 4	O	07/16/82	S/700	-	Carries MSS and TM
NOAA 8	O	03/28/83	S/815	06/12/84	Had search and rescue capability.
GOES 6	O	04/28/83	G/35,791	-	Alternates between 98°W and 108°W. Only spacecraft with operating VAS.
NOAA 9	O	12/12/84	S/815	-	Has search and rescue capability and sensors for ozone and earth radiation budget.

1 R-Research; O-Operations; R/O-Operational Prototype
2 S-Sun-Synchronous; G-Geosynchronous
3 APT-Automatic Picture Transmission; AVCS-Advanced Vidicon Camera System; WEFAX-Weather Facsimile; VTPR-Vertical Temperature Profile Radiometer; VISSR-Visible Infrared Spin-Scan Radiometer; VAS-VISSR Atmospheric Sounder; MSS-Multi Spectral Scanner; TM-Thematic Mapper; DCS-Data Collection System.

GOES WEFAX SCHEDULES

Key to Abbreviations:

CHNL	Channel	PRES	Pressure
DIR	Daytime Infrar-red (image)	PROG	Prognostic (chart)
EU	Europe	PS	Pacific South
FL	Flight Level	QUADS	Quadrants
IR	Infra-red	SA	South Atlantic
MERC	Mercator (map/chart projection)	SH/NH	Southern/Northern Hemisphere
MOIST	Moisture	SH	Southern Hemisphere
MSL	Mean Sea Level	SIG	Significant
NIR	Nighttime Infra-red	THK	Thickness
NE/SE	Northeast/Southeast	TROP	Troposphere
NH	Northern hemisphere	VIS	Visible
NW/SW	Northwest/Southwest	VT	Valid Time
NT	North Atlantic	WEA	Weather
PN	Pacific North	W/E	West/East
POLAR	Polar (map/chart projection)		

GOES-East Wefax Broadcast Schedule

(Revised October 23, 1985)

Chart #	Time (Z)	Product
T01	0010	METEOSAT (D2) 0000Z IR
T02	0014	METEOSAT (D1) 0000Z IR
T03	0018	METEOSAT (D3) 0000Z IR
T04	0022	METEOSAT (D4) 0000Z IR
T05	0026	METEOSAT (D5) 0000Z IR
T06	0030	METEOSAT (D6) 0000Z IR
T07	0034	METEOSAT (D7) 0000Z IR
GE01	0038	NOAA-9(R)100E-10E SH POLAR VIS, 10E-80W SH POLAR DIR
GE02	0050	NOAA-9(N)10E-80W NH POLAR DIR, 100E-10E NHPOLAR DIR
GE03	0102	NOAA-9(P)170W-10E NH/SH POLAR NIR
GE04	0117	GOES 0000Z TROPICAL W/E IR
GE05	0130	GOES 0000Z NW/SW QUADS IR
GE06	0142	GOES 0000Z NE/SE QUADS IR
GE07	0157	GOES 0000Z TROPICAL W/E VIS
GE08	0210	GOES 0000Z NW/SW QUADS VIS
GE09	0222	GOES 0000Z NE/SE QUADS VIS
GE10	0237	GOES 2330Z TROPICAL W/E (MOIST CHNL)
GE11	0250	NOAA-7(Q)10E-60W MERC VIS, 160W-130E MERC NIR
T08	0310	METEOSAT 0300Z (D2) IR
T09	0314	METEOSAT 0300Z (D1) IR
T10	0318	METEOSAT 0300Z (D3) IR
T11	0322	METEOSAT 0300Z (D4) IR
T12	0326	METEOSAT 0300Z (D5) IR
T13	0330	METEOSAT 0300Z (D6) IR
T14	0334	METEOSAT 0300Z (D7) IR
GE12	0338	GOES 2330Z NW/SW QUADS (MOIST CHNL)
GE13	0350	GOES 2330Z NE/SE QUADS (MOIST CHNL) ECLIPSE
GE14	0600	NOAA-9(C)100W-170W, 40W-110W MERC VIS
T15	0610	METEOSAT 0600Z (D2) IR
T16	0614	METEOSAT 0600Z (CO2) VIS
T17	0618	METEOSAT 0600Z (CO3) VIS
T18	0622	METEOSAT 0600Z (D1) IR
T19	0626	METEOSAT 0600Z (D3) IR
T20	0630	METEOSAT 0600Z (D4) IR
GE15	0635	NOAA-9(S)NORTH POLE DIR, 80W-170W NH DIR
GE16	0650	NOAA-9(B)100E-10E NH/SH POLAR NIR
GE17	0705	GOES 0600Z TROPICAL W/E IR (05Z DURING ECLIPSE)
GE18	0720	GOES 0600Z NW/SW QUADS IR (05Z DURING ECLIPSE)
GE19	0735	GOES 0600Z NE/SE QUADS IR (05Z DURING ECLIPSE)
GE20	0750	NOAA-9(A)80W-170W NH/SH POLAR VIS
GE21	0805	NOAA-9(D)80W-10E, 140E-70E MERC NIR
GE22	0820	NOAA-9(T)80W-170W SH POLAR DIR
T21	0922	METEOSAT 0900Z (D1) IR
T22	0926	METEOSAT 0900Z (D3) IR
T23	0930	METEOSAT 0900Z (D4) IR
GE23	1035	GOES 0900Z TROPICAL W/E IR
GE24	1050	GOES 0900Z NW/SW QUADS IR
GE25	1105	GOES 0900Z NE/SE QUADS IR
GE26	1115	TBUS NOAA-9
GE27	1120	OPERATIONAL MESSAGE
T24	1154	METEOSAT 1130Z(E1)MOIST CHNL
T25	1158	METEOSAT 1130Z(E2)MOIST CHNL
T26	1202	METEOSAT 1130Z(E3)MOIST CHNL

T27 1210 METEOSAT 1200Z (D2) IR
T28 1214 METEOSAT 1200Z (C2) VIS
T29 1218 METEOSAT 1200Z (C03) VIS
T30 1222 METEOSAT 1200Z (D1) IR
T31 1226 METEOSAT 1200Z (D3) IR
T32 1230 METEOSAT 1200Z (D4) IR
GE28 1235 GOES 1130Z TROPICAL W/E (MOIST CHNL)
GE29 1248 GOES 1130Z NW/SW QUADS (MOIST CHNL)
GE30 1300 GOES 1130Z NE/SE QUADS (MOIST CHNL)
GE31 1315 GOES 1200Z TROPICAL W/E VIS
GE32 1328 GOES 1200Z NE/SE QUADS VIS
GE33 1343 GOES 1200Z TROPICAL W/E IR
GE34 1355 GOES 1200Z NW/SW QUADS IR
GE35 1407 GOES 1200Z NE/SE QUADS IR
GE36*1430 TBUS NOAA 8
GE37 1435 NOAA-9(H)170W-120E MERC VIS, 20E-50W MERC VIS
GE38 1450 NOAA-9(G) 10E-80W NH/SH POLAR NIR
T33 1510 METEOSAT 1500Z (D2) IR
T34 1514 METEOSAT 1500Z (CO2) VIS
T35 1518 METEOSAT 1500Z (CO3) VIS
T36 1522 METEOSAT 1500Z (D1) IR
T37 1526 METEOSAT 1500Z (D3) IR
GE39 1530 NOAA-9(E) 170W-100E NH/SH POLAR VIS
GE40 1545 GOES 1500Z TROPICAL W/E IR
GE41 1558 GOES 1500Z NW/SW QUADS IR
GE42 1610 GOES 1500Z NE/SE QUADS IR
GE43 1625 GOES 1500Z TROPICAL W/E VIS
GE44 1638 GOES 1500Z NE/SE QUADS VIS
GE45 1650 NOAA-9(J)80W-170W NH/SH POLAR NIR
GE46 1705 NOAA-9(F)170W-100E POLAR DIR, 100E-10E NH POLAR VIS
T38 1810 METEOSAT 1800Z (D2) IR
T39 1814 METEOSAT 1800Z (D1) IR
T40 1818 METEOSAT 1800Z (D3) IR
T41 1822 METEOSAT 1800Z (D4) IR
T42 1826 METEOSAT 1800Z (D5) IR
T43 1830 METEOSAT 1800Z (D6) IR
T44 1834 METEOSAT 1800Z (D7) IR
GE47 1840 GOES 1800Z TROPICAL W/E IR
GE48 1853 GOES 1800Z NW/SW QUADS IR
GE49 1905 GOES 1800Z NE/SE QUADS IR
T45 1922 METEOSAT 1900Z(E1)MOIST CHNL
T46 1926 METEOSAT 1900Z(E2)MOIST CHNL
T47 1930 METEOSAT 1900Z(E3)MOIST CHNL
T48 1934 METEOSAT 1900Z(E4)MOIST CHNL
GE50 1940 TBUS NOAA-6
GE51 1945 GOES 1800Z TROPICAL W/E VIS
GE52 1958 GOES 1800Z NW/SW QUADS VIS
GE53 2010 GOES 1800Z NE/SE QUADS VIS
GE54 2025 NOAA-9(K) 130E-60E MERC VIS, 70E-00W MERC VIS
GE55 2040 NOAA-9(L)40W-110W MERC VIR, 100W-170W MERC NIR
T49 2110 METEOSAT 2100Z (D2) IR
T50 2114 METEOSAT 2110Z (D1) IR
T51 2118 METEOSAT 2100Z (D3) IR
T52 2122 METEOSAT 2100Z (D4) IR
T53 2126 METEOSAT 2100Z (D5) IR
T54 2130 METEOSAT 2100Z (D6) IR
T55 2134 METEOSAT 2100Z (D7) IR
GE56 2208 GOES 2100Z TROPICAL W/E VIS
GE57 2223 GOES 2100Z NW/SW QUADS VIS
GE58 2236 GOES 2100Z TROPICAL W/E IR
GE59 2251 GOES 2100Z NW/SW QUADS IR
GE60 2303 GOES 2100Z NE/SE QUADS IR
GE61 2315 GOES 2100 TROPICAL W/E VIS
GE62 2342 NOAA-9(M)10E-80W NH/SH POLAR VIS
GE63 2358 NOAA-9(I)170W-100E, 100E-10E SH POLAR DIR

GOES-Central Wefax Broadcast Schedule

(Revised October 23, 1985)

Chart #	Time (Z)	Product
T01	0010	METEOSAT 0000Z (D2) IR
T02	0014	METEOSAT 0000Z (D1) IR
T03	0018	METEOSAT 0000Z (D3) IR
T04	0022	METEOSAT 0000Z (D4) IR
T05	0026	METEOSAT 0000Z (D5) IR
T06	0030	METEOSAT 0000Z (D6) IR
T07	0034	METEOSAT 0000Z (D7) IR
GC01	0038	NOAA-9(L)40W-110W MERC NIR, 100W-70W MERC VIS
GC02	0050	NOAA-9(N)10E-80W NH POLAR DIR, 100E-10E NH POLAR DIR
GC03	0102	NOAA-9(P)170W-100E NH/SH POLAR NIR
GC04	0117	GOES 0000Z TROPICAL W/E IR
GC05	0130	GOES 0000Z NW/SW QUADS IR
GC06	0142	GOES 0000Z NE/SE QUADS IR
GC07	0157	GOES 0000Z TROPICAL W/E VIS
GC08	0210	GOES 0000Z NW/SW QUADS VIS
GC09	0222	GOES 0000Z NE/SE QUADS VIS
GC10	0237	GOES 2330Z TROPICAL W/E (MOIST CHNL)
GC11	0250	NOAA-9(Q)10E-60W MERC VIS,
WO64	0300	SIG WEATHER PROG FL250-600 VT12Z NT-24
WO65	0305	SIG WEATHER PROG FL250-600 VT12Z EU-6
T08	0310	METEOSAT 0300Z (D2) IR
T09	0314	METEOSAT 0300Z (D1) IR
T10	0318	METEOSAT 0300Z (D3) IR
T11	0322	METEOSAT 0300Z (D4) IR
T12	0326	METEOSAT 0300Z (D5) IR
T13	0330	METEOSAT 0300Z (D6) IR
T14	0334	METEOSAT 0300Z (D7) IR
GC12	0338	GOES 2330Z NW/SW QUADS (MOIST CHNL)
GC13	0350	GOES 2330Z NE/SE QUADS

(MOIST CHNL)
GC14 0407 GOES 0300Z TROPICAL W/E VIS
GC15 0420 GOES 0300Z NW/SW QUADS IR
GC16 0432 GOES 0300Z NE/SE QUADS IR
GC17 0447 GOES 0300Z TROPICAL W/E IR
GC18 0500 NOAA-9(A)80W-170W NH/SH POLAR VIS
GC19 0512 NOAA-9(B)100E-10E NH/SH POLAR NIR
GC20 0525 NOAA-9(C)100W-170W, 40W-110W MERC VIS
GC21 0540 NOAA-9(D)80E-10E, 140E-70E MERC NIR
GC22 0550 NOAA-9(S)80W-170W NH POLAR DIR ECLIPSE
GC23 0820 NOAA-9(T)SOUTH POLE DIR, 80W-170W SH POLAR DIR
W014 0835 24HR 250 HEIGHT/ISOTACHS PN26
W015 0840 24HR 250 HEIGHT/ISOTACHS NT21
W016 0845 24HR TROP PRESSURE/VWS PN26
W017 0850 24HR TROP PRESSURE/VWS NT21
W018 0855 24HR 300 STREAM/ISOTACHS PS2
W019 0900 24HR 300 STREAM/ISOTACHS SA8
W020 0905 24HR 200 STREAM/ISOTACHS PS2
W021 0910 24HR 200 STREAM/ISOTACHS SA8
T15 0922 METEOSAT 0900Z (D1) IR
T16 0926 METEOSAT 0900Z (D3) IR
T17 0930 METEOSAT 0900Z (D4) IR
W022 0935 24HR 500 STREAM/ISOTACHS PS2
W023 0940 24HR 500 STREAM/ISOTACHS SA8
W010 0945 24HR 850 HEIGHT/ISOTACHS PN26
W011 0950 24HR 850 HEIGHT/ISOTACHS NT21
W012 0955 24HR 700 HEIGHT/ISOTACHS PN26
W013 1000 24HR 700 HEIGHT/ISOTACHS NT21
W100 1005 00Z MSL PRES/1000-500 THK PN26
W101 1010 00Z MSL PRES/1000-500 THK NT21
W102 1015 24HR MSL PRES/1000-500 THK PN26
W103 1020 24HR MSL PRES/1000-500 THK NT21
W104 1025 48HR MSL PRES/1000-500 THK PN26
W105 1030 48HR MSL PRES/1000-500 THK NT21
GC24 1035 GOES 0900Z TROPICAL W/E IR
GC25 1050 GOES 0900Z NW/SW QUADS IR
GC26 1105 GOES 0900Z NE/SE QUADS IR
GC27 1115 TBUS NOAA-9
GC28 1120 OPERATIONAL MESSAGE
T18* 1130 ICE CHART-NORTHERN LIMIT (PART 1)
T19* 1140 ICE CHART-NORTHERN LIMIT (PART 2)
T20 1154 METEOSAT 1130Z(E1)MOIST CHNL
T21 1158 METEOSAT 1130Z(E2)MOIST CHNL
T22 1202 METEOSAT 1130Z(E2)MOIST CHNL
T23 1210 METEOSAT 1200Z (D2) IR
T24 1214 METEOSAT 1200Z (C1) VIS
T25 1218 METEOSAT 1200Z (C3) VIS
T26 1222 METEOSAT 1200Z (D1) IR
T27 1226 METEOSAT 1200Z (D3) IR
T28 1230 METEOSAT 1200Z (D4) IR
GC29 1235 GOES 1130 TROPICAL W/E (MOIST CHNL)
GC30 1248 GOES 1130Z NW/SW QUADS (MOIST CHNL)
GC31 1300 GOES 1130Z NE/SE QUADS (MOIST CHNL)
GC32 1315 GOES 1200Z TROPICAL W/E VIS
GC33 1327 GOES 1200Z NE/SE QUADS VIS
GC34 1342 GOES 1200Z TROPICAL W/E IR
GC35 1355 GOES 1200Z NW/SW QUADS IR
GC36 1407 GOES 1200Z NE/SE QUADS IR
W066 1420 SIG WEA PROG FL250-600 VT00Z NT-24
W067 1425 SIG WEA PROG FL250-600 VT 00Z EU-6
GC37 1430 TBUS NOAA 8 (SEE MSG NOAA-8 STATUS CHANGE)
GC38 1435 NOAA-9(H)170W-120E MERC VIS, 203-50W MERC NIR
GC39 1450 NOAA-9(G)10E-80W NH/SH POLAR NIR
GC40 1840 GOES 1800Z TOPICAL W/E IR
GC41 1853 GOES 1800Z NW/SW QUADS IR
GC42 1905 GOES 1800Z NE/SE QUADS IR
T29 1922 METEOSAT 1900Z(E1)MOIST CHNL
T30 1926 METEOSAT 1900Z(E2)MOIST CHNL
T31 1930 METEOSAT 1900Z(E3)MOIST CHNL
T32 1934 METEOSAT 1900Z(E4)MOIST CHNL
GC43 1940 TBUS NOAA-6
GC44 1945 GOES 1800Z TROPICAL W/E VIS
GC45 1958 GOES 1800Z NW/SW QUADS VIS
GC46 2010 GOES 1800Z NE/SE QUADS VIS
W054 2020 24HR 250 HEIGHT/ISOTACHS PN26
W055 2025 24HR 250 HEIGHT/ISOTACHS NT21
W056 2030 24HR TROP PRESSURE/VWS PN26
W057 2035 24HR TROP PRESSURE/VWS NT21
W058 2040 24HR 300 STREAM/ISOTACHS PS2
W059 2045 24HR 300 STREAM/ISOTACHS

Chart #	Time (Z)	Product
		SA8
W060	2050	24HR 200 STREAM/ISOTACHS PS2
W061	2055	24HR 200 STREAM/ISOTACHS SA8
W062	2100	24HR 500 STREAM/ISOTACHS PS2
W063	2105	24HR 500 STREAM/ISOTACHS SA8
W050	2110	24HR 850 HEIGHT/ISOTACHS PN26
W051	2115	W24HR 850 HEIGHT/ISOTACHS NT21
W052	2120	24HR 700 HEIGHT/ISOTACHS PN26
W053	2125	24HR 700 HEIGHT/ISOTACHS NT21
W150	2130	00Z MSL PRES/1000-500 THK PN26
W151	2135	00Z MSL PRES/1000-500 THK NT21
W152	2140	24HR MSL PRES/1000-500 THK PN26
W153	2145	24HR MSL PRES/1000-500 THK NT21
W154	2150	48HR MSL PRES/1000-500 THK PN26
W155	2155	48HR MSL PRES/1000-500 THK NT21
GC47	2208	GOES 2100Z TROPICAL W/E VIS
GC48	2223	GOES 2100Z NW/SW QUADS VIS
GC49	2236	GOES 2100Z TROPICAL W/E IR
GC50	2251	GOES 2100Z NW/SW QUADS IR
GC51	2303	GOES 2100Z NE/SE QUADS IR
GC52	2315	GOES 2100Z NE/SE QUADS VIS
GC53	2330	NOAA-9(K)130E-60E MERC VIS, 70E-00W MERC VIS
GC54	2342	NOAA-9(M)10E-80W NH/SH POLAR VIS
GC55	2358	NOAA-9(I)170W-100E, 100E-10E SH POLAR DIR

GOES-Central Areas

PN-6		PS-2
21N/169W	87N/10E	70W TO 150W
1S/1	17N/78W	26N TO 28S

NT-21	SA-8	
73N/155W	30N/4E	0 TO 80W
14N/89W	1N/43W	26N TO 48S

NT-24	EU-6
28N/145W	78N/135E
78N/135E	25N/54.5E
2.5N/83.5W	16.5N/45W
16.5N/45W	1.5N/5W

GOES-West Wefax Broadcast Schedule

(Revised October 23, 1985)

Chart #	Time (Z)	Product
W48W	0010	SIG WEA 400-70MB VT 12Z PN34
W483	0015	SIG WEA 400-70MB VT 12Z PN35
W273	0020	24HR 500 HEIGHT/ISOTACHS PN26
W473	0025	24HR 500 HEIGHT/ISOTACHS PN27
W480	0030	24HR 1000 STREAM/ISOTACHS PS3
W481	0035	24HR 1000 STREAM/ISOTACHS PS2
W484	0040	24HR 700 STREAM/ISOTACHS PS3
W485	0045	24HR 700 STREAM/ISOTACHS PS2
W486	0050	48HR 1000 STREAM/ISOTACHS PS3
W487	0055	48HR 1000 STREAM/ISOTACHS PS2
GW01	0102	NOAA-9(P)170W-100E NH/SH POLAR NIR
GW02	0117	GOES 0000Z TROPICAL W/E IR
GW03	0130	GOES 0000Z NW/SW QUADS IR
GW04	0142	GOES 0000Z NE/SE QUADS IR
GW05	0157	GOES-1 2345Z TROPICAL W/E VIS
GW06	0210	GOES-1 2345Z NW/SW QUADS VIS
GW07	0222	GOES-1 2345Z NE/SE QUADS VIS
GW08	0237	GOES 2330Z TROPICAL W/E (MOIST CHNL)
GW09	0250	NOAA-9(Q)10E-60W MERC VIS, 160W-130E MERC NIR
T001	0302	ICE CHART-NORTHERN LIMIT (PART 1)
T002*	0312	ICE CHART-NORTHERN LIMIT (PART 2)
GW10	0327	NOAA-9(R)100E-10E SH POLAR VIS, 10-80W SH POLAR DIR
GW11	0338	GOES 2330Z NW/SW QUADS (MOIST CHNL)
GW12	0350	GOES 2330Z NE/SE QUADS (MOIST CHNL)
GW13	0407	OPEN
GW14	0420	OPEN
GW15	0432	OPEN
GW16	0447	GOES 0300Z TROPICAL W/E IR
GW17	0500	GOES 0300Z NW/SW QUADS IR
GW18	0512	GOES 0300Z NE/SE QUADS IR
GW19	0525	NOAA-9(C)100W-170W, 40W-110 MERC VIS
GW20	0540	NOAA-9(D)80E-10E, 140E-70E MERC NIR
GW21	0600	NOAA-9(A)802-170W NH/SH POLAR VIS
GW22	0615	NOAA-9(B)100E-10E NH/SH POLAR NIR

GW23 0635 NOAA-9(S)NORTH POLE DIR, 80W-170W NH DIR
W410 0650 SIG WEA 400-70MB VT18Z PN34
W211 0655 SIG WEA 400-70MB VT18Z PN35
W300 0700 ANAL 1000 STREAMS/WINDS SFC-1
W301 0705 ANAL 100 STREAMS/WINDS SFC-2
W302 0710 ANAL 1000 STREAM/WINDS SFC-3
W303 0715 ANAL 1000 STREAMS/WINDS SFC-4
W304 0720 ANAL 1000 STREAMS/WINDS SFC-5
W305 0725 ANAL 1000 STREAMS/WINDS SFC-6
W307 0730 ANAL 1000 STREAMS/WINDS SFC-8
W308 0740 ANAL 1000 STREAMS/WINDS SFC-9
W309 0745 ANAL 1000 STREAMS/WINDS SFC-10
W310 0750 ANAL 1000 STREAMS/WINDS SFC-11
W311 0755 ANAL 1000 STREAMS/WINDS SFC-12
W210 0800 ANAL MSL PRES/1000-500 THK PN26
ECLIPSE
W411 1005 24HR MSL PRES/1000-500 THK PN27
W212 1010 24HR 250 HEIGHT/ISOTACHS PN26
W412 1015 24HR 250 HEIGHT/ISOTACHS PN27
W213 1020 24HR TROP PRESSURE/VWS PN26
W413 1025 24HR TROP PRESSURE/VWS PN27
W214 1030 24HR 300 STREAMS/ISOTACHS PS3
W215 1035 24HR 300 STREAMS/ISOTACHS PS2
W216 1040 24HR 200 STREAMS/ISOTACHS PS3
W217 1045 24HR 200 STREAMS/ISOTACHS PS2
W218 1050 24HR 500 STREAMS/ISOTACHS PS3
W219 1055 24HR 500 STREAMS/ISOTACHS PS2
W220 1100 24HR 850 HEIGHTS/ISOTACHS PN26
W420 1105 24HR 850 HEIGHTS/ISOTACHS PN27
W221 1110 24HR 700 HEIGHTS/ISOTACHS PN27
GW24 1115 TBUS NOAA-9
W421 1120 24HR 700 HEIGHTS/ISOTACHS PN26
W222 1125 48HR MSL PRES/1000-500 THK PN26
W422 1130 48HR MSL PRES/1000-500 THK PN27
W223 1135 24HR 500 HEIGHTS/ISOTACHS PN26
W423 1140 24HR 500 HEIGHTS/ISOTACHS PN27
W430 1145 24HR 1000 STREAMS/ISOTACHS PS-3
W431 1150 24HR 1000 STREAMS/ISOTACHS PS2
W432 1155 24HR 850 STREAMS/ISOTACHS PS3
W433 1200 24HR 850 STREAMS/ISOTACHS PS2
W434 1205 24HR 700 STREAMS/ISOTACHS PS3
W435 1210 24HR 700 STREAMS/ISOTACHS PS2
W440 1215 SIG WEA 400-70MB VT00Z PN34
W441 1220 SIG WEA 400-70MB VT00Z PN35
GW25 1225 OPERATIONAL MESSAGE
GW26 1235 GOES 1130Z TROPICAL W/E (MOIST CHNL)
GW27 1247 GOES 1130Z NW/SW QUADS (MOIST CHNL)
GW28 1302 GOES 1130Z NE/SE QUADS (MOIST CHNL)
GW29 1315 NOAA-9(T)80W-170W SH POLAR DIR
GW30 1327 GOES 1200Z NE/SE QUADS VIS
GW31 1342 GOES 1200Z TROPICAL W/E IR
GW32 1355 GOES 1200Z NW/SW QUADS IR
GW33 1407 GOES 1200Z NE/SE QUADS IR
GW34 1420 GOES 1200Z TROPICAL W/E VIS
GW35 1430 TBUS NOAA 8
GW36 1435 NOAA-9(H)170W-120E MERC VIS, 20E-50W MERC NIR
GW37 1450 NOAA-9(G)10E-80W NH/SH POLAR NIR
W436 1505 48HR 1000 STREAM/ISOTACH PS-3
W437 1510 48HR 1000 STREAM/ISOTACH PS-2
T003 1518 METEOSAT 1200Z (C3) VIS
T004 1522 METEOSAT 1200Z (D1) IR
GW38 1530 NOAA-7(E)170W-100E NH/SH POLAR VIS
GW39 1545 GOES 1500Z TROPICAL W/E IR
GW40 1558 GOES 1500Z NW/SW QUADS IR
GW41 1610 GOES 1500Z NE/SE QUADS IR
GW42 1625 GOES 1500Z TROPICAL W/E VIS
GW43 1638 GOES 1500Z NE/SE QUADS VIS
GW44 1650 NOAA-9(J)80W-170W NH/SH POLAR NIR
GW45 1705 NOAA-9(F) 170W-100E NH POLAR DIR, 100E-10E NH POLAR VIS
W263 1735 SIG WEA 400-70MB VT06Z PN34
W463 1740 SIG WEA 400-70MB VT06Z PN35
W350 1735 ANAL 1000 STREAMS/WINDS SFC1
W351 1750 ANAL 1000 STREAMS/WINDS SFC2
W352 1755 ANAL 1000 STREAMS/WINDS

		SFC3
W353	1800	ANAL 1000 STREAMS/WINDS SFC4
W354	1820	ANAL 1000 STREAMS/WINDS SFC5
W355	1825	ANAL 1000 STREAMS/WINDS SFC6
W356	1830	ANAL 1000 STREAMS/WINDS SFC7
W357	1835	ANAL 1000 STREAMS/WINDS SFC8
W358	1840	ANAL 1000 STREAMS/WINDS SFC9
W359	1845	ANAL 1000 STREAMS/WINDS SFC10
W360	1850	ANAL 1000 STREAMS/WINDS SFC11
W361	1855	ANAL 1000 STREAMS/WINDS SFC12
W260	1900	ANAL MSL PRES/1000-500 THK PN26
W460	1905	ANAL MSL PRES/1000-500 THK PN27
W261	1910	24HR MSL PRES/1000-500 THK PN26
W461	1915	24HR MSL PRES/1000-500 THK PN27
W262	1920	24HR 250 HEIGHT/ISOTACHS PN26
W462	1925	24HR 250 HEIGHT/ISOTACHS PN27
W264	1930	24HR 300 STREAMS/ISOTACHS PS3
W265	1935	24HR 300 STREAMS/ISOTACHS PS2
GW46	1940	TBUS NOAA-6
W266	1945	24HR 200 STREAMS/ISOTACHS PS3
GW47	1950	GOES-1 1845Z TROPICAL W/E VIS
GW48	2003	GOES-1 1845Z NW/SW QUADS VIS
GW49	2015	GOES-1 1845Z NE/SE QUADS VIS
GW50	2030	GOES 1800Z TROPICAL W/E IR
GW51	2043	GOES 1800Z NW/SW QUADS IR
GW52	2055	GOES 1800Z NE/SE QUADS IR
GW53	2105	NOAA-9(L)40W-110W MERC NIR, 100W-170W MERC NIR
GW54	2117	GOES 2030Z TROPICAL W/E VIS
GW55	2132	GOES 2030Z NW/SW QUADS VIS
W267	2150	24HR 200 STREAMS/ISOTACHS PS2
W268	2155	24HR 500 STREAMS/ISOTACHS PS3
W269	2200	24HR 500 STREAMS/ISOTACHS PS2
W270	2205	24HR 850 HEIGHTS/ISOTACHS PN26
W470	2210	24HR 850 HEIGHTS/ISOTACHS PN27
W271	2215	24HR 700 HEIGHTS/ISOTACHS PN26
W471	2220	24HR 700 HEIGHTS/ISOTACHS PN27
W272	2225	48HR MSL PRES/1000-500 THK PN26
W474	2230	48HR MSL PRES/1000-500 THK PN27
GW56	2236	GOES 2100Z TROPICAL W/E IR
GW57	2251	GOES 2100Z NW/SW QUADS IR
GW58	2303	GOES 2100Z NE/SE QUADS IR
GW59	2315	NOAA-9(N)10E-80W POLAR DIR, 100E-10E NH POLAR DIR
GW60	2330	NOAA-9(K)130E-60E MERC VIS, 70E-00W MERC VIS
GW60	2342	NOAA-9(M)10E-80W NH/SH POLAR VIS
GW61	2358	NOAA-9(I)170W-100E, 1003-10E SH POLAR IR

GOES West Areas

PS-2	PS-3
70W TO 150W	140W TO 140E
26N TO 48S	26N TO 48S

PN-26		PN-27	
19N/169W	87N/10E	19N/168W	19N/103E
1S/122W	17N/78W	87N/160W	1S/146E

PN-34	PN-35
170E TO 110W	50N TO 20S
50N TO 20	100E TO 180W
180W TO 130W (SH)	45N TO EQUATOR

SFC-1	50N TO 17N	90W TO 132W
SFC-2	50N TO 17N	130W TO 172W
SFC-3	50N TO 17N	170W TO 148E
SFC-4	50N TO 17N	150E TO 118E
SFC-5	20N TO 20S	90W TO 132W
SFC-6	20N TO 20S	130W TO 172W
SFC-7	20N TO 20S	170W TO 148E
SFC-8	20N TO 20S	150E TO 118E
SFC-9	17S TO 50S	90W TO 132W
SFC-10	17S TO 50S	130W TO 172W
SFC-11	17S TO 50S	170W TO 148E
SFC-12	17S TO 50S	150E TO 118E

Appendix E

FREQUENCY CROSS REFERENCE

(Mode designators at end of this appendix; other abbreviations such as country and agency appear in the Glossary)

Frequency (MHz)	Country/ Agency	Use	Mode	Notes
1.500	US	Nongeostationary		Explorer 20
1.610	NASA	Ground Network	AM	Kennedy Space Center grandstand shuttle audio
2.000	US	Nongeostationary		Explorer 20
2.0175	NASA	Ground Network		Hampton, VA
2.2074	NASA	Ground Network		New Orleans, LA
2.357	NASA	Ground Network		Malabar
2.383	NASA	Ground Network		New Orleans, LA
2.4314	NASA	Ground Network		Mobile, AL
2.576	USN	Ground Network		Antigua
2.622	USN	Ground Network		Malabar, Booster recovery vessels
2.624	NASA	Ground Network		Houston, Texas
2.638	USN	Ground Network		Malabar
2.678		Ground Network	USB	Range Control
2.717	USN	Ground Network		Malabar
2.744	USN	Ground Network		Antigua
2.764	USN	Ground Network		Malabar
2.800	USN	Ground Network		Malabar
2.8215	USN	Ground Network		Malabar
2.8375	USN	Ground Network		Malabar
2.850	US	Nongeostationary		Explorer 20
3.023	USCG	Ground Network		Search & Rescue, primary
3.039	USN	Ground Network		Antigua
3.099	USN	Ground Network		Antigua
3.120	USN	Ground Network		Malabar
3.130	USN	Ground Network		Malabar
3.162	USAF	Ground Network	USB	Pacific Missile Range
3.187	USN	Ground Network	USB	Malabar, Booster recovery vessels
3.219	USN	Ground Network		Antigua
3.365	USN	Ground Network		Malabar
3.380	NASA	Ground Network		Nationwide Tracking Net
3.385	NASA	Ground Network		Nationwide Tracking Net
3.395	NASA	Ground Network		Nationwide Tracking Net
3.720	US	Nongeostationary		Explorer 20
3.860	ARO	Ground Network	LSB	WA3NAN shuttle audio
3.940	ARO	Ground Network	LSB	W5RRR shuttle audio (occ.)
4.376	USCG	Ground Network	USB	Search & Rescue ops
4.486	USAF	Ground Network	USB	Pacific Missile Range
4.500	USAF	Ground Network		Ascension Island
4.510	USN	Ground Network	USB	Booster recovery vessels
4.521	USN	Ground Network		Malabar
4.6835	NASA	Ground Network		Fairbanks, Alaska
4.704	USN	Ground Network	USB	Malabar
4.714	USN	Ground Network		Antigua

Frequency (MHz)	Country/ Agency	Use	Mode	Notes
4.760	USAF	Ground Network	USB	Pacific Missile Range
4.765	USN	Ground Network		Malabar
4.825	USN	Ground Network		Malabar
4.8565	USN	Ground Network	USB	Malabar, Cape Radio/Cape Leader
4.860	USN	Ground Network		Malabar
4.864	USN	Ground Network		Malabar
4.900	USAF	Ground Network		Malabar
4.992	USN	Ground Network	USB	Malabar, Cape Radio
5.060	USN	Ground Network		Antigua
5.180	USN	Ground Network		Malabar
5.1885	USN	Ground Network	USB	Malabar, tracking ships
5.246	USN	Ground Network		Malabar
5.297	USAF	Ground Network	USB	NORAD channel
5.350	USN	Ground Network	USB	Malabar, support aircraft
5.436	USN	Ground Network		Antigua
5.470	US	Nongeostationary		Explorer 20
5.4705	NASA	Ground Network		Fairbanks, Alaska
5.680	USN	Ground Network	USB	Launch support ships
5.692	USCG	Ground Network	USB	SAR helicopters
5.696	USCG	Ground Network	USB	SAR helicopters
5.700	USAF	Ground Network	USB	Pacific Missile Range
5.710	USAF	Ground Network	USB	Malabar
5.7195	USAF	Ground Network	USB	Malabar, support aircraft
5.724	USN	Ground Network	Data	Antigua
5.775	USAF	Ground Network		Ascension Island
5.810	USAF	Ground Network	USB	Malabar, Pri night calling freq ETR
5.822	USAF	Ground Network		Ascension Island
	USAF	Ground Network	USB	Pacific Missile Range
6.6945	USAF	Ground Network	USB	Malabar, support aircraft
6.701	USN	Ground Network	USB	Antigua
6.7095	USAF	Ground Network	USB	Malabar, support aircraft
6.720	USN	Ground Network	USB	SAR primary Atlantic
6.723	USN	Ground Network		Malabar
6.7545	USN	Ground Network		Malabar
6.810	USN	Ground Network		Antigua
6.820	USAF	Ground Network		Pacific Missile Range
6.880	USN	Ground Network	USB	Antigua
6.889	USAF	Ground Network	USB	Pacific Missile Range
6.897	USN	Ground Network	USB	Malabar, Cape Radio
6.919	USN	Ground Network		Antigua
6.937	USN	Ground Network	USB	Malabar, Cape Radio
6.9825	NASA	Ground Network	USB	Nationwide Tracking Net/Emergency radio net
7.050	ARS	Nongeostationary	CW	Oscar 9(UOSAT 1)/Oscar II(UOSAT 2)
7.185	ARO	Ground Network	LSB	WA3NAN-shuttle audio
7.220	US	Nongeostationary		Explorer 20
7.265	ARO	Ground Network	LSB	W5RRR-shuttle audio (occ.)
7.313	USAF	Ground Network		Ascension Island
	USN	Ground Network		Antigua
7.332	USAF	Ground Network		Ascension Island
7.355	USN	Ground Network		Malabar
7.387	USAF	Ground Network		Malabar
	USN	Ground Network		Malabar
7.412	USN	Ground Network	USB	Malabar, comms w/Bahamas
7.461	USN	Ground Network	USB	Malabar, Cape Radio/support aircraft
7.512	USN	Ground Network		Malabar
7.525	USN	Ground Network	USB	Malabar, booster recovery vessels
7.605	USAF	Ground Network		Ascension Island
7.676	USN	Ground Network	USB	Malabar, support aircraft
7.697	USAF	Ground Network		Ascension Island
7.705	USAF	Ground Network	USB	Pacific Missile Range
7.715	USAF	Ground Network		Ascension Island
7.742	USAF	Ground Network		Ascension Island
7.765	USN	Ground Network	USB	Malabar, booster recovery vessels

Frequency (MHz)	Country/ Agency	Use	Mode	Notes
7.804	USAF	Ground Network		Ascension Island
7.833	USN	Ground Network		Malabar
7.860	USN	Ground Network		Antigua
7.910	USAF	Ground Network	USB	Ascension Island
7.919	USN	Ground Network		Malabar
7.985	USN	Ground Network	Data	Malabar
8.035	URS	Nongeostationary	AM Voice	Soyuz 1(manned spacecraft)
8.7792	USAF	Ground Network	USB	Malabar
8.964	USN	Ground Network		Antigua
8.972	USN	Ground Network	USB	Antigua, Atlantic support
8.981	USAF	Ground Network	USB	Support aircraft
8.9825	USAF	Nongeostationary		OV1-17A
8.984	USCG	Ground Network	USB	SAR helicopters
8.991	USN	Ground Network		Antigua
9.0075	USAF	Ground Network	USB	Malabar, shuttle launch support, duplex w/11.2065
9.019	URS	Nongeostationary	AM Voice	Vostok 1(manned spacecraft)
9.022	USN	Ground Network		Malabar
9.023	USAF	Ground Network	USB	NORAD channel
9.029	USAF	Ground Network	USB	Pacific Missile Range
9.043	USN	Ground Network	USB	Malabar, support aircraft
9.115	USN	Ground Network	Data	Antigua
9.132	USN	Ground Network	USB	Malabar, support aircraft
9.138	USN	Ground Network		Antigua
9.170	USAF	Ground Network		Ascension Island
	USN	Ground Network		Antigua
9.212	USAF	Ground Network	USB	Pacific Missile Range
9.793	USAF	Ground Network	USB	NORAD channel
10.000	US-DOD	Nongeostationary		Orbis
10.004	USAF	Nongeostationary		OV2-5
10.013	URS	Nongeostationary	PDM TM's	Cosmos 138 (Soyuz precursor)
10.147	USAF	Ground Network		Ascension Island
10.159	USAF	Ground Network		Ascension Island
10.215	USAF	Ground Network		Ascension Island
10.230	USAF	Ground Network		Ascension Island
10.270	USN	Ground Network		Antigua
10.272	USAF	Ground Network	USB	Pacific Missile Range
10.301	USN	Ground Network		Antigua
10.305	USN	Ground Network	USB	Malabar, support aircraft
10.310	USN	Ground Network	LSB	Malabar, Cape Radio
10.327	USN	Ground Network		Antigua
10.352	USAF	Ground Network	USB	Pacific Missile Range
10.475	USN	Ground Network		Antigua
10.510	USAF	Ground Network	USB	Pacific Missile Range
10.660	USAF	Ground Network	USB	Pacific Missile Range
10.780	USN	Ground Network	USB	Malabar, Pri calling channel ETR
10.804	USAF	Ground Network	USB	Pacific Missile Range
10.850	USAF	Ground Network		Ascension Island
10.880	USAF	Ground Network		Ascension Island
11.104	USN	Ground Network	USB/LSB	Malabar, shuttle launch support
11.183	USN	Ground Network		Malabar
11.201	USCG	Ground Network	USB	SAR helicopters
11.2065	USAF	Ground Network	USB	Malabar, shuttle launch support, duplex w/9.0075
11.252	USN	Ground Network	USB	Malabar, support operations
11.407	USN	Ground Network	USB	Malabar, booster recovery vessels
11.414	USN	Ground Network	USB	Malabar, Cape Radio
11.441	USAF	Ground Network	USB	NORAD channel
11.510	USAF	Ground Network	USB	Pacific Missile Range
11.548	USN	Ground Network	USB	Malabar, Cape Radio
11.6225	USN	Ground Network	USB	Malabar, support ships
11.634	USAF	Ground Network		Ascension Island
11.984	USN	Ground Network		Antigua
11.988	USAF	Ground Network		Ascension Island

Frequency (MHz)	Country/ Agency	Use	Mode	Notes
12.107	USN	Ground Network		Malabar
12.160	USN	Ground Network		Antigua
12.277	USN	Ground Network		Malabar
12.287	USN	Ground Network		Antigua
13.1735	USAF	Ground Network	USB	Malabar, support aircraft
13.204	USN	Ground Network		Malabar
13.210	USN	Ground Network		Malabar
13.218	USAF	Ground Network	USB	Pacific Missile Range
13.2285	USAF	Ground Network	LSB	Malabar, support aircraft
13.237	USN	Ground Network	Data	Malabar
13.2525	USAF	Nongeostationary		OV1-17A
13.468	USAF	Ground Network		Ascension Island
13.495	USAF	Ground Network	Data	Ascension Island
13.600	USN	Ground Network		Malabar
13.635	USN	Ground Network		Malabar
13.735	USN	Ground Network		Malabar
13.742	USAF	Ground Network	LSB	Ascension Island
13.756	USAF	Ground Network	USB	Pacific Missile Range, ground-duplex
13.878	USN	Ground Network	USB	Malabar, support aircraft
13.900	USAF	Ground Network	USB	Pacific Missile Range, duplex
14.002	ARS	Nongeostationary	CW	Oscar 9(UOSAT 1)/Oscar II(UOSAT 2)
14.280	ARO	Ground Network	USB	WA3NAN shuttle audio
14.295	ARO	Ground Network	USB	W5RRR shuttle audio (occ.)
14.397	USN	Ground Network		Malabar
14.432	USN	Ground Network		Antigua
14.452	NASA	Ground Network		Wallops Island
14.455	NSAS	Ground Network		Nationwide tracking net/vessels
14.497	USAF	Ground Network		Ascension Island
14.547	USAF	Ground Network		Malabar
14.585	USAF	Ground Network		Ascension Island
136.100	NASA	Nongeostationary		Explorer 15
136.110	NASA	Nongeostationary		Explorer 35
136.111	US	Nongeostationary		Explorer 18 (IMP 1)
136.112	FRANCE/	Nongeostationary		Ayame 2/ETS4/ATS 6/ECS-1,2/
	JAPAN/US	Geostationary		
	JAPAN	Nongeostationary		MOS-1/MABES
136.1121	ESA	Geostationary		ETS-2
136.125	NASA	Nongeostationary		Explorer 28 (IMP-3), 49
136.140	US/ITALY	Geostationary		Sirio Comsat/ Relay 1
136.141	US	Nongeostationary		Relay 2/ Explorer 34 (IMP-4)
136.142	US	Nongeostationary	TM	Explorer 21 (IMP-2)
136.145	NASA	Nongeostationary		Explorer series
136.159	JAPAN	Nongeostationary		Ohsumi-1
136.170	NASA	Nongeostationary	30F9	HCMM/Explorer 42
	US	Nongeostationary		Echo 2
	ESRO	Nongeostationary		Aurorae (ESRO-1A)/Boreas (ESRO-1B)
136.171	US	Nongeostationary		Explorer 22 (BE-2)
136.200	NASA	Nongeostationary	40F9	
	US	Nongeostationary		Cameo-1/Nimbus 2/OGO-1,2,3,5,6/ Injun-SR3/Explorer 13, 16
136.220	US	Nongeostationary		OAO-1
136.230	NASA	Geostationary		ATS-6
	US	Nongeostationary		Tiros 4,5,6,10/ESSA-1/SERT-2
136.231	US	Nongeostationary		Tiros 9
136.233	US	Nongeostationary		Tiros 8
136.234	US	Nongeostationary		Tiros 7
136.250	FRANCE	Nongeostationary		D 5B Castor
136.260	NASA/USAF	Nongeostationary	30F9	OAO 3,C(Copernicus)/ERS 20(OV5-3)
136.273	NASA	Nongeostationary		Explorer series satellites
136.275	US	Nongeostationary		Explorer 26
136.290	NASA	Nongeostationary		Explorer 40/Hawkeye/OSO-3,5,7
136.293	US	Nongeostationary		Explorer 25
136.300	NASA	Nongeostationary		SMS 1
136.319	USAF	Nongeostationary		Gravity Gradient 1 (GGSE-1)

Frequency (MHz)	Country/ Agency	Use	Mode	Notes
136.320	NASA/USAF	Nongeostationary	30F9/90F9	Explorer 17,32,46(MTS)/Ferret/ GEOS-3/PAC-A
136.34827	AUS	Nongeostationary		WRESAT-1
136.350	ESA/USAF/ US	Nongeostationary		FR1/D5 B2/CAS-A/Eole 1/SR 11B/ Explorer 20,38
136.350	F	Nongeostationary		Peole
136.380	NASA/USAF/ C	Geostationary	100F9/90F9	Weather satellites/ERS 27/OV5-6/ SMS/GOES/Explorer 44(Solradio)
	CANADA	Geostationary		Alouette 2
136.408	ESA	Nongeostationary		Ariel 1
136.410	CANADA/ITSO	Nongeostationary		ISIS 1/ISIS 2/Intelsat 1/Expl 37
	US	Nongeostationary		Pegasus 1,2,3
136.4138	ITSO	Nongeostationary		Intelsat 1 satellite
136.415	USAF	Nongeostationary		ERS-6 (TRS-3)
136.430	INDIA	Nongeostationary		Bhaskara 2
136.440	NASA	Nongeostationary	0A0	Intelsat/ERS 15/OAO 1,2,3/Early Bird/ OAO-C(Copernicus)/Pegasus 1
136.468	NASA	Geostationary	30F9	ATS 1,3,5/Syncom 2
136.470	US	Geostationary		Syncom 1,3/ATS-2,3,4,5
136.499	US	Nongeostationary		Nimbus-1
136.500	NASA	Nongeostationary	30F9	ATS series/Injun/SR 3/Nimbus 3-6
136.521	NASA	Nongeostationary		Explorer 37/OVS 9/Solrad 11-B
136.530	USAF/US	Nongeostationary		Explorer 30,37/Vela Hotel 8/OV 5-9/ Solrad 11B/ERS-18/Cannonball-2
	FRANCE	Nongeostationary		A-1
136.536	ITALY	Nongeostationary	TM	San Marco-1
136.557	ESA	Nongeostationary		Ariel 2
136.560	UK/FRG/US	Nongeostationary		UK 6(Ariel 6)/GRS-A(Azur)/ Explorer 17,32
	G	Nongeostationary		Ariel 3
	D	Nongeostationary		WIKA
136.563	USAF	Nongeostationary		Int'l Designator 1964-83C
136.590	CANADA	Nongeostationary		ISIS 1&2/Alouette-1&2
136.610	ESA	Nongeostationary		CAT-1,3/Ariane LO3/D-5A Pollux
136.620	ITALY/US/	Geostationary		Sirio 1(6W TT&C)/Explorer 39/Relay 1
	USAF	Nongeostationary		OV2-5
	NASA	Nongeostationary		TETR-4
136.621	US	Nongeostationary		Relay 2
136.630	FRANCE	Nongeostationary		Signe 3/Tounesol
136.650	USAF	Nongeostationary		OV5-5(ERS- 29)/SN 39/TRAAC/ Transit 5B5/Saturn SA-6
	ESRO	Nongeostationary		HEOS-A1
	NASA	Nongeostationary		Explorer-46(MTS)
136.651	USAF	Nongeostationary		SN-39,43
136.678	US	Nongeostationary		SAS 3/Explorer 53/GOES/SMS
136.680	US,NASA	Nongeostationary		Explorer 20,42,48(SAS-2),53/ Biosatellite 1,3
	ESRO	Nongeostationary		HEOS-A2
136.694	JAPAN	Nongeostationary		Shinsei/Jiki'ken telemetry
136.695	JAPAN	Nongeostationary		Denpa(REXS)
136.6952	JAPAN	Nongeostationary		Tansei 1
136.709	US	Nongeostationary		Explorer 24 (Air Density 1)
136.710	US	Nongeostationary		OSO-4,6
	E	Nongeostationary		Intasat-1
136.712	US	Nongeostationary		OGO-2
136.713	NASA	Nongeostationary		OSO-2 scientific satellites
136.725	JAPAN	Nongeostationary		Corsa B tracking/TM beacon/MS-T4/ Tansei-2(MS-T2),3(MST-3)TT&C/ Exos-1(Kyokko)/SRATS(Taiyo)
136.738	ITALY	Nongeostationary	TM	San Marco-1
136.739	USAF	Nongeostationary		Gravity Gradient 2
136.740	NASA/FRG	Nongeostationary	250F9	Explorer 27/ERS-A/San Marco-2/Azur
	ESA	Nongeostationary		D-2B Aura
	ITALY	Nongeostationary		San Marco-4

Frequency (MHz)	Country/ Agency	Use	Mode	Notes
136.768	USAF	Nongeostationary		ERS-17 (ORS-3)
136.770	C/NASA/USAF	Nongeostationary	30F9	
	US	Nongeostationary		ESSA 2-9/EGRS-7/ITOS-1/NOAA 1-9/ DOD weather sats/DSB channel
136.771	USAF	Nongeostationary		ERS-13
136.774	US	Nongeostationary		OSO-1
136.776	USAF	Nongeostationary		Gravity Gradient 3
136.797	NASA	Nongeostationary	30F9	Nimbus-4
136.800	NASA/USAF/	Nongeostationary	30F9	Expl 50/Dodge/EGRS 1,3,7,13/FR-1
	ESA	Geos		Solrad 1&2/Symphonie 2&3/ Transit 4B/Ferret 24
	JAPAN	Nongeostationary		JISS
136.801	USAF	Nongeostationary		Solrad 7B
136.804	USAF	Nongeostationary		EGRS-1 (Secor 1B)
136.8097	JAPAN	Nongeostationary		UME 1 & 2
136.81	JAPAN	Nongeostationary		ETS-1(Kiku)
136.830	NASA/USAF	Nongeostationary		ERS 28/EGRS 8 (SECOR 8)/Expl 29,45
136.840	USAF/US	Nongeostationary		Nimbus 4/EGRS 2,3,9/TOPO 1
136.857	US	Nongeostationary		Explorer 23
136.860	NASA	Nongeostationary		Landsat 2/Expl 13,16,25/IUE TETR 2/ ERS-21/OV5-4/RMS
	USAF	Nongeostationary		Cannonball-2
		Geostationary		GOES-2
136.870	US	Nongeostationary		Injun 3
136.886	USAF	Nongeostationary		Solrad 1 (GREB)
136.887	USAF	Nongeostationary		Solrad 7A
136.890	NASA/	Nongeostationary	30F9	Explorer 47/Solrad 6/GMS 1,2
	JAPAN	Geostationary		Geostationary satellite/GMS 1(7.4W)
	US	Nongeostationary		Pegasus 1,2,3
	ESRO	Nongeostationary		IRIS (ESRO II)
136.891	USAF	Nongeostationary		ERS-9
136.892	USAF	Nongeostationary		ERS-5(TRS-2)
136.919	US	Nongeostationary		Tiros 9
136.920	NASA/USAF	Nongeostationary		
	US	Nongeostationary		Tiros 4-10/ESSA-1/SERT-2/OSO-8
136.922	US	Nongeostationary		Tiros 7
136.923	US	Nongeostationary		Tiros 8
136.928	USAF	Nongeostationary		SERT 2
136.950	US	Nongeostationary		Tiros 8/Nimbus 1,2,3,4/Cos-B1
	ESRO	Nongeostationary		Aurorae(ESRO-1A)/Boreas(ESRO-1B)
136.980	CANADA	Nongeostationary		Alouette 1,2/Syncom 1,2,3
	FRANCE/US	Nongeostationary		Early Bird/D-1A/Diademe 1,2
136.992	US	Nongeostationary		Tiros 7
136.995	US	Nongeostationary		Saturn SA-5,SA-7
136.980	ITSO	Geostationary		Intelsat II F-4
137.020	USN/ESA	Nongeostationary	33F9	
		Geostationary		Symphonie
137.050	ESA/NASA/ USAF/G	Nongeostationary		OTS 2/(USAF) 71-100F Ferret/ Ariel 4
137.080	ESA/USAF	Geostationary		Meteosat
		Nongeostationary		71-100C(USAF) Ferret
137.100	NASA/UK	Geostationary		ATS series/UK-X4
137.110	US	Geostationary		ATS-6
137.120	URS	Nongeostationary		Meteor class 1 satellites
137.130	URS	Nongeostationary		Meteor class 1 satellites
137.140	NASA	Nongeostationary	30F9	U.S. weather satellites/Aurora-1
	ETSO	Geostationary	TM	Eutelsat I F1,2,3 beacon
137.150	URS /CZECH	Nongeostationary		Meteor class 1/Magion/Cosmos 1066
137.170	FRANCE	Nongeostationary		Marots/HCMM/Marecs-1
	NASA	Nongeostationary		Explorer 42
137.180	INDIA	Nongeostationary	AM/FM	Bhaskara 1,2
137.190	US	Geostationary		Goes-3
137.192	INDIA	Nongeostationary		Apple 1

Frequency (MHz)	Country/ Agency	Use	Mode	Notes
137.195	NASA	Geostationary	70F9	Geostationary weather sats
137.200	ESA	Geostationary		SMS-1 weather telemetry/ GOES(6W) 1,2/ Marecs 1
	ESA	Nongeostationary		ESRO-4
137.230	NASA/INDIA	Nongeostationary	90F9	Bhaskara 1/NOAA-61
	NASA	Nongeostationary		Explorer 51(AE-3),54(AE-4),55(AE-5)
137.260	US	Nongeostationary		OAO-A2
	G/NASA	Nongeostationary		Aeros-A
137.290	NASA	Nongeostationary		Explorer 38
	D/NASA	Nongeostationary		Heros-B
137.300	URS	Nongeostationary		Meteor weather satellites
137.350	NASA	Geostationary	30F9	ATS 2,3,4,5
137.380	USN	Nongeostationary		Timation 2 Navsat
137.400	URS /US	Nongeostationary		Meteor class 2-APT/Ocean resources/ SMS-2/Rohini-4(India)
137.400	URS	Nongeostationary	APT	Meteor 3 class/Cosmos 1602,1766,1500
137.410	USAF/US	Nongeostationary		USAF satellite 1969-82E/Explorer 30
137.420	INDIA	Nongeostationary		Rohini 3, 4
137.440	INDIA/US	Nongeostationary		Aryabhata/Bhaskari/Explorer 30
	USN	Nongeostationary		Solrad-11A
137.500	USAF/C	Nongeostationary	30F9	DOD Block weather satellites/ NOAA weather satellite APT/ESSA
	FRANCE	Nongeostationary		MIKU
137.530	FRANCE	Nongeostationary		SRET 2
137.560	UK	Nongeostationary		UK 6/Prospero
137.570	NASA	Nongeostationary	40F9	Explorer series
137.580	NASA	Nongeostationary	40F9	Explorer series
137.590	US	Nongeostationary		Explorer 37
137.620	USAF/C	Nongeostationary	30F9	DOD Block weather satellites/ NOAA APT channel/(TIROS-N)/ESSA
137.676	USN	Nongeostationary	0A0	P76-5 satellite
137.680	UK	Nongeostationary		UK-5(Ariel 5)
137.710	NASA	Nongeostationary		Explorer 44 (Solrad-10)
137.725	JAPAN	Nongeostationary		MS-T4 (Tansei 4)
137.740	USAF	Nongeostationary		Gravity Gradient 4
	ITALY	Nongeostationary		San Marco 3
	ESRO	Nongeostationary		TD-1A
137.770	USAF/C	Nongeostationary	30F9	DOD weather sats/NOAA (TIROS-N)
137.800	USAF	Nongeostationary		Solrad 11
137.850	URS	Nongeostationary	APT	Meteor class 2, class 3 wx satellite/ Interkosmos series (18)
137.860	NASA	Nongeostationary	90F9	Earth Resources telemetry channel
137.890	NASA	Nongeostationary		RMS
	US	Nongeostationary		ANS-1
137.920	NASA	Nongeostationary	30F9	Explorer 47, 50
137.950	CANADA	Nongeostationary		ISIS telemetry channel 1 & 2
	NASA	Nongeostationary		Explorer 45(SSS-1)
137.980	NASA	Nongeostationary	30F9	Explorer 50/Gravity Gradient 5
138.000	USAF	Nongeostationary		Hilat-1
138.350	USAF	Space		Unknown
138.360	USAF	Space		Unknown
138.375	USAF	Space		Unknown
139.056	US	Geostationary		ATS-6
140.056	US	Geostationary		ATS-6
141.056	US	Geostationary		ATS-6
142.400	URS	Nongeostationary		Military Salyut voice channel
142.417	URS	Nongeostationary	FM	Civilian Salyut 7 voice channel
142.600	URS	Nongeostationary	FM	Soyuz 19 voice channel
143.144	URS	Nongeostationary	FM	Soyuz voice channel
143.625	URS	Nongeostationary		Common Russian manned voice channel Vostok 1,2,3,4,5,6/Voskhod 1,2
143.825	URS	Nongeostationary		Military Salyut coded channel
144.050	ARS	Nongeostationary		OSCAR 5 telemetry beacon

Frequency (MHz)	Country/ Agency	Use	Mode	Notes
144.100	ARS	Ground Station		OSCAR 4 Uplink to 431.935 MHz
144.3-145.5	URS	Nongeostationary		Cosmos ASAT targets, "squiggers" every 4 kHz
144.983	ARS	Nongeostationary		OSCAR 1 & 2 beacon downlink
145.8-145.9	ARS	Ground Station		RS-1,2 uplink
145.810	ARS	Nongeostationary		OSCAR 10 Mode B beacon
145.825	ARS	Nongeostationary		OSCAR 9,11 telemetry beacon
145.825-145.977	ARS	Nongeostationary		OSCAR 10 Mode B downlink transponder
145.826	ARS	Ground Station		Radio Sputnik 5 transponder uplink
145.835	ARS	Ground Station		Radio Sputnik 7 transponder uplink
145.850	ARS	Nongeostationary		OSCAR telemetry (early shots)
145.875-145.925	ARS	Nongeostationary		Oscar 3 downlink transponder
145.950	ARS	Ground Station		OSCAR 7,8 Mode A uplink
145.900-146.000	ARS	Ground Station		OSCAR 6 uplink/OSCAR8 Mode J uplink
145.910-145.950	ARS	Ground Station		Radio Sputnik 5 & 6 uplink
145.925-145.975	ARS	Nongeostationary		OSCAR 7, Mode B downlink
145.950	ARS	Nongeostationary		OSCAR Phase 1 beacons
145.960-146.000	ARS	Ground Station		Radio Sputnik 7 & 8 uplinks
145.975	ARS	Nongeostationary		OSCAR 7 telemetry beacon
145.987	ARS	Nongeostationary		OSCAR 10 engineering beacon
148.250	Various	Ground Station		Telecommand uplink channel
	ESA	Ground Station		Cameo C&C
148.260	ESA	Ground Station		Sirio Command & Control
148.270	ESA	Ground Station		ECS satellite series C&C
148.300	US	Ground Station		ATS C&C; ISIS, UK series; SMS
148.560	C	Ground Station		NOAA Telecommand uplink (weather)
148.600	US	Ground Station		SMS-1 C&C
	US	Ground Station		Tiros-N C&C
148.980	NASA	Ground Station		Telecommand
149.000	US	Ground Station		Explorer series C&C
149.175	NASA	Ground Station		ATS satellites channel 1 uplink
149.195	NASA	Ground Station		ATS satellites channel 2 uplink
149.220	NASA	Ground Station		ATS satellites channel 3 uplink
149.245	NASA	Ground Station		ATS satellites channel 4 uplink
149.265	NASA	Ground Station		ATS satellites channel 5 uplink
149.282	C	Ground Station		Telecommand
149.300	ESA	Ground Station		Meteosat C&C
149.340	ETSO	Geostationary		Eutelsat-I F1,2,3 TT&C
149.480	NASA	Ground Station		Telecommand
149.500	US	Ground Station		OAO/HCMM/Nimbus 5-6 C&C
149.520	US	Ground Station		OSO uplink channel
149.600	US	Ground Station		Explorer 53 C&C
149.700	FRANCE	Nongeostationary		Diademe 1(D-1C),II(D-1D)
149.910	URS	Nongeostationary		Military Cosmos NAVSAT ch.
149.940	URS	Nongeostationary		Military Cosmos NAVSAT ch.
149.970	FRANCE	Nongeostationary		D-1A Diapason
	URS	Nongeostationary		Military Cosmos NAVSAT ch.
149.978	US	Nongeostationary		US NAVSAT maintenance ch.
149.981	USAF	Nongeostationary		Polar Bear
149.982	US	Nongeostationary		US Improved Transit NAVSAT ch.
149.988	US	Nongeostationary		US NAVSAT operational ch/Hilat-1
150.000	URS	Nongeostationary		Civilian Cosmos NAVSAT ch.
150.300	URS	Nongeostationary	PCM	Cosmos geodetic
150.012	US	Nongeostationary		Geosat downlink channel
	Unknown	Nongeostationary		Unknown carrier downlink
150.030	URS	Nongeostationary		Military Cosmos NAVSAT ch.
	US	Nongeostationary		Score Comsat
153.000	URS	Nongeostationary	CW TM	Cosmos 1151-3rd harmonic of 51.0

Frequency (MHz)	Country/ Agency	Use	Mode	Notes
153.420	URS	Nongeostationary	CW	Cosmos Heavy Elint, 3rd harmonic
153.480	URS	Nongeostationary	CW	Cosmos Heavy Elint, 3rd harmonic
153.600	URS	Nongeostationary	CW	Cosmos Store-Dump Comsat, 3rd harmnc
154.200	NASA/USAF	Ground Station		Telemetry telecommand channel
160.00	CHINA	Nongeostationary		China 9B
161.992	NASA	Nongeostationary		
162.000	NASA/USN	Nongeostationary	0A0	Scientific, geodetic satellites
165.000	URS	Nongeostationary	PPM-AM	Salyut 7
166.000	URS	Nongeostationary		Wideband Cosmos/Soyuz TM/Rorsat
179.985	CHINA	Nongeostationary	CW	China 17,19
180.000	CHINA	Nongeostationary		China 4-7 sat. telemetry, China 9C
183.538	URS	Space		Lunar probes-narrow band telemetry
183.600	URS	Space	TM	Lunik 1,2,3/Mars probes
192.000	URS	Nongeostationary	PPM-AM	Salyut 7
192.040	URS	Nongeostationary		Soyuz 19/Civilian Salyut telemetry
204.00	URS	Nongeostationary	CW TM	Cosmos 1151-4th harmonic of 51.0
204.560	URS	Nongeostationary	CW	Cosmos Heavy Elint; 4th harmonic
204.640	URS	Nongeostationary	CW	Cosmos Heavy Elint; 4th harmonic
204.800	URS	Nongeostationary	CW	Cosmos Store-Dump comsat; 4th harmnc
216.000	USN	Nongeostationary		Military geodetic sats/(SURCAL)
216.980	USN	Ground Station		NAVSPASUR space surveillance sys
225-400	USAF	Nongeostationary		NAVSTAR Global Burst Detection
225.0000	USAF	Unknown	500F9	Possible LES 8/9 224.75-225.250
225.025	USAF	Ground Station		Uplink to missile silos
225.700	US	Nongeostationary		Atlas-Centaur 4 (beacon)
226.200	US	Nongeostationary		Mercury-Atlas manned program
227.50	NASA	Nongeostationary		Apollo 6
228.5138	US	Nongeostationary		Apollo 4
230.400	US	Nongeostationary		Skylab 1/Mercury-Atlas manned prog/ Gemini capsules
230.9	NASA	Nongeostationary		Apollo 5
231.500	URS	Nongeostationary	FM TM	Cosmos photo recons
231.900	US	Nongeostationary		Skylab 1
232.000	URS	Nongeostationary	PPM-AM	Cosmos photo recon, 3rd generation
235.0000	US	Nongeostationary		TV xmit Tiros & ESSA series wx/Skylab
			TM	Atlas-Centaur 4
	FRANCE	Nongeostationary		MIKU
235.8500		Unknown; Down		WBFM reported here-new downlink?
237.000	US	Nongeostationary		Skylab 1
237.8000	US	Nongeostationary		Tiros 2,3,4 wx
	NASA	Nongeostationary		Apollo 5,6
240.00	USSR	Nongeostationary	FM TM	4th generation photo recon
240.20	US	Nongeostationary		Biosatellite 2
241.5	NASA	Nongeostationary		Apollo 5
243.695-243.7325	USAF	Nongeostationary		SDS AFSATCOM transponder
243.7200	USAF	Nongeostationary	25F9	AFSATCOM on SDS? (see next)
243.7275	USAF	Nongeostationary	70F9	SDS relay 243.7075-243.7325
	USAF	Geostationary		TACSAT-1
243.730	USAF	Geostationary		TACSAT-1
243.800	US	Nongeostationary	TM	Atlas-Centaur 4
243.8525-243.9225	USAF	Nongostationary		SDS AFSTACOM trasponder
243.8550	USN	Geostationary		Leasat Whiskey ch9 243.8425-243.8675
243.8600	USN	Geostationary		Leasat Whiskey ch10 243.8475-243.8725
243.8750	USN	Geostationary		Leasat Whiskey ch11 243.8625-243.8875
243.8875	USAF	Nongeostationary	70F9	SDS relay 243.8525-243.9225
	USAF	Geostationary		TACSAT-1
243.9000	USN	Geostationary		Leasat Whiskey ch12 243.8875-243.9125
243.9100	USN	Geostationary		Leasat Whiskey ch13 243.8975-243.9225
243.9450	USN	Geostationary	25F9	Alpha ch12 243.9325-243.9575
243.9550	USN	Geostationary	25F9	Alpha ch13/Xray ch.9 243.9425-243.9675
243.9600	USN	Geostationary	25F9	Alpha ch14/Xray ch10 243.9475-243.9725
243.9650	USN	Geostationary	25F9	Alpha ch15 243.9525-243.9775

Frequency (MHz)	Country/ Agency	Use	Mode	Notes
243.9700	USN	Geostationary	25F9	Alpha ch16 243.9575-243.9825
243.9750	USN	Geostationary	25F9	Alpha ch17/Xray ch11 243.9625-243.9875
243.9800	USN	Geostationary	25F9	Alpha ch18 243.9675-243.9925
243.9850	USN	Geostationary	25F9	Alpha ch19 243.9735-243.9985
243.9900	USN	Geostationary	25F9	Alpha ch20 243.9775-244.0025
243.9950	USN	Geostationary	25F9	Alpha ch21 243.9825-244.0075
244.0000	USN	Geostationary	25F9	Alpha ch.22/Xray ch.12 243.9875-244.0125
244.0100	USN	Geostationary	25F9	Alpha ch.23/Xray ch.13 243.9975-244.0225
244.0450	USN	Geostationary	25F9	Bravo ch12 244.0325-244.0575
244.0550	USN	Geostationary	25F9	Bravo ch.13/Yankee ch.9 244.0425-244.0675
244.0600	USN	Geostationary	25F9	Bravo ch.14/Yankee ch.10 244.0475-244.0725
244.0650	USN	Geostationary	25F9	Bravo ch15 244.0525-244.0775
244.0700	USN	Geostationary	25F9	Bravo ch16 244.0575-244.0825
244.0750	USN	Geostationary	25F9	Bravo ch17/Yankee ch1 244.0625-244.0875
244.0775	USAF	Geostationary	70F9	244.0425-244.1125
244.0800	USN	Geostationary	24F9	Bravo ch18 244.0675-244.0925
244.0850	USN	Geostationary	25F9	Bravo ch19 244.0725-244.0975
244.0900	USN	Geostationary	25F9	Bravo ch20 244.0775-244.1025
244.0950	USN	Geostationary	25F9	Bravo ch21 244.0825-244.1075
244.1000	USN	Geostationary	25F9	Bravo ch22/Yankee ch12 244.0875-244.1125
	USN	Geostationary	5F9	
244.1100	USN	Geostationary	25F9	Bravo ch23/Yankee ch13 244.0975-244.1225
244.1450	USN	Geostationary	25F9	Charlie ch12 244.1325-244.1575
244.1550	USN	Geostationary	25F9	Charlie ch13/Zulu ch9 244.1425-244.1675
244.1600	USN	Geostationary	25F9	Charlie ch14/Zulu ch10 244.1475-244.1725
244.1650	USN	Geostationary	25F9	Charlie ch15 244.1525-244.1775
244.1700	USN	Geostationary	25F9	Charlie ch16 244.1575-244.1825
244.1750	USN	Geostationary	25F9	Charlie ch17/Zulu ch11 244.1625-244.1875
244.1775	USAF	Geostationary	70F9	244.1425-244.2125
	USAF	Geostationary	5F9	
244.1800	USN	Geostationary	25F9	Charlie ch18 244.1675-244.1925
244.1850	USN	Geostationary	25F9	Charlie ch19 244.1725-244.1975
244.1900	USN	Geostationary	25F9	Charlie ch20 244.1775-244.2025
244.1950	USN	Geostationary	25F9	Charlie ch21 244.1825-244.2075
244.2000	USN	Geostationary	25F9	Charlie ch22/Zulu ch12 244.1875-244.2125
244.2100	USN	Geostationary	25F9	Charlie ch23/Zulu ch13 244.1975-244.2225
244.300	NASA	Nongeostationary		Gemini 6,7
245.400	US-DOD			Satellite reserve
245.600	US-DOD			Satellite reserve
246.300	US	Nongeostationary		Skylab 1/Mercury-Atlas manned prog
246.400	US-DOD			Satellite reserve
246.600	US-DOD			Satellite reserve
246.900	US-DOD			Satellite reserve
247.000	URS	Nongeostationary	FM TM	Cosmos 1686 (See 19.954)
247.100	US-DOD			Satellite reserve
247.300	NASA	Nongeostationary		Apollo 5
	US-DOD			Satellite reserve
247.800	US-DOD			Satellite reserve
248.000	URS	Nongeostationary	PPM-AM	Cosmos 1689 Sun-synchronous earth resource sats
248.8500	USN	Geostationary		Marisat wideband ch.1
248.8750	USN	Geostationary		Marisat wideband ch.2
248.9000	USN	Geostationary		Marisat wideband ch.3
248.9250	USN	Geostationary		Marisat wideband ch.4
248.9500	USN	Geostationary		Marisat wideband ch.5
248.9750	USN	Geostationary		Marisat wideband ch.6
249.0000	USN	Geostationary		Marisat wideband ch.7
249.0250	USN	Geostationary		Marisat wideband ch.8
249.0500	USN	Geostationary		Marisat wideband ch.9
249.0750	USN	Geostationary		Marisat wideband ch.10

Frequency (MHz)	Country/ Agency	Use	Mode	Notes
249.1000	USN	Geostationary	500F9	Marisat wideband ch.11 (center)
	USAF	Geostationary		LES-6 wideband
	USAF	Nongeostationary	500F9	
	USAF	Nongeostationary	1A/0	
	USAF	Nongeostationary	10F3	
	USAF	Nongeostationary	400F1	
	USAF	Nongeostationary	400F9	
	USAF	Nongeostationary	400P/0	
	USAF	Nongeostationary	6A3	
	USAF	Geostationary	500F9	
249.1250	USN	Geostationary		Marisat wideband ch.12
249.1500	USN	Geostationary		Marisat wideband ch.13
249.1750	USN	Geostationary		Marisat wideband ch.14
249.2000	USN	Geostationary		Marisat wideband ch.15
	NASA	Nongeostationary		Gemini 6
249.2250	USN	Geostationary		Marisat wideband ch.16
249.2500	USN	Geostationary		Marisat wideband ch.17
249.2750	USN	Geostationary		Marisat wideband ch.18
249.3000	USN	Geostationary		Marisat wideband ch.19
249.3250	USN	Geostationary		Marisat wideband ch.20
249.3500	USN	Geostationary		Marisat wideband ch.21
	USAF	Geostationary		LES 8/9 wideband ch.1
249.3750	USAF	Geostationary		LES 8/9 wideband ch.2
249.3875	USAF	Nongeostationary		TACSAT
249.4000	USAF	Geostationary		LES 8/9 wideband ch.3
249.4250	USAF	Geostationary		LES 8/9 wideband ch.4
249.4500	USAF	Geostationary		LES 8/9 wideband ch.5
249.4750	USAF	Geostationary		LES 8/9 wideband ch.6
249.5000	USAF	Geostationary		LES 8/9 wideband ch.7
249.5250	USAF	Geostationary		LES 8/9 wideband ch.8
249.5500	USAF	Geostationary		LES 8/9 wideband ch.9
249.5750	USAF	Geostationary		LES 8/9 wideband ch.10
249.6000	USAF	Geostationary	500F9	LES 8/9 wideband ch.11 (center)
	USAF	Nongeostationary	500F9	
	USAF	Nongeostationary	1A/0	
	USAF	Nongeostationary	10F3	
	USAF	Nongeostationary	400F1	
	USAF	Nongeostationary	400F9	
	USAF	Nongeostationary	400P/0	
	USAF	Nongeostationary	400P9	
	USAF	Nongeostationary	6A3	
	USN	Geostationary	500F9	
	USAF/USN	Nongeostationary		6 USAF/ 2 USN wideband sats listed
249.6250	USAF	Geostationary		LES 8/9 wideband ch.12
249.6500	USAF	Geostationary		LES 8/9 wideband ch.13
249.6750	USAF	Geostationary		LES 8/9 wideband ch.14
249.7000	USAF	Geostationary		LES 8/9 wideband ch.15
	USAF/USN	Nongeostationary		6 USAF/2 USN wideband sats listed
249.7250	USAF	Geostationary		LES 8/9 wideband ch.16
249.7500	USAF	Geostationary		LES 8/9 wideband ch.17
249.7750	USAF	Geostationary		LES 8/9 wideband ch.18
249.8000	USAF	Geostationary		LES 8/9 wideband ch.19
249.8250	USAF	Geostationary		LES 8/9 wideband ch.20
249.8500	USAF	Geostationary		LES 8/9 wideband ch.21
250.0000	USAF	Geostationary		Rhyolite Early Warning satellites (wideband)
250.3500	USN	Geostationary		Fleet Broadcast Whiskey ch.1
250.4500	USN	Geostationary	25F9	Fleet Brdcst Alpha/Xray ch.1
250.5500	USN	Geostationary	25F9	Fleet Brdcst Bravo/Yankee ch.1
250.6500	USN	Geostationary	25F9	Fleet Brdcst Charlie/Zulu ch.1
251.000	US	Nongeostationary	TM	Atlas-Centaur 4
251.8500	USN	Geostationary		Fleet Relay Whiskey ch.3
251.900	USN	Geostationary	72F9	
	USN	Geostationary	6A3	

Frequency (MHz)	Country/ Agency	Use	Mode	Notes
251.9500	USN	Geostationary	25F9	Fleet Relay Alpha/Xray ch.3
252.0500	USN	Geostationary	25F9	Fleet Relay Bravo/Yankee ch.3
252.1500	USN	Geostationary	25F9	Fleet Relay Charlie/Zulu ch.3
253.5500	USN	Geostationary		Fleet Relay Whiskey ch.4
253.6500	USN	Geostationary	25F9	Fleet Relay Alpha/Xray ch.4
253.7500	USN	Geostationary	25F9	Fleet Relay Bravo/Yankee ch.4
253.8500	USN	Geostationary	25F9	Fleet Relay Charlie/Zulu ch.4
254.100	USAF	Geostationary		LES-6 narrowband
254.140	USAF	Geostationary		LES-6 tracking
254.1500	USN	Geostationary	25F9	Marisat (Gapfiller) ch.1
255.2500	USN	Geostationary		Fleet Relay Whiskey ch.5
255.3500	USN	Geostationary	25F9	Fleet Relay Alpha/Xray ch.5
255.4500	USN	Geostationary	25F9	Fleet Relay Bravo/Yankee ch.5
255.5500	USN	Geostationary	25F9	Fleet Relay Charlie/Zulu ch.5
256.8500	USN	Geostationary		Fleet Relay Whiskey ch.6
256.9500	USN	Geostationary	25F9	Fleet Relay Alpha/Xray ch.6
257.0500	USN	Geostationary	25F9	Fleet Relay Bravo/Yankee ch.6
257.1500	USN	Geostationary	25F9	Fleet Relay Charlie/Zulu ch.6
257.300	NASA	Nongeostationary		Apollo 5
257.5500	USN	Geostationary	25F9	Marisat (Gapfiller) ch.2
258.3500	USN	Geostationary		Fleet Relay Whiskey ch.7
258.4500	USN	Geostationary	25F9	Fleet Relay Alpha/Xray ch.7
258.5500	USN	Geostationary	25F9	Fleet Relay Bravo/Yankee ch.7
258.6500	USN	Geostationary	25F9	Fleet Relay Charlie/Zulu ch.7
259.500	URS	Nongeostationary		Soyuz 19/Apollo (ASTP)
259.7000	NASA	Nongeostationary	100F9	
	NASA	Nongeostationary	55A9	Shuttle-secnd/Mercury-Atlas manned prog/Gemini capsules/Skylab 2,3,4/
	NASA	Nongeostationary		TETR-4
260.000	US			Wideband allocation
260.3500	USAF/USN	Geostationary		Alpha/Xray wideband ch.1
260.3750	USAF/USN	Geostationary		Alpha/Xray wideband ch.2
260.4000	USAF/USN	Geostationary		Alpha/Xray wideband ch.3
260.4250	USAF/USN	Geostationary		Alpha/Xray wideband ch.4
260.4500	USAF/USN	Geostationary		Alpha/Xray wideband ch.5
260.4750	USAF/USN	Geostationary		Alpha/Xray wideband ch.6
260.5000	USAF/USN	Geostationary		Alpha/Xray wideband ch.7
260.5250	USAF/USN	Geostationary		Alpha/Xray wideband ch.8
260.5500	USAF/USN	Geostationary		Alpha/Xray wideband ch.9
260.5750	USAF/USN	Geostationary		Alpha/Xray wideband ch.10
260.6000	USAF/USN	Geostationary	500F9	Alpha/Xray wideband ch.11 (center)
	US-DOD	Geostationary		4 USAF/5 USN sats listed
260.6250	USAF/USN	Geostationary		Alpha/Xray wideband ch.12
260.6500	USAF/USN	Geostationary		Alpha/Xray wideband ch.13
260.6750	USAF/USN	Geostationary		Alpha/Xray wideband ch.14
260.7000	USAF/USN	Geostationary		Alpha/Xray wideband ch.15
260.7250	USAF/USN	Geostationary		Alpha/Xray wideband ch.16
260.7500	USAF/USN	Geostationary		Alpha/Xray wideband ch.17
260.7750	USAF/USN	Geostationary		Alpha/Xray wideband ch.18
260.8000	USAF/USN	Geostationary		Alpha/Xray wideband ch.19
260.8250	USAF/USN	Geostationary		Alpha/Xray wideband ch.20
260.8500	USAF/USN	Geostationary		Alpha/Xray wideband ch.21
261.4500	USAF/USN	Geostationary		Bravo/Yankee wideband ch.1
261.4750	USAF/USN	Geostationary		Bravo/Yankee wideband ch.2
261.5000	USAF/USN	Geostationary		Bravo/Yankee wideband ch.3
261.5250	USAF/USN	Geostationary		Bravo/Yankee wideband ch.4
261.5500	USAF/USN	Geostationary		Bravo/Yankee wideband ch.5
261.5750	USAF/USN	Geostationary		Bravo/Yankee wideband ch.6
261.6000	USAF/USN	Geostationary		Bravo/Yankee wideband ch.7
261.6250	USAF/USN	Geostationary		Bravo/Yankee wideband ch.8
261.6500	USAF/USN	Geostationary		Bravo/Yankee wideband ch.9
261.6750	USAF/USN	Geostationary		Bravo/Yankee wideband ch.10
261.7000	USAF/USN	Geostationary	500F9	Bravo/Yankee wideband ch.11 (center)
	US-DOD	Geostationary		5 sats total listed

Frequency (MHz)	Country/ Agency	Use	Mode	Notes
261.7250	USAF/USN	Geostationary		Bravo/Yankee wideband ch.12
261.7500	USAF/USN	Geostationary		Bravo/Yankee wideband ch.13
261.7750	USAF/USN	Geostationary		Bravo/Yankee wideband ch.14
261.8000	USAF/USN	Geostationary		Bravo/Yankee wideband ch.15
261.8250	USAF/USN	Geostationary		Bravo/Yankee wideband ch.16
261.8500	USAF/USN	Geostationary		Bravo/Yankee wideband ch.17
261.8750	USAF/USN	Geostationary		Bravo/Yankee wideband ch.18
261.9000	USAF/USN	Geostationary		Bravo/Yankee wideband ch.19
261.9250	USAF/USN	Geostationary		Bravo/Yankee wideband ch.20
261.9500	USAF/USN	Geostationary		Bravo/Yankee wideband ch.21
261.975	US-DOD	Geostationary		Noted SAM A/C activity
262.025	US-DOD	Geostationary		Noted SAM A/C activity
262.0500	USAF/USN	Geostationary		Charlie/Zulu wideband ch.1
262.0750	USAF/USN	Geostationary		Charlie/Zulu wideband ch.2
262.1000	USAF/USN	Geostationary		Charlie/Zulu wideband ch.3
262.1250	USAF/USN	Geostationary		Charlie/Zulu wideband ch.4
262.1500	USAF/USN	Geostationary		Charlie/Zulu wideband ch.5
262.1750	USAF/USN	Geostationary		Charlie/Zulu wideband ch.6
262.2000	USAF/USN	Geostationary		Charlie/Zulu wideband ch.7
262.2250	USAF/USN	Geostationary		Charlie/Zulu wideband ch.8
262.2500	USAF/USN	Geostationary		Charlie/Zulu wideband ch.9
262.2750	USAF/USN	Geostationary		Charlie/Zulu wideband ch.10
262.3000	USAF/USN	Geostationary		Charlie/Zulu wideband ch.11(center)
262.3250	USAF/USN	Geostationary		Charlie/Zulu wideband ch.12
262.3500	USAF/USN	Geostationary		Charlie/Zulu wideband ch.13
262.3750	USAF/USN	Geostationary		Charlie/Zulu wideband ch.14
262.4000	USAF/USN	Geostationary		Charlie/Zulu wideband ch.15
262.4250	USAF/USN	Geostationary		Charlie/Zulu wideband ch.16
262.4500	USAF/USN	Geostationary		Charlie/Zulu wideband ch.17
262.4750	USAF/USN	Geostationary		Charlie/Zulu wideband ch.18
262.5000	USAF/USN	Geostationary		Charlie/Zulu wideband ch.19
262.5250	USAF/USN	Geostationary		Charlie/Zulu wideband ch.20
262.5500	USAF/USN	Geostationary		Charlie/Zulu wideband ch.21
263.0500	USAF	Geostationary		Ch.1 wideband
263.0750	USAF	Geostationary		Ch.2 wideband
263.1000	USAF	Geostationary		Ch.3 wideband
263.1250	USAF	Geostationary		Ch.4 wideband
263.1500	USAF	Geostationary		Ch.5 wideband
263.1750	USAF	Geostationary		Ch.6 wideband
263.2000	USAF	Geostationary		Ch.7 wideband
263.2250	USAF	Geostationary		Ch.8 wideband
263.2500	USAF	Geostationary		Ch.9 wideband
263.2750	USAF	Geostationary		Ch.10 wideband
263.3000	USAF	Geostationary	500F9	Ch.11 Unknown wideband (center)
	US-DOD	Geostationary		2 USAF sats listed
263.3250	USAF	Geostationary		Ch.12 wideband
263.3500	USAF	Geostationary		Ch.13 wideband
263.3750	USAF	Geostationary		Ch.14 wideband
263.4000	USAF	Geostationary		Ch.15 wideband
263.4250	USAF	Geostationary		Ch.16 wideband
263.4500	USAF	Geostationary		Ch.17 wideband
263.4750	USAF	Geostationary		Ch.18 wideband
263.5000	USAF	Geostationary		Ch.19 wideband
263.5250	USAF	Geostationary		Ch.20 wideband
263.5500	USAF	Geostationary		Ch.21 wideband
263.5500	USN/USAF	Geostationary		Whiskey wideband ch.1
263.5750	USN/USAF	Geostationary		Whiskey wideband ch.2
263.6000	USN/USAF	Geostationary		Whiskey wideband ch.3
263.6250	USN/USAF	Geostationary		Whiskey wideband ch.4
263.6500	USN/USAF	Geostationary		Whiskey wideband ch.5
263.6750	USN/USAF	Geostationary		Whiskey wideband ch.6
263.7000	USN/USAF	Geostationary		Whiskey wideband ch.7
263.7250	USN/USAF	Geostationary		Whiskey wideband ch.8
263.7500	USN/USAF	Geostationary		Whiskey wideband ch.9 (center Alpha)

Frequency (MHz)	Country/ Agency	Use	Mode	Notes
263.7750	USN/USAF	Geostationary		Whiskey wideband ch.10
263.8000	USN/USAF	Geostationary		Whiskey wideband ch.11 (center Bravo)
263.8250	USN/USAF	Geostationary		Whiskey wideband ch.12
263.8500	USN/USAF	Geostationary		Whiskey wideband ch.13
263.8750	USN/USAF	Geostationary		Whiskey wideband ch.14
263.9000	USN/USAF	Geostationary		Whiskey wideband ch.15
263.9250	USN/USAF	Geostationary		Whiskey wideband ch.16
263.9500	USN/USAF	Geostationary		Whiskey wideband ch.17
263.9750	USN/USAF	Geostationary		Whiskey wideband ch.18
264.0000	USN/USAF	Geostationary		Whiskey wideband ch.19
264.0250	USN/USAF	Geostationary		Whiskey wideband ch.20
264.0500	USN/USAF	Geostationary		Whiskey wideband ch.21
265.2500	USN	Geostationary		Fleet Relay Whiskey ch.8
265.3500	USN	Geostationary	25F9	Fleet Relay Alpha/Xray ch.8
265.4500	USN	Geostationary	25F9	Fleet Relay Bravo/Yankee ch.8
265.5500	USN	Geostationary	25F9	Fleet Relay Charlie/Zulu ch.8
266.8500	USN	Geostationary	25F9	Fleet Relay Alpha ch.9
266.950	USN	Geostationary	25F9	Fleet Relay Bravo ch.9
267.050	USN	Geostationary	25F9	Fleet Relay Charlie ch.9
268.2500	USN	Geostationary	25F9	Fleet Relay Alpha ch.10
268.3500	USN	Geostationary	25F9	Fleet Relay Bravo ch.10
268.4500	USN	Geostationary	25F9	Fleet Relay Charlie ch.10
269.7500	USN	Geostationary	25F9	Fleet Relay Alpha ch.11
269.8500	USN	Geostationary	25F9	Fleet Relay Bravo ch.11
269.9500	USN	Geostationary	25F9	Fleet Relay Charlie ch.11
279.000	NASA	Nongeos; Nongeos	100F9	
	NASA	Nongeos; Nongeos	55A9	STS EVA link
292.8500	USN	Ground Station		Uplink Whiskey ch.3
292.9500	USN	Ground Station	25F9	Uplink Alpha/Xray ch.3
293.050	USN	Ground Station	25F9	Uplink Bravo/Yankee ch.3
293.0750	USN	Ground Station	25F9	Uplink Charlie/Zulu ch.3
293.9500	USAF/USN	Ground Station		Uplink Alpha/Xray wideband ch.1
293.9750	USAF/USN	Ground Station		Uplink Alpha/Xray wideband ch.2
294.0000	USAF/USN	Ground Station		Uplink Alpha/Xray wideband ch.3
294.0250	USAF/USN	Ground Station		Uplink Alpha/Xray wideband ch.4
294.0500	USAF/USN	Ground Station		Uplink Alpha/Xray wideband ch.5
294.0750	USAF/USN	Ground Station		Uplink Alpha/Xray wideband ch.6
294.1000	USAF/USN	Ground Station		Uplink Alpha/Xray wideband ch.7
294.1250	USAF/USN	Ground Station		Uplink Alpha/Xray wideband ch.8
294.1500	USAF/USN	Ground Station		Uplink Alpha/Xray wideband ch.9
294.1750	USAF/USN	Ground Station		Uplink Alpha/Xray wideband ch.10
294.2000	USAF/USN	Ground Station	500F9	Uplink Alpha/Xray wideband ch.11(ctr)
294.2250	USAF/USN	Ground Station		Uplink Alpha/Xray wideband ch.12
294.2500	USAF/USN	Ground Station		Uplink Alpha/Xray wideband ch.13
294.2750	USAF/USN	Ground Station		Uplink Alpha/Xray wideband ch.14
294.3000	USAF/USN	Ground Station		Uplink Alpha/Xray wideband ch.15
294.3250	USAF/USN	Ground Station		Uplink Alpha/Xray wideband ch.16
294.3500	USAF/USN	Ground Station		Uplink Alpha/Xray wideband ch.17
294.3750	USAF/USN	Ground Station		Uplink Alpha/Xray wideband ch.18
294.4000	USAF/USN	Ground Station		Uplink Alpha/Xray wideband ch.19
294.4250	USAF/USN	Ground Station		Uplink Alpha/Xray wideband ch.20
294.4500	USAF/USN	Ground Station		Uplink Alpha/Xray wideband ch.21
294.5400	USN	Ground Station	25F9	
294.5500	USN	Ground Station		Uplink Whiskey ch.4
294.6500	USN	Ground Station	25F9	Uplink Alpha/Xray ch.4
294.825	Unknown	Nongeostationary		
294.8500	USN	Ground Station	25F9	Uplink Bravo/Yankee ch.4
295.0000	USN	Ground Station	25F9	
295.0500	USAF/USN	Ground Station		Uplink Bravo/Yankee wideband ch.1
295.0750	USAF/USN	Ground Station		Uplink Bravo/Yankee wideband ch.2
295.1000	USAF/USN	Ground Station		Uplink Bravo/Yankee wideband ch.3
295.1250	USAF/USN	Ground Station		Uplink Bravo/Yankee wideband ch.4
295.1500	USAF/USN	Ground Station		Uplink Bravo/Yankee wideband ch.5
295.1750	USAF/USN	Ground Station		Uplink Bravo/Yankee wideband ch.6

Frequency (MHz)	Country/ Agency	Use	Mode	Notes
295.2000	USAF/USN	Ground Station		Uplink Bravo/Yankee wideband ch.7
295.2250	USAF/USN	Ground Station		Uplink Bravo/Yankee wideband ch.8
295.2250	USAF/USN	Ground Station		Uplink Bravo/Yankee wideband ch.9
295.2250	USAF/USN	Ground Station		Uplink Bravo/Yankee wideband ch.10
295.2500	USAF/USN	Ground Station	500F9	Uplink Bravo/Yankee wideband ch11 (Center frequency)
295.2750	USAF/USN	Ground Station		Uplink Bravo/Yankee wideband ch.12
295.3000	USAF/USN	Ground Station		Uplink Bravo/Yankee wideband ch.13
295.3250	USAF/USN	Ground Station		Uplink Bravo/Yankee wideband ch.14
295.3500	USAF/USN	Ground Station		Uplink Bravo/Yankee wideband ch.15
295.3500	USN	Ground Station	25F9	
295.3750	USAF/USN	Ground Station		Uplink Bravo/Yankee wideband ch.16
295.4000	USAF/USN	Ground Station		Uplink Bravo/Yankee wideband ch.17
295.4250	USAF/USN	Ground Station		Uplink Bravo/Yankee wideband ch.18
295.4500	USAF/USN	Ground Station		Uplink Bravo/Yankee wideband ch.19
295.4750	USAF/USN	Ground Station		Uplink Bravo/Yankee wideband ch.20
295.5000	USAF/USN	Ground Station		Uplink Bravo/Yankee wideband ch.21
295.6500	USAF/USN	Ground Station		Uplink Charlie/Zulu wideband ch.1
295.6750	USAF/USN	Ground Station		Uplink Charlie/Zulu wideband ch.2
295.7000	USAF/USN	Ground Station		Uplink Charlie/Zulu wideband ch.3
295.7250	USAF/USN	Ground Station		Uplink Charlie/Zulu wideband ch.4
295.7500	USAF/USN	Ground Station		Uplink Charlie/Zulu wideband ch.5
295.7750	USAF/USN	Ground Station		Uplink Charlie/Zulu wideband ch.6
295.8000	USAF/USN	Ground Station		Uplink Charlie/Zulu wideband ch.7
295.8250	USAF/USN	Ground Station		Uplink Charlie/Zulu wideband ch.8
295.8500	USAF/USN	Ground Station		Uplink Charlie/Zulu wideband ch.9
295.8750	USAF/USN	Ground Station		Uplink Charlie/Zulu wideband ch.10
295.9000	USAF/USN	Ground Station	500F9	Uplink Charlie/Zulu wideband ch.11 (Center frequency)
295.9250	USAF/USN	Ground Station		Uplink Charlie/Zulu wideband ch.12
295.9500	USAF/USN	Ground Station		Uplink Charlie/Zulu wideband ch.13
295.9750	USAF/USN	Ground Station		Uplink Charlie/Zulu wideband ch.14
296.0000	USAF/USN	Ground Station		Uplink Charlie/Zulu wideband ch.15
296.0250	USAF/USN	Ground Station		Uplink Charlie/Zulu wideband ch.16
296.0500	USAF/USN	Ground Station		Uplink Charlie/Zulu wideband ch.17
296.0750	USAF/USN	Ground Station		Uplink Charlie/Zulu wideband ch.18
296.1000	USAF/USN	Ground Station		Uplink Charlie/Zulu wideband ch.19
296.1250	USAF/USN	Ground Station		Uplink Charlie/Zulu wideband ch.20
296.1500	USAF/USN	Ground Station		Uplink Charlie/Zulu wideband ch.21
296.2500	USN	Ground Station		Uplink Whiskey ch.5
296.3500	USN	Ground Station	25F9	Uplink Alpha/Xray ch.5
296.4500	USN	Ground Station	25F9	Uplink Bravo/Yankee ch.5
296.5500	USN	Ground Station	25F9	Uplink Charlie/Zulu ch.5
296.8000	NASA	Ground Station	6A3	Space Shuttle UHF primary
	NASA	Nongeostationary	100F9	Mercury-Atlas manned prog/Gemini caps Skylab 2,3,4
	NASA	Nongeostationary	55A9	
	URS	Nongeostationary		Soyuz 19/Apollo (ASTP)
297.1500	USAF/USN	Ground Station		Uplink Whiskey wideband ch.1
297.1750	USAF/USN	Ground Station		Uplink Whiskey wideband ch.2
297.2000	USAF/USN	Ground Station		Uplink Whiskey wideband ch.3
297.2250	USAF/USN	Ground Station		Uplink Whiskey wideband ch.4
297.2500	USAF/USN	Ground Station		Uplink Whiskey wideband ch.5
297.2750	USAF/USN	Ground Station		Uplink Whiskey wideband ch.6
297.3000	USAF/USN	Ground Station		Uplink Whiskey wideband ch.7
297.3250	USAF/USN	Ground Station		Uplink Whiskey wideband ch.8
297.3500	USAF/USN	Ground Station		Uplink Whiskey wideband ch.9
297.3750	USAF/USN	Ground Station		Uplink Whiskey wideband ch.10
297.4000	USAF/USN	Ground Station		Uplink Whiskey wideband ch.11(Center)
297.4250	USAF/USN	Ground Station		Uplink Whiskey wideband ch.12
297.4500	USAF/USN	Ground Station		Uplink Whiskey wideband ch.13
297.4750	USAF/USN	Ground Station		Uplink Whiskey wideband ch.14
297.5000	USAF/USN	Ground Station		Uplink Whiskey wideband ch.15
297.5250	USAF/USN	Ground Station		Uplink Whiskey wideband ch.16

Frequency (MHz)	Country/ Agency	Use	Mode	Notes
297.5500	USAF/USN	Ground Station		Uplink Whiskey wideband ch.17
297.5750	USAF/USN	Ground Station		Uplink Whiskey wideband ch.18
297.6000	USAF/USN	Ground Station		Uplink Whiskey wideband ch.19
297.6250	USAF/USN	Ground Station		Uplink Whiskey wideband ch.20
297.6500	USAF/USN	Ground Station		Uplink Whiskey wideband ch.21
297.8500	USN	Ground Station		Uplink Whiskey ch.6
297.9500	USN	Ground Station	25F9	Uplink Alpha/Xray ch.6
298.0500	USN	Ground Station	25F9	Uplink Bravo/Yankee ch.6
298.1500	USN	Ground Station	25F9	Uplink Charlie/Zulu ch.6
299.3500	USN	Ground Station		Uplink Whiskey ch.7
299.4500	USN	Ground Station	25F9	Uplink Alpha/Xray ch.7
299.5500	USN	Ground Station	25F9	Uplink Charlie/Zulu ch.7
302.4500	USN/USAF	Ground Station		Uplink Marisat wideband ch.1
302.4750	USN/USAF	Ground Station		Uplink Marisat wideband ch.2
302.5-312.5	USAF	Ground Station		TACSAT
302.5000	USN/USAF	Ground Station		Uplink Marisat wideband ch.3
302.5250	USN/USAF	Ground Station		Uplink Marisat wideband ch.4
302.5500	USN/USAF	Ground Station		Uplink Marisat wideband ch.5
302.5750	USN/USAF	Ground Station		Uplink Marisat wideband ch.6
302.6000	USN/USAF	Ground Station		Uplink Marisat wideband ch.7
302.6250	USN/USAF	Ground Station		Uplink Marisat wideband ch.8
302.6500	USN/USAF	Ground Station		Uplink Marisat wideband ch.9
302.6750	USN/USAF	Ground Station		Uplink Marisat wideband ch.10
302.7000	USN/USAF	Ground Station	500F9	Uplink Marisat wideband ch.11(Center)
	USAF	Ground Station	500F9	
	USAF	Ground Station	0A0	
	USAF	Ground Station	10F3	
	USAF	Ground Station	400F1	
	USAF	Ground Station	400F9	
	USAF	Ground Station	400P0	
	USAF	Ground Station	6A3	
	USAF	Ground Station		LES-6 wideband
302.7250	USN/USAF	Ground Station		Uplink Marisat wideband ch.12
302.7500	USN/USAF	Ground Station		Uplink Marisat wideband ch.13
302.7750	USN/USAF	Ground Station		Uplink Marisat wideband ch.14
302.8000	USN/USAF	Ground Station		Uplink Marisat wideband ch.15
302.8250	USN/USAF	Ground Station		Uplink Marisat wideband ch.16
302.8500	USN/USAF	Ground Station		Uplink Marisat wideband ch.17
302.8750	USN/USAF	Ground Station		Uplink Marisat wideband ch.18
302.9000	USN/USAF	Ground Station		Uplink Marisat wideband ch.19
302.9250	USN/USAF	Ground Station		Uplink Marisat wideband ch.20
302.9500	USN/USAF	Ground Station		Uplink Marisat wideband ch.21
303.1500	USN/USAF	Ground Station		Uplink LES 8/9 wideband ch.1
303.1750	USN/USAF	Ground Station		Uplink LES 8/9 wideband ch.2
303.2000	USN/USAF	Ground Station		Uplink LES 8/9 wideband ch.3
303.2250	USN/USAF	Ground Station		Uplink LES 8/9 wideband ch.4
303.2500	USN/USAF	Ground Station		Uplink LES 8/9 wideband ch.5
303.2750	USN/USAF	Ground Station		Uplink LES 8/9 wideband ch.6
303.3000	USN/USAF	Ground Station		Uplink LES 8/9 wideband ch.7
303.3250	USN/USAF	Ground Station		Uplink LES 8/9 wideband ch.8
303.3500	USN/USAF	Ground Station		LES 8/9 wideband ch.9
303.3750	USN/USAF	Ground Station		Uplink LES 8/9 wideband ch.10
303.4000	USN/USAF	Ground Station	500F9	Uplink LES 8/9 wideband ch.11(Center)
303.4250	USN/USAF	Ground Station		Uplink LES 8/9 wideband ch.12
303.4500	USN/USAF	Ground Station		Uplink LES 8/9 wideband ch.13
303.4750	USN/USAF	Ground Station		Uplink LES 8/9 wideband ch.14
303.5000	USN/USAF	Ground Station		Uplink LES 8/9 wideband ch.15
303.5250	USN/USAF	Ground Station		Uplink LES 8/9 wideband ch.16
303.5500	USN/USAF	Ground Station		Uplink LES 8/9 wideband ch.17
303.5750	USN/USAF	Ground Station		Uplink LES 8/9 wideband ch.18
303.6000	USN/USAF	Ground Station		Uplink LES 8/9 wideband ch.19
303.6250	USN/USAF	Ground Station		Uplink LES 8/9 wideband ch.20
303.6500	USN/USAF	Ground Station		Uplink LES 8/9 wideband ch.21
305.5500				MPX-WBFM & SSB Foxtrot BC reported

Frequency (MHz)	Country/ Agency	Use	Mode	Notes
306.2500	USN	Ground Station		Uplink Whiskey Fleet Relay ch.2
306.3500	USN	Ground Station	25F9	Uplink Alpha/Xray ch.8
306.4500	USN	Ground Station	25F9	Uplink Bravo/Yankee ch.8
306.5500	USN	Ground Station	25F9	Uplink Charlie/Zulu ch.8
307.7500	USN	Ground Station	25F9	Uplink Marisat F1
307.8500	USN	Ground Station	25F9	Uplink Alpha ch.9
307.9500	USN	Ground Station	25F9	Uplink Bravo ch.9
308.0500	USN	Ground Station	25F9	Uplink Charlie ch.9
309.2500	USN	Ground Station	25F9	Uplink Alpha ch.10
309.3500	USN	Ground Station	25F9	Uplink Bravo ch.10
309.4500	USN	Ground Station	25F9	Uplink Charlie ch.10
310.7500	USN	Ground Station	25F9	Uplink Alpha ch.11
310.8500	USN	Ground Station	25F9	Uplink Bravo ch.11
310.9500	USN	Ground Station	25F9	Uplink Charlie ch.11
311.1500	USN	Ground Station	25F9	Uplink Marisat F2
316.6200	USAF	Ground Station	25F9	Uplink to 243.7200
316.6375	USAF	Ground Station	70F9	Uplink SDS 243.7275
316.7875	USAF	Ground Station	70F9	Uplink SDS 242.8875
316.9550	USN	Ground Station		Uplink Whiskey ch.9
316.9600	USN	Ground Station		Uplink Whiskey ch.10
316.9750	USN	Ground Station		Uplink Whiskey ch.11
317.0000	USN	Ground Station		Uplink Whiskey ch.12
317.0100	USN	Ground Station		Uplink Whiskey ch.13
317.0450	USN	Ground Station	25F9	Uplink Alpha ch.12
317.0550	USN	Ground Station	25F9	Uplink Alpha ch.13/Xray ch.9
317.0600	USN	Ground Station	25F9	Uplink Alpha ch.14/Xray ch.10
317.0650	USN	Ground Station	25F9	Uplink Alpha ch.15
317.0700	USN	Ground Station	25F9	Uplink Alpha ch.16
317.0750	USN	Ground Station	25F9	Uplink Alpha ch.17/Xray ch.11
317.0800	USN	Ground Station	25F9	Uplink Alpha ch.18
317.0850	USN	Ground Station	25F9	Uplink Alpha ch.19
317.0900	USN	Ground Station	25F9	Uplink Alpha ch.20
317.0950	USN	Ground Station	25F9	Uplink Alpha ch.21
317.1000	USN	Ground Station	25F9	Uplink Alpha ch.22/Xray ch.12
317.1100	USN	Ground Station	25F9	Uplink Alpha ch.23/Xray ch.13
317.1450	USN	Ground Station	25F9	Uplink Bravo ch.12
317.1550	USN	Ground Station	25F9	Uplink Bravo ch.13/Yankee ch.9
317.1600	USN	Ground Station	25F9	Uplink Bravo ch.14/Yankee ch.10
317.1650	USN	Ground Station	25F9	Uplink Bravo ch.15
317.1700	USN	Ground Station	25F9	Uplink Bravo ch.16
317.1750	USN	Ground Station	25F9	Uplink Bravo ch.17/Yankee ch.11
317.1775	USAF	Ground Station	70F9	Uplink to 244.0775
317.1800	USN	Ground Station	25F9	Uplink Bravo ch.18
317.1850	USN	Ground Station	25F9	Uplink Bravo ch.19
317.1900	USN	Ground Station	25F9	Uplink Bravo ch.20
317.1950	USN	Ground Station	25F9	Uplink Bravo ch.21
317.2000	USN	Ground Station	25F9	Uplink Bravo ch.22/Yankee ch.12
317.2100	USN	Ground Station	25F9	Uplink Bravo ch.23/Yankee ch.13
317.2450	USN	Ground Station	25F9	Uplink Charlie ch.12
317.2550	USN	Ground Station	25F9	Uplink Charlie ch.13/Zulu ch.9
317.2600	USN	Ground Station	25F9	Uplink Charlie ch.14/Zulu ch.10
317.2650	USN	Ground Station	25F9	Uplink Charlie ch.15
317.2700	USN	Ground Station	25F9	Uplink Charlie ch.16
317.2750	USN	Ground Station	25F9	Uplink Charlie ch.17/Zulu ch.11
317.2775	USAF	Ground Station	70F9	Uplink to 244.1775
	USAF	Ground Station	5F1	
317.2800	USN	Ground Station	25F9	Uplink Charlie ch.18
317.2850	USN	Ground Station	25F9	Uplink Marisat wideband ch.19
	USN	Ground Station	5F9	
317.3100	USN	Ground Station	25F9	Uplink Charlie ch.23/Zulu ch.13
	USN	Ground Station	5F9	
317.3500	USN	Ground Station	25F9	
318.3000	USAF	Ground Station		
323.9840	NASA	Nongeostationary		

Frequency (MHz)	Country/ Agency	Use	Mode	Notes
324.0000	NASA	Nongeostationary	5F9	Scientific satellites
	USN	Nongeostationary	0A0	Navy NAVSATs
328.625	USN	Nongeostationary		
328.725	USN	Nongeostationary		
328.825	USN	Nongeostationary		
328.925	USN	Nongeostationary		
328.9250	USN	Nongeostationary	0A0	
333.7250	USN	Nongeostationary	0A0	
334.3250	USN	Nongeostationary	0A0	
334.9250	USN	Nongeostationary	0A0	Timation Naval Technology Sats 1&2
335.0000	USN	Nongeostationary	0A0	Timation Naval Technology Sats 1&2
335.2250	USN	Nongeostationary	0A0	Timation Naval Technology Sats 1&2
335.26105	USN	Nongeostationary		
335.2650	USN	Nongeostationary	0A0	Timation Naval Technology Sats 1&2
335.30905	USN	Nongeostationary		
335.3100	USN	Nongeostationary	0A0	Timation Naval Technology Sats 1&2
335.3200	USN	Nongeostationary	0A0	Timation Naval Technology Sats 1&2
335.32105	USN	Nongeostationary		
335.3240	USN	Nongeostationary	0A0	Timation Naval Technology Sats 1&2
335.3247	USN	Nongeostationary	0A0	Timation Naval Technology Sats 1&2
335.32475	USN	Nongeostationary		
335.3249	USN	Nongeostationary	0A0	Timation Naval Technology Sats 1&2
335.3250	USN	Nongeostationary	0A0	Timation Naval Technology Sats 1&2
335.33495	USN	Nongeostationary		
335.3550	USN	Nongeostationary	0A0	Timation Naval Technology Sats 1&2
335.3700	USN	Nongeostationary	0A0	Timation Naval Technology Sats 1&2
335.3850	USN	Nongeostationary	0A0	Timation Naval Technology Sats 1&2
335.409	USN	Nongeostationary		
335.421	USN	Nongeostationary		
335.42405	USN	Nongeostationary		
335.42475	USN	Nongeostationary		
335.42495	USN	Nongeostationary		
335.425	USN	Nongeostationary		
335.6650	USN	Nongeostationary	0A0	Timation Naval Technology Sats 1&2
344.825	USN	Nongeostatinary		
344.925	USN	Nongeostationary		
346.300	NASA	Nongeostationary		Gemini-5
359.144	US	Geostationary		ATS-6
360.0000	NASA	Nongeostationary		Explorer 27/OGO-3
360.0450	NASA	Geostationary		ATS 6
360.0900	NASA	Nongeostationary		Explorer 22
360.1450	NASA	Geostationary		ATS 6
360.2450	NASA	Geostationary		ATS 6
361.1450	NASA	Geostationary		ATS 6
366.4000	US-DOD	Geostationary		Multiplexed downlink
366.4150	US-DOD	Geostationary		FM downlink
366.4415	US-DOD	Geostationary		USB (typical HF SAC broadcast)
367.1360	USN	Nongeostationary	0A0	VIP support & MPX down/BMEWS/AF-GEOS
367.1365	USAF	Geostationary		BMEWS AFSATCOM link
369.050	USN	Ground Station		NAVSAT telecommand chan
374.53105	USAF	Geostationary	100F9	DSP
	USAF	Nongeostationary	100F9	Recon sats
374.53105	USAF	Geostationary		DSP
		Nongeostationary		Recon sats
374.63105	USAF	Geostationary		DSP
		Nongeostationary		Recon sats
374.73105	USAF	Geostationary		DSP
		Nongeostationary		Recon sats
375-400	UK/NATO	Geostationary		Skynet 2/NATO-1 Comsats
375.3600	USAF	Geostationary	100F9	DSP
	USAF	Ground Station	100F9	Recon sat
	USAF	Ground Station	36F9	
375.8560	USAF	Ground Station	100F9	DSP

Frequency (MHz)	Country/ Agency	Use	Mode	Notes
	USAF	Ground Station	100F9	Recon sat
375.900	US-DOD	Ground Station		
378.6090	USAF	Ground Station	0A0	P76-5 satellite
378.60905	USN	Ground Station		
380.9633	NASA	Space	290F9	Viking Landers 1&2 spacecraft
382.300	USAF	Unknown		Possible SAC
382.3500		Unknown		USB downlink
382.3570		Unknown		USB (Norfolk, Langley) FM
390.000	USAF	Nongeostationary		Hilat-1
390.0820	USN	Nongeostationary	0A0	P76-5 satellite
393.0000	CHINA	Nongeostationary		China 4-7 telemetry
393.675	CHINA	Nongeostationary	FM TM	China 17, 19
393.8550	USN	Ground Station	1F9	Missiles
395.25	USAF	Ground Station		Missile command center
397.0000	CHINA	Nongeostationary		China 4-7 telemetry
396.250	USAF	Ground Station		Missile Command Chan
397.050	USAF	Ground Station		
398.775	USN	Ground Station		
399.050	USAF	Nongeostationary		SAC channel
399.650	USA	Nongeostationary		
399.750	USA	Nongeostationary		
399.760	URS	Nongeostationary		Military Cosmos NAVSAT ch.
399.840	URS	Nongeostationary		Military Cosmos NAVSAT ch.
399.850	USA	Nongeostationary		
399.9200	FRANCE	Nongeostationary		D-1A/Diademe 1(D1-C), 2(D1-D)
	URS	Nongeostationary		Military Cosmos NAVSAT ch.
399.9413	US	Nongeostationary		US NAVSAT maintenance ch.
399.9434	USN	Nongeostationary	4.94A3	
399.952	US	Nongeostationary		US Improved Transit NAVSAT channel
399.9680	USN	Nongeostationary	40F9	US NAVSAT operational ch.
400.0000	USN	Nongeostationary	40F9	Navy Transit NAVSATs
	URS	Nongeostationary		Civilian Cosmos NAVSAT
	NASA	Nongeostationary		OGO-3
400.249	NASA	Nongeostationary		OGO-5, 6
400.032	US	Nongeostationary		Geosat downlink
	Unknown	Nongeostationary		Unknown carrier
400.080	URS	Nongeostationary		Military Cosmos NAVSAT
400.190	FRANCE	Nongeostationary		Peole satellite
400.2500	NASA	Nongeostationary		OGO 1 & 2
400.350	US	Nongeostationary		OAO-A2 scientific satellite
400.3840	USAF	Nongeostationary	36F9	
400.4500	JAPAN	Nongeostationary	TM	Tansei-2/SRATS(Taiyo)
400.4971	JAPAN	Nongeostationary		Tansei 1
400.5000	JAPAN	Nongeostationary	TM	Tansei 3(MST-3)/Denpa(REXS)
	NASA	Nongeostationary		OFO-1
400.5500	NASA	Nongeostationary	300F9	0A0 1,2,3,C(Copernicus)
400.6500	NASA	Nongeostationary	300F9	Explorer 40
400.695	JAPAN	Nongeostationary		Shinsei
400.800	URS	Nongeostationary	CW BCN	Cosmos geodetic sats
400.848	US	Nongeostationary		OGO-6 scientific satellite
400.8490	NASA	Nongeostationary		0G0-5
400.850	US	Nongeostationary		OGO-1,2
400.900	JAPAN	Nongeostationary		JISS
400.9008	JAPAN	Nongeostationary	TM	UME1 & 2
401.0125	USAF	Geostationary	4F9	
	USAF	Nongeostationary	4F9	IDCSP BCN 9,17,25
401.0375	USAF	Geostationary	4F9	
	USAF	Nongeostationary	4F9	IDCSP BCN 10,18,26
401.0625	USAF	Geostationary	4F9	
	USAF	Nongeostationary	4F9	IDCSP BCN 11,19
401.0875	USAF	Geostationary	4F9	
	USAF	Nongeostationary	4F9	IDCSP BCN 12,20
401.1000	Interior	Ground Station	1.5F9	Geos DCS uplinks
401.1125	USAF	Geostationary	4F9	

Frequency (MHz)	Country/ Agency	Use	Mode	Notes
	USAF	Nongeostationary	4F9	IDCSP BCN 13,21
401.1375	USAF	Geostationary	4F9	
	USAF	Nongeostationary	4F9	IDCSP BCN 14,22
401.1625	USAF	Geostationary	4F9	
	USAF	Nongeostationary	4F9	IDCSP BCN 15,23
401.1875	USAF	Geostationary	4F9	
	USAF	Nongeostationary	4F9	IDCSP 8,16,24
401.2000	C	Ground Station	6F9	Weather satellite uplink, Nimbus 6
	NASA	Ground Station	30F9	
	NSF	Ground Station	100F9	
401.2125	USAF	Geostationary	4F9	
	USAF	Nongeostationary	4F9	IDCSP Telemetry satellites 8-26
401.2375	USAF	Geostationary	4F9	
	USAF	Nongeostationary	4F9	IDCSP Telemetry satellites 8-26
401.5000		Nongeostationary		Weather satellite downlink, Nimbus 3
	NASA	Nongeostationary		OFO-1
401.5500	USN	Nongeostationary	0A0	P76-5 satellite
	NASA	Ground Station	100F9	Landsat satellites uplink
401.5515	USN	Nongeostationary	0A0	P76-5 satellite
401.6000	C	Ground Station	3F9	
401.6500	C	Ground Station	.10F9	
	NASA	Ground Station	1.60F9	NOAA DCS uplinks
	NG	Ground Station	2.4F9	
401.7000	C	Ground Station	1.5F9	DCS uplinks
	C	Ground Station	3F9	
	C	Ground Station	12F9	
401.7039	Agri/Int	Ground Station	1.5F9	DCS uplinks
401.7099	USA	Ground Station	1.5F9	DCS uplinks
401.7100	USA	Ground Station	1.5F9	DCS uplinks
401.7144	DOE/Int/NG	Ground Station	1.5F9	DCS uplinks
401.7145	NG	Ground Station	1.5F9	DCS uplinks
401.7189	Int	Ground Station	1.5F9	DCS uplinks
401.7190	Int	Ground Station	1.5F9	DCS uplinks
401.7250	USA	Ground Station	1.5F9	DCS uplinks
401.7279	C	Ground Station	1.5F9	DCS uplinks
401.7280	Int	Ground Station	1.5F9	DCS uplinks
401.7325	Int	Ground Station	1.5F9	DCS uplinks
401.7355	Int	Ground Station	1.5F9	DCS uplinks
401.7399	C	Ground Station	1.5F9	DCS uplinks
401.7414	DOE	Ground Station	1.5F9	DCS uplinks
401.7415	Agri/Int	Ground Station	1.5F9	DCS uplinks
401.7445	C/EPA/ /Int/NG	Ground Station	1.5F9	DCS uplinks
401.7489	C	Ground Station	1.5F9	DCS uplinks
401.7490	C	Ground Station	1.5F9	DCS uplinks
401.7500	NASA	Nongeostationary		Canadian ISIS 1 & 2
401.7519	USA	Ground Station	1.5F9	DCS uplinks
401.7520	USA	Ground Station	1.5F9	DCS uplinks
401.7565	Int	Ground Station	1.5F9	DCS uplinks
401.7624	Int	Ground Station	1.5F9	DCS uplinks
401.7625	Int	Ground Station	1.5F9	DCS uplinks
401.7654	Int	Ground Station	1.5F9	DCS uplinks
401.7655	Int	Ground Station	1.5F9	DCS uplinks
401.7669	C	Ground Station	1.5F9	DCS uplinks
401.7699	NASA	Ground Station	1.5F9	DCS uplinks
401.7700	Int	Ground Station	1.5F9	DCS uplinks
401.7774	USA	Ground Station	1.5F9	DCS uplinks
401.7819	USA	Ground Station	1.5F9	DCS uplinks
401.7820	USA	Ground Station	1.5F9	DCS uplinks
401.7895	USA	Ground Station	1.5F9	DCS uplinks
401.7990	USA	Ground Station	1.5F9	DCS uplinks
401.7999	USA	Ground Station	1.5F9	DCS uplinks
401.8000	USA	Ground Station	1.5F9	DCS uplinks
401.8029	C/NASA	Ground Station	1.5F9	DCS uplinks

Frequency (MHz)	Country/ Agency	Use	Mode	Notes
401.8149	Agri/Int	Ground Station	1.5F9	DCS uplinks
401.8150	Int	Ground Station	1.5F9	DCS uplinks
401.8179	C	Ground Station	1.5F9	DCS uplinks
401.8180	C	Ground Station	1.5F9	DCS uplinks
401.8239	C	Ground Station	1.5F9	DCS uplinks
401.8359	NASA/NSF	Ground Station	1.5F9	DCS uplinks
401.8389	C	Ground Station	1.5F9	DCS uplinks
401.8419	C	Ground Station	1.5F9	DCS uplinks
401.8420	C	Ground Station	1.5F9	DCS uplinks
401.8420	C	Ground Station	1.5F9	DCS uplinks
401.8585	DOE	Ground Station	1.5F9	DCS uplinks
	NG	Ground Station	.2F9	
401.8630	C	Ground Station	1.5F9	DCS uplinks
401.8765	Int	Ground Station	1.5F9	DCS uplinks
401.8850	C	Ground Station	1.5F9	DCS uplinks
401.8855	C	Ground Station	1.5F9	DCS uplinks
401.8930	USA	Ground Station	1.5F9	DCS uplinks
401.9020	C	Ground Station	1.5F9	DCS uplinks
401.9095	Int	Ground Station	1.5F9	DCS uplinks
401.9140	C	Ground Station	1.5F9	DCS uplinks
401.9185	Int	Ground Station	1.5F9	DCS uplinks
401.9260	NG	Ground Station	1.5F9	DCS uplinks
401.9335	Int	Ground Station	1.5F9	DCS uplinks
401.9410	C	Ground Station	1.5F9	DCS uplinks
401.9470	USA	Ground Station	1.5F9	DCS uplinks
401.9471	USA	Ground Station	1.5F9	DCS uplinks
402.0505	NASA	Ground Station	3F9	
402.1000	ESA	Ground Station		ESA Meteosat uplink
403.0000	NASA	Balloon	200F9	
406.0250	Worldwide	Ground Station	-	New emergency location transmitter ch
408.0000	Canada	- -	-	Radio Astronomy frequency
412.0500	NASA	Geostationary	90F9	ATS-1 BCN
412.800	US	Geostationary		ATS-3
413.000	USAF	Nongeostationay		Hilat-1
413.0280	USN	Nongeostationary	0A0	P76-5 satellites
415.0000	NSF	Balloon	5P/0	
416.5000	USAF	Ground Station	100F0/	
		Ground Station	360F9	
416.6000	DOE	Ground Station	500F9	
423.3000	NASA	Ground Station	0A0	
		Ground Station	100F2	
424.5010	USN	Nongeostationary	0A0	P76-5 satellite
432.925	ARS	Nongeostationary		Oscar 4 beacon
435.9740	USN	Nongeostationary	0A0	P76-5 satellite
447.4470	USN	Nongeostationary	0A0	P76-5 satellite
458.9200	USN	Nongeostationary	0A0	P76-5 satellite
431.935	ARS	Nongeostationary	TM	OSCAR 4 beacon
432.150-432.157	ARS	Nongeostationary	Transp	OSCAR 7 Mode B uplink
435.025	ARS	Nongeostationary	TM	OSCAR 9 & 11 beacon
435.027-435.179	ARS	Ground Station		OSCAR 10 Mode B uplink
435.095	ARS	Nongeostationary	TM	OSCAR 8 Mode J beacon
435.100	ARS	Nongeostationary	TM	OSCAR 7 beacon
435.100-435.200	ARS	Nongeostationary	Transp	OSCAR 8 Mode J
435.795	JAPAN	Nongeostationary		JAS-1
435.910	JAPAN	Nongeostationary		JAS-1
436.020	ARS	Nongeostationary		OSCAR 10 Mode L engineering beacon
436.040	ARS	Nongeostationary		OSCAR 10 Mode L general beacon
436.150-436.950	ARS	Nongeostationary	Transp	OSCAR 10 Mode L
447.4431	USAF	Nongeostationary		P76-5
461.150	URS	Nongeostationary		Meteosat

Frequency (MHz)	Country/ Agency	Use	Mode	Notes
461.500	URS	Nongeostationary		Cosmos 144-weather precursor
463.000	URS	Nongeostationary		Soyuz 19 downlink
464.000	URS	Nongeostationary		Meteosat/Cosmos 156-weather precursor
466.000	US	Nongeostationary		Nimbus 3 weather satellite
466.500	URS	Nongeostationary		Meteosat/Cosmos ocean resources sats/ Cosmos 184
468.500	URS	Nongeostationary		Meteosat
468.750	C	Nongeostationary	.40F9	
468.825	C	Geostationary	150F9	Geostationary weather (Time signals) GOES-West 10W or 20W
468.850	C	Geostationary	200F9	Geostationary weather (Time signals) GOES-East 10W
468.875	ESA/J	Geostationary		Meteosat geostationary/ GMS-1(J) 5.5W
468.883	JAPAN	Geostationary		GMS-1 weather satellite (5.5W)
468.900	JAPAN	Geostationary	GMS	Geostationary weather satellites
468.924	JAPAN	Geostationary		GMS-1 weather satellite (5.5W)
468.925	ESA	Geostationary		Meteosat geostationary satellites
469.000	Japan	Geostationary		GMS geostationary weather satellites
470.393	USN	Nongeostationary	0A0	
479.970	CHINA	Nongeostationary	CW BCN	China 19
480.000	CHINA	Nongeostationary		China 4-7 telemetry/China 9C
536.000	USAF	Nongeostationary		Hilat-1
648.000	USAF	Nongeostationary		5E3(SN-41)
702-726	URS	Geostationary		Ekran UHF DBS TV satellite
800-1000	URS	Nongeostationary		Molniya class 1 COMSATs
821-825	US/CANADA	Ground Station		Mobilesat uplink
845-851	US/CANADA	Ground Station		Mobilesat uplink
860.000	US	Geostationary		ATS-6 satellite
866-870	US/CANADA	Geostationary		Mobilesat downlink
890-896	US	Geostationary		Mobilesat downlink
919.760	URS	Nongeostationary		Soyuz 19 PM command system
922.750	URS	Nongeostationary	TM CW	Manned spcraft/Cosmos/Progress/Salyut
922.760	URS	Space	TM	Mars-1 planetary spacecraft
922.763	URS	Space	TM	Venera planetary spacecraft
922.800	URS	Space	TM	Venera planetary spacecraft-1
925.240	URS	Nongeostationary	TM	Salyut space station
926.060	URS	Nongeostationary	TM	Salyut space station
926.070	URS	Nongeostationary	CW BCN	Salyut 7
928.400	URS	Space	TM	Mars/Venera/Prognoz/Vega 1 & 2
928.429	URS	Space	TM	Venera spacecraft
959.520	US	Space		Ranger 6,7,8,9
960.000	US	Space	TM	Ranger-4,5 spacecraft
960.050	US	Space		Mariner-2/Ranger/Pioneer-3,4
960.100	US	Space		Ranger 1,2,3,6,7,8,9
960.580	US	Space		Ranger 6,7,8,9
972.000	US	Nongeostationary		Explorer 29,36
974.000-985.000	URS	Nongeostationary		Molniya class 1
1000.000	URS	Nongeostationary		Meteosat
1227.600	USAF	Nongeostationary	24F9	NAVSTAR satellite system L2 channel
1228.000	USN	Nongeostationary		NTS-2 NAVSAT
1239.000	USAF	Nongeostationary		Hilat-1
1239.084	USN	Nongeostationary	36F9	P76-5 satellite
1246.875	URS	Nongeostationary	(SS)	Glonass 1520
1247.750	URS	Nongeostationary	(SS)	Glonass 1710
1249.0625	URS	Nongeostationary	(SS)	Glonass 1650
1249.9375	URS	Nongeostationary	(SS)	Glonass 1554,1651
1250.000	URS	Nongeostationary		Glonass NAVSAT system
1253.4375	URS	Nongeostationary	(SS)	Glonass 1595
1254.3125	URS	Nongeostationary	(SS)	Glonass 1711
1256.5000	URS	Nongeostationary	(SS)	Glonass 1555
1269.05-1269.850	ARS	Gnd Stn; Up-Ngeos		Oscar 10 Mode L uplink
1381.08	USAF	Nongeostationary		NAVSTAR Global Burst Detection

Frequency (MHz)	Country/ Agency	Use	Mode	Notes
1427.230	USN	Nongeostationary	NBTM	Project Whitecloud-Ocean surveillance
1427.430	USN	Nongeostationary	NBTM	Project Whitecloud-Ocean surveillance
1427.630	USN	Nongeostationary	NBTM	Project Whitecloud-Ocean surveillance
1430.200	USN	Nongeos; Down	WBTM	Project Whitecloud-Ocean surveillance
1432.200	USN	Nongeos; Down	WBTM	Project Whitecloud-Ocean surveillance
1434.200	USN	Nongeos; Down	WBTM	Project Whitecloud-Ocean surveillance
1535-1542.5	US/ITSO	Geostationary		Intelsat 5-Maritime transponder
1535-1542	URS	Geostationary		Volna maritime transponder
1537-1541	US	Geostationary		Marisat commercial maritime downlink
1537.75-1542.5	ESA	Geostationary		MARECS maritime transponder
1540-1542.5	ESA	Geostationary		MAROTS maritime transponder
1543-1558	URS	Geostationary		Volna aeronautical transponder
1543.5-1558.500	ESA	Geostationary		Proposed aerosat aeronautical transp.
1543.000	NASA	Nongeostationary	800F9	
1544.500	NASA/URS	Nongeostationary	800F9	SARSAT downlink frequency
1545-1559	CANADA	Geostationary		M-Sat
1550.000	US	Geostationary		ATS-6
1575.000	USN	Nongeostationary		Timation NAVSAT
1575.420	USAF	Nongeostationary	24F9	NAVSTAR satellite system L1 Channel
1580.000	USN	Nongeostationary		Timation NAVSAT
1603.000	URS	Nongeostationary		Glonass NAVSAT system
1603.125	URS	Nongeostationary	(SS)	Glonass 1520
1603.6875	URS	Nongeostationary	(SS)	Glonass 1490
1604.250	URS	Nongeostationary	(SS)	Glonass 1710
1605.9375	URS	Nongeostationary	(SS)	Glonass 1650
1607.0625	URS	Nongeostationary	(SS)	Glonass 1554
1607.625	URS	Nongeostationary	(SS)	Glonass 1651
1611.5625	URS	Nongeostationary	(SS)	Glonass 1595
1612.125	URS	Nongeostationary	(SS)	Glonass 1519
1612.6875	URS	Nongeostationary	(SS)	Glonass 1711
1636-1644	URS	Ground Station		Volna maritime service uplink
1638.500-1642.500	US	Ground Station		Marisat ship to shore
1636.500	US	Ground Station	40F1/40F3/40F9/115F9	
1636.5-1644	ITSO	Ground Station		Intelsat V MCS ship-to-shore
1638.500	US	Ground Station	6.3F9/30F3	
1638.5-1644	ESA	Ground Station		Marecs A & B ship-to-shore
1639.750	US	Ground Station	9.6F9	
1639.975	US	Ground Station	9.6F9	
1641.500-1644.500	USA	Ground Station		MAROTS maritime uplink
1645-1660	ESA/URS /US	Ground Station		Aerosat uplink/Volna aeronautical sevice uplink/Mobilesat-air and land mobile to satellite uplink
1646.5-1660.5	CANADA	Ground Station		M-Sat
1668.00	URS	Deep Space		Vega 1 & 2 tracking beacon
1675.281	ESA	Geostationary		Meteosat geostationary satellite
1675.929	ESA	Geostationary		Meteosat geostationary satellite
1681.600	C/Japan	Geostationary		GOES/GMS weather satellites
1682.500	C	Geostationary	25F9	GOES weather satellites (20W)/SMS-1,2
1684.000	C/Japan	Geostationary		GOES/GMS weather satellites
1686.833	ESA	Geostationary		Meteosat weather satellites
1687.100	C/Japan	Geostationary		SMS/GOES/GMS weather satellites Stretched VISSR
1688.200	C/Japan	Geostationary		GOES/GMS weather satellites
1690.200	C/Japan	Geostationary		GOES/GMS weather satellites
1691.000	C/Japan/ESA	Geostationary		SMS/GOES/GMS/Meteosat weather sats. WEFAX channel
	URS	Nongeostationary		Meteor 30

Frequency (MHz)	Country/ Agency	Use	Mode	Notes
1694.000	C/Japan	Geostationary		GOES/GMS weather satellites
	URS	Nongeostationary		Meteor 30
1694.300	Japan	Geostationary		GMS 20W TT&C weather satellites
1694.500	ESA	Geostationary		Meteosat weather sats. WEFAX chan.
1696.000	US	Geostationary		GOES Wefax
1697.500	C	Space	3F9	GOES-to-GOES satellite link
1698.000	C/USAF	Nongeostationary		NOAA/DOD Block weather sats.HRPT chan.
1700.000	US	Nongeostationary		Relay 1 & 2 satellites
1702.470	NASA	Nongeostationary		Nimbus weather satellites
1702.480	JAPAN	Nongeostationary		MOS-1
1702.486	US	Nongeostationary		Nimbus 4
1702.500	C/USAF/ NASA	Nongeos; Down	15F9/3F9	NOAA/DOD Block/Nimbus wx sats HRPT channel
1705.000	Japan	Geostationary		ETS satellite series
1707.000	C/USAF	Nongeostationary		NOAA/DOD Block/TIROS wx satellites HRPT channel
1707.500	US	Nongeostationary		Nimbus 2,5
1775.700	USAF	Ground Station		DOD Space Shuttle uplink (secondary)PM
1779.736	USAF	Ground Station		Scatha 1
1782.500	NASA	Nongeostationary		Nimbus weather satellites
1783.74	USAF	Nongeostationary		NAVSTAR L3 channel
1784.740	USAF	Ground Station	5F9	
	USAF	Ground Station	5F9	
1787.744	USAF	Ground Station	5F9	
	USAF	Ground Station	5F9	
1791.748	USAF	Ground Station	5F9	
	USAF	Ground Station	5F9	
1795.752	USAF	Ground Station	5F9	
	USAF	Ground Station	5F9	
1799.756	USAF	Ground Station	5F9	
	USAF	Ground Station	5F9	
1803.760	USAF	Ground Station	5F9	
	USAF	Ground Station	5F9	
1807.764	USAF	Ground Station	5F9	
	USAF	Ground Station	5F9	
1811.768	USAF	Ground Station	5F9	
	USAF	Ground Station	5F9	
1815.000	US	Geostationary		Syncom 1,2,3
1815.771	USAF	Ground Station	5F9	
	USAF	Ground Station	5F9	
1819.775	USAF	Ground Station	5F9	
	USAF	Ground Station	5F9	
1820.000	US	Geostationary		Syncom 1,2,3
1823.779	USAF	Ground Station	5F9	
	USAF	Ground Station	5F9	
1827.783	USAF	Ground Station	5F9	
	USAF	Ground Station	5F9	
1831.787	USAF	Ground Station	5F9	
	USAF	Ground Station	5F9	
1831.800	USAF	Ground Station	5F9	DOD Space Shuttle uplink (primary)
1835.791	USAF	Ground Station	5F9	
1839.795	USAF	Ground Station	5F9	
1900-2100	US	Geostationary		ATS-6
2025-2100	FRANCE	Ground Station		F-SAT 1,2
2025.000	NASA	Geostationary	5F9	
	NASA	Nongeostationary	4F9	
2028.000	D	Ground Station		DGS 1,2 command chan
	ITALY	Ground Station		Italsat TT&C
2029.100	USAF	Ground Station	8.2F9	Weather satellite uplink
2030.000	C	Ground Station	10F9	
2030.200	C	Ground Station	1F9	
2035.960	NASA	Geostationary		TDRSS A satellite
2038.725	NASA	Ground Station	1.5F9	

Frequency (MHz)	Country/ Agency	Use	Mode	Notes
2040.100	NASA	Ground Station	1F9	ISEE 1
2041.900	NASA	Ground Station	3.6F9	Space Shuttle telemetry&control uplnk
	NASA	Geostationary	6F9	
2041.950	NASA	Ground Station	1F9	ISEE 3 satellite
	US	Ground Station	3.6F9	
2042.870	NASA	Ground Station	1F9	SAGE
2060.3646	NASA	Nongeostationary	5F9	SPAS-01A to Space Shuttle channel
2062.850	NASA	Ground Station	1F9	
2062.900	US	Ground Station		Nimbus 6
2068.190	NASA	Ground Station	1F9	
2069.1125	NASA	Ground Station	1F9/3.6F9	Head-2/Head-1/GEOS-1
2074.6375	NASA	Ground Station	3.6F9	
2074.638	NASA	Ground Station	40F9	Head-2/Head-1
2075.000	US	Geostationary		ATS-6
2081.820	NASA	Ground Station	1F9	ISEE 2
2085.500	NASA	Ground Station	1F9	ISEE 1
2087.990	NASA	Ground Station	5F9	
	NASA	Geostationary	5F9	
2090.660	NASA	Ground Station	1F9	ISEE 3 satellite
2092.594	NASA	Ground Station	5F9	
	NASA	Geostationary	5F9	
2093.515	NASA	Ground Station	1F9	Weather satellites
2094.896	NASA	Ground Station	1.5F9	
2097.1975	CANADA	Ground Station		CTS-1 Comsat
2106.400	NASA	Ground Station	1F9/3.6F9/ 2F9/3F9	Space Shuttle command and voice ranging, Landsat satellites
	US	Ground Station	3.6F9	Ground station to shuttle
	NASA	Geostationary	6F9	STA/Landsat to TDRSS
	NASA	Geostationary	6F9	Ground station to TDRSS
	NASA	Nongeostationary		Apollo 7
2106.406	NASA	Ground Station	3F9	Space Shuttle uplink command voice and ranging/Landsat/Seasat
	NASA	Geostationary	6F9	TDRSS to Space Shuttle
2108.250	NASA	Ground Station	3F9	Explorer 55
	USAF	Ground Station	3F9	
2110.000	NASA	Ground Station	5F9	
	NASA	Ground Station		Pioneer 10,11
2110.5841	NASA	Ground Station	3F9	Pioneer 7
2110.9252	NASA	Ground Station	170F9	
2112.2894	NASA	Ground Station	170F9	
2112.6304	NASA	Ground Station	170F9	
2112.9715	NASA	Ground Station	3F9	
2113.3125	NASA	Ground Station	3.3F9	Voyager-1 telecommand
2113.6536	NASA	Ground Station	2.7F9	Mariner
2114.6767	NASA	Ground Station	3.3F9	Voyager-2 telecommand
2115.6998	NASA	Ground Station	2.7F9	ISEE 1,2
2116.7229	ESA	Ground Station		Giotto Comet Halley satellite
2119.000	NASA	Ground Station		Alsep package
2200-2290	FRANCE	Geostationary		F-SAT 1,2
2200.0000	USAF	Space		
2202.350	D	Geostationary	TM BCN	DFS 1,2
2202.500	USAF	Geostationary	5F9	Fleetsatcom telemetry/tracking chan.
	USAF	Nongeostationary	5F9	
2205.000	NASA	Nongeostationary	10F9	Space Shuttle DFI downlink
2205.900	ESA	Nongeostationary		Spot-1
2207.500	USAF	Geostationary	5F9	
	USAF	Nongeostationary	5F9	
2208.500	NASA	Nongeostationary		Nimbus 5 weather satellite
2209.086	NASA	Nongeostationary	350F9	Geos weather satellites
2211.000	NASA	Geostationary	2.5F9	TDRSS downlink
	NASA	Nongeostationary	2.5F9/3F9	Weather satellite downlink
	CANADA	Nongeostationary		Nimbus 7 4W (PCM/FM)
2212.500	USAF	Geostationary	5F9	NATO 3C

Frequency (MHz)	Country/ Agency	Use	Mode	Notes
	USAF	Nongeostationary	5F9	OSO-8
2214.000	NASA	Nongeostationary	3F9/300F9	Weather satellite downlink, DE1 satellite downlink
2215.000	NASA	Nongeostationary		Space Shuttle TV downlink
2215.400	US	Space		Viking Orbiter 2
2215.500	NASA	Nongeostationary	1F9	ISEE 3 satellite downlink/ISEE 1(2.5W)
2217.500	USAF	Geostationary	5F9	
	USAF	Nongeostationary	5F9	
	NASA	Nongeostationary	5F9	Space Shuttle TT&C/ISEE-3
2218.500	NASA	Nongeostationary	2F9	Sage satellite downlink
2220.00	JAPAN	Nongeostationary		MOS-1
2222.500	USAF	Geostationary	5F9	
	USAF	Nongeostationary	5F9	Scatha 1 satellite
2227.500	USAF	Geostationary	5F9	NATO 3C
	USAF	Nongeostationary	5F9	NAVSTAR NAVSAT series
2229.500	NASA	Nongeostationary	20F9	Landsat sat series(MSS data)/Seasat
2232.500	USAF	Geostationary	5F9	
	USAF	Nongeostationary	5F9	
	NASA	Geostationary		TDRSS A downlink
2237.500	USAF	Geostationary	5F9	
	USAF	Nongeostationary	5F9	
	NASA	Nongeostationary	4.1F9	
	NASA	Nongeostationar	100F9	SPAS-01A from Orbiter (Shuttle)
2389.140	US	Space		Pioneer Venus Orbiter
2240.000	USAF	Nongeostationary	20F9	
2241.500	USAF	Nongeostationary	20F9	Scatha 1 (Stage 2 TM)/Anik-B
2242.500	USAF	Geostationary	5F9	
	USAF	Nongeostationary	5F9	Solwind 1
2244.500	USAF	Nongeostationary	20F9	Scatha 1 (Stage 1 TM)/Anik-B
2246.000	NASA	Nongeostationary	2F9	
2247.000	ESA/NASA	Nongeostationary	4F9/400F9	GEOS ESA satellite/Head-1 (1W),2,3
2247.500	USAF	Geostationary	5F9	
	USAF	Nongeostationary	5F9	Solwind 1
2248.000	NASA	Nongeostationary	3F9/4F9	Sage satellite/HCMM (2W) TM's
2249.800	NASA	Space		International Ultraviolet Explorer
2250.000	USAF	Nongeostationary	20F9	Scatha 1 (Stage 3 TM)
	NASA	Nongeostationary	10F9	Space Shuttle wideband FM/Sirio
2250.500	US	Geostationary		Telstar 3 TT&C channel/Scatha 1
2252.500	USAF	Geostationary	5F9	Fleetsatcom TT&C
	USAF	Nongeostationary	5F9	Solwind 1
2253.000	NASA	Nongeostationary	3F9	IRAS 1/weather sat downlink/Head-1,2
	NASA	Nongeostationary	4F9	
	NASA	Nongeostationary	6F9	
2255.500	NASA	Nongeostationary	15F9	
2256.00	US	Nongeostationary		Apollo (ASTP)
2257.500	USAF	Geostationary	5F9	
	USAF	Nongeostationary	1F9	ISEE 2/EXOSAT satellites
2260.800	ESA	Nongeostationary		ISEE 2 3W TT&C/ISEE 3
	US	Nongeostationary		EXOSAT
2262.500	USAF	Geostationary	5F9	Fleetsatcom satellites
	USAF	Nongeostationary	5F9	
2264.800	NASA	Nongeostationary	1F9	ISEE 1 satellite
2265.100	US	Nongeostationary		Seasat-1 5W TM's
2265.500	NASA	Nongeostationary	20F9	Landsat series (video data)/Seasat
2267.500	USAF	Geostationary	5F9	
	USAF	Nongeostationary	5F9	
	NASA	Nongeostationary	5F9/40F9	
	NASA	Nongeostationary	5F9/40F9	
2270.000	USN	Nongeostationary	.01A0	
2270.400	NASA	Nongeostationary	1F9	ISEE 3
2270.855	USN	Nongeostationary	6F9	
2272.500	USAF	Geostationary	5F9	
	USAF	Nongeostationary	5F9	
	NASA	Nongeostationary	5F9	Skylab 2,3,4/Apollo (ASTP)/Apollo CM

Frequency (MHz)	Country/ Agency	Use	Mode	Notes
	NASA	Nongeostationary	5F9	TDRSS
2273.420	USN	Nongeostationary	6P9	
2273.500	NASA	Nongeostationary	5F9	Weather satellite downlink
2274.800	NASA	Nongeostationary		ISEE 1 2.5W TT&C
2275.000	NASA	Nongeostationary	3F9/6F9	DE 2 satellite
2276.990	JAPAN	Geostationary	5F9	BS 2A/2B TT&C
2277.500	USAF	Geostationary	5F9	
	USAF	Nongeostationary	5F9	
	CANADA	Geostationary		CTS-1 Comsat tracking beacon
2280-2304	URS	Nongeostationary		Cosmos Early Warning satellites
2280.500	JAPAN	Nongeostationary		Telemetry channel (Tansei-4)
2280.72	JAPAN	Geostationary		BS-2B
2282.070	US	Nongeostationary		Pioneer 9
2282.500	USAF	Geostationary	5F9	
	USAF	Nongeostationary	5F9	
	NASA	Space		Apollo Lunar Module/Pioneer 8/ Apollo CM & LM
	NASA	Nongeostationary		TETR-4/Skylab 2,3,4
2285.000	JAPAN	Geostationary		BSE
2286.500	JAPAN	Geostationary		TM channel/Sakura/Yuri(BSE)
2287.500	NASA	Geostationary	6F9	
	USAF	Geostationary	5F9	
	USAF	Nongeostationary	5F9	
	NASA	Nongeostationary	5F9/6F9	Space Shuttle TT&C/Landsat/SMM/SME/ Seasat 1/Apollo(ASTP)
	NASA	Nongeostationary	5F9/3F9/ 6F9	Above listed spacecraft to TDRSS Explorer 55
	NASA	Space		Apollo CM
2289.500	NASA	Nongeostationary	3F9	Explorer 51,54,55
2292.000	URS	Nongeostationary		Cosmos Early Warning satellite center frequency/Pioneer-Venus 2(10/20W TT&C)
2292.0	NASA	Space		Pioneer Venus Multiprobe
2292.037	NASA	Space	20F9/ 200F9	Pioneer 6,7,8,9,10,11
2292.4074	NASA	Space	20F9/ 200F9	Pioneer 6,7,8,9
2293.148	NASA	Space		Viking Orbiter 1
2293.150	JAPAN	Nongeostationary		MS T5 scientific satellite
2293.8889	NASA	Space	170F9	Pioneer Venus obiter
2293.890	JAPAN	Space		Planet A (Suisei)
2294.2593	NASA	Space	170F9	Pioneer satellite series
2294.6296	NASA	Space	3F9	
2295.0000	NASA	Space	3.3F9	Voyager 1 (9.4/28.3W TM's)/Mariner 3,4 Surveyor 1,2,6,7
	NASA	Nongeostationary		Atlas-Centaur 6,8
2295.240	NASA	Space		Viking Orbiter 2
2295.3703	NASA	Space	3F0	Helios 2
2295.740	NASA	Space		Viking Orbiter 1,2
2296.11	NASA	Space		Mariner 9
2296.4815	NASA	Space	3.3F9	Voyager 2
2296.851	NASA	Space		Mariner 6
2297.2205	NASA	Space		Mariner 7
2297.5925	NASA	Space		GOES weather satellites/Helios 1
2297.720	NASA	Space		Viking Orbiter 2
2298.000	NASA	Space		Mariner 5
	URS	Nongeostationary		Cosmos Early Warning Sats
2298.300	NASA	Space		Lunar Orbiter 1,2,3,4
2298.7037	NASA	Space		Giotto Comet Halley satellite
2299.500	ESA	Geostationary		BW-on/off TM's/GEOS-112
2304.000	ARS	Nongeostationary		OSCAR 7 telemetry beacon
	URS	Nongeostationary		Cosmos Early Warning Sats
2400.000	US	Nongeostationary		Mercury-Atlas manned program
2401.000	ARS	Nongeostationary		OSCAR 9 & 11 telemetry beacon
2544-2577	ARB	Geostationary		Arabsat 1A,1B

Frequency (MHz)	Country/ Agency	Use	Mode	Notes	
2560-2590	India	Geostationary		INSAT 1B TV channel 1	
2569.200	US	Geostationary		ATS-6	
2600-2630	India	Geostationary		INSAT 1B TV channel 2	
2618-2651	ARB	Geostationary		Arabsat 1A,1B	
2670.000	US	Geostationary		ATS-6	
2802.000	URS	Nongeostationary		Salyut 6 space station	
2891.171	USN	Nongeostationary		P76-5 satellite	
2910.000	NASA	Nongeostationary		Gemini Agena Target vehicle	
3400-4100	URS	Nongeostationary		Molniya class 1 satellites	
3408-3442	URS	Geostationary		Raduga TR 1(3425 center freq)	
3458-3492	URS	Geostationary		Raduga TR 2(3475 center freq)	
3500.000	URS	Geostationary	TM	AMS-1,2	
3508-3542	URS	Geostationary		Raduga TR 3(3525 center freq)	
3512-3548	ISR	Geostationary		AMS-1,2 Ch1(3530 center freq)	
3552-3588	ISR	Geostationary		AMS-1,2 Ch2(3570 center freq)	
3592-3628	ISR	Geostationary		AMS-1,2 Ch3(3610 center freq)	
3558-3592	URS	Geostationary		Raduga TR 4(3575 center freq)	
3608-3642	URS	Geostationary		Raduga TR 5(3625 center freq)	
3632-3668	ISR	Geostationary		AMS-1,2 Ch4(3650 center freq)	
3658-3692	URS	Geostationary		Raduga TR 6(3675 center freq)	
3672-3708	ISR	Geostationary		AMS-1,2 Ch4(3690 center freq)	
3700.000	URS	Space		Mars 1	
3700-3740	FRANCE	Geostationary		Telecom F3,F1 Ch.F1	
3700-4200	Various	Geostationary		ASC 1-5	CCC 1-2
				Digisat 1B	Digisat 2
				Equastar 102	Ford 1-3
				Spotnet C1-C2	AMS 1-2
				Pacific Area Satellite System A/B	

Domestic C-Band Satellites

Group A Satellites		Group B Satellites
Comstar D1-4	Galaxy 1-4	Spacenet 1-3
Satcom 1-4	Anik D1-D2	Anik A1-A3
Telstar 301/302/303	Westar 4-5	Anik B
Aurora 1-3		Westar 1-3
		Palapa A1-A2
		F-Sat 1

Frequency (MHz)	Country/ Agency	Use	Mode	Notes
3700-4195	France	Geostationary		Telecom 1A/1B
3700.5	US	Geostationary		Satcom 5 (Aurora 1) TT&C
3701	US	Geostationary		Satcom 1-4/Westar 4-5 TT&C
3702-3738	--	Geostationary		Group A satellite channel 1
	--	Geostationary		Group B satellite channel 1
3704-3781	ITSO	Geostationary		Intelsat 5/5A channel 1-2
3705-3930	ITSO	Geostationary		Intelsat 3 channel 1
3705-3741	MEXICO	Geostationary		Morelos 1 & 2
3707-3743	ITSO	Geostationary		Intelsat 4/4A channel 1
3708-3742	URS	Geostationary		Gorizont/Raduga TR 7(3725 center freq)
3712-3748	INDIA	Geostationary		Insat 1B channel 1
	ISR	Geostationary		AMS-1,2 Ch6(3730 center freq)
3715-3805	ESA	Geostationary		Symphonie
3718.75-3751.75	ARB	Geostationary		Arabsat 1A,1B
3720-4180	Japan	Geostationary		CS (Sakura)
3722-3758	--	Geostationary		Group A satellites channel 2
3722-3798	US	Geostationary		Spacenet 1-3 channel 13
3729-3800	MEXICO	Geostationary		Morelos 1 & 2
3735-3915	Japan	Geostationary		CS-2A/2B channel 1
3737.25-3770.25	ARB	Geostationary		Arabsat 1A,1B
3742-3778	--	Geostationary		Group A satellite channel 3
	--	Geostationary		Group B satellite channel 2

Frequency (MHz)	Country/ Agency	Use	Mode	Notes
3745-3781	MEXICO	Geostationary		Morelos 1 & 2
3747-3783	ITSO	Geostationary		Intelsat 4/4A channel 2
3750.000	US	Geostationary		ATS-6
3752-3788	INDIA	Geostationary		Insat 1B channel 2
	ISR	Geostationary		AMS-1,2 Ch7(3770 center freq)
3755-3875	FRANCE	Geostationary		Telecom F3,F1 Ch.F2
3755.75-3788.75	ARB	Geostationary		Arabsat 1A,1B
3758-3792	URS	Geostationary		Gorizont/Molniya class 3/Raduga TR 8(3775 center freq)
3762-3798	--	Geostationary		Group A satellites channel 4
3772.000	USA	Nongeostationary	60F9	
3774.25-3807.25	ARB	Geostationary		Arabsat 1A,1B
3774.7200	USA	Nongeostationary	8F9	
3782-3818	--	Geostationary		Group A satellites channel 5
	--	Geostationary		Group B stallites channel 3
3783.75-3862.75	ARB	Geostationary		Arabsat 1A,1B
3785-3821	MEXICO	Geostationary		Morelos 1 & 2
3787-3823	ITSO	Geostationary		Intelsat 4/4A channel 3
3789-3861	ITSO	Geostationary		Intelsat 5/5A channel 3-4
3789.600	USA	Nongeostationary	60F9	
3792-3823	INDIA	Geostationary		Insat 1B channel 3
	ISR	Geostationary		AMS-1,2 Ch8(3810 center freq)
3792.75-3819.75	ARB	Geostationary		Arabsat 1A,1B
3802-3838	--	Geostationary		Spacenet 1-3 channel 14
3808.000	ESA	Geostationary		Symphonie 2
3808-3842	URS	Geostationary		Gorizont/Raduga TR 9(3825 center freq)
3809-3880	MEXICO	Geostationary		Morelos 1 & 2
3811.25-3844.25	ARB	Geostationary		Arabsat 1A,1B
3820.000	Japan	Geostationary		CS 2A/2B TT&C
3822-3858	--	Geostationary		Group A satellites channel 7
	--	Geostationary		Group B satellites channel 4
3825-3861	MEXICO	Geostationary		Morelos 1 & 2
3827-3863	ITSO	Geostationary		Intelsat 4/4A channel 4
3832-3868	INDIA	Geostationary		Insat 1B channel 4
	ISR	Geostationary		AMS-1,2 Ch9(3850 center freq)
3837.000	ESA	Geostationary		Symphonie 1
3840-3930	ESA	Geostationary		Symphonie 1
3842-3878	--	Geostationary		Group A satellites channel 8
3848.25-3881.25	ARB	Geostationary		Arabsat 1A,1B
3858-3892	URS	Geostationary		Gorizont/Molniya class 3/Raduga TR10(3875 center freq)
3854.500	ITSO	Geostationary		Marisat 2
3862-3898	--	Geostationary		Group A satellites channel 9
	--	Geostationary		Group B satellites channel 5
3865-3901	MEXICO	Geostationary		Morelos 1 & 2
3866.75-3899.25	ARB	Geostationary		Arabsat 1A,1B
3867-3903	ITSO	Geostationary		Intelsat 4/4A channel 5
3869-3941	ITSO	Geostationary		Intelsat 5/5A channel 5-6
3872-3908	INDIA	Geostationary		Insat 1B channel 5
	ISR	Geostationary		AMS-1,2 Ch10(3890 center freq)
3882-3918	--	Geostationary		Group A satellites channel 10
3882-3958	US	Geostationary		Spacenet 1-3 channel 15
3885.25-3918.25	ARB	Geostationary		Arabsat 1A,1B
3889-3960	MEXICO	Geostationary		Morelos 1 & 2
3890-3930	FRANCE	Geostationary		Telecom F3,F1 Ch.F3
3902-3938	--	Geostationary		Group A satellites channel 11

Frequency (MHz)	Country/ Agency	Use	Mode	Notes
	--	Geostationary		Group B satellites channel 6
3903.75-3936.75	ARB	Geostationary		Arabsat 1A,1B
3905-3941	MEXICO	Geostationary		Morelos 1 & 2
3907-3943	ITSO	Geostationary		Intelsat 4/4A channel 6
3908-3442	URS	Geostationary		Gorizont/Raduga TR 11(3925 center)
3912-3948	INDIA	Geostationary		Insat 1B channel 6
	ISR	Geostationary		AMS-1,2 Ch11(3930 center freq)
3922-3958	--	Geostationary		Group A satellites channel 12
3922.25-3955.25	ARB	Geostationary		Arabsat 1A,1B
3933-3967	ITSO	Geostationary		Intelsat 3 TT&C
3940.000	JAPAN	Geostationary		Ayame 2
3940.75-3973.75	ARB	Geostationary		Arabsat 1A,1B
3942-3978	--	Geostationary		Group A satellites channel 13
	--	Geostationary		Group B satellites channel 7
3945-3981	MEXICO	Geostationary		Morelos 1 & 2
3945.500	ITSP	Geostationary		Marisat 1-3
3947.500	ITSO	Geostationary		Intelsat 4/4A/5/5A TT&C
	US	Geostationary		Comstar D1-4 TT&C
3950.000	JAPAN	Geostationary		CS 2A/2B TT&C
	US	Geostationary		ATS-6
3952.000	ISR	Geostationary		AMS-1,2 Ch12(3970 center freq)
3952.500	ITSO	Geostationary		Intelsat 4/4A/5/5A TT&C
	US	Geostationary		Comstar D1-4
3954.500	ITSO	Geostationary		Marisat 1-3
3957-3993	ITSO	Geostationary		Intelsat 4/4A channel 7
3959-4031	ITSO	Geostationary		Intelsat 5/5A channel 7-8
3959.25-3992.25	ARB	Geostationary		Arabsat 1A,1B
3962-6998	--	Geostationary		Group A satellites channel 14
	INDIA	Geostationary		Insat 1B channel 7
3962-4038	US	Geostationary		Spacenet 1-3 channel 16
3969-4040	MEXICO	Geostationary		Morelos 1 & 2
3970.000	FRANCE	Geostationary	TM BCN	Telecom F1,2,3 beacon
3970-4060	ESA	Geostationary		Symphonie 2
3970-4195	ITSO	Geostationary		Intelsat 3 channel
3975.000	FRANCE	Geostationary	TM BCN	Telecom F1,2,3 beacon
3977.75-4010.75	ARB	Geostationary		Arabsat 1A,1B
3982-4018	--	Geostationary		Group A satellites channel 15
	--	Geostationary		Group B satellites channel 8
3985-4021	MEXICO	Geostationary		Morelos 1 & 2
3985-4165	JAPAN	Geostationary		CS 2A/2B channel 2
3992-4028	ISR	Geostationary		AMS-1,2 Ch13 (4010 center freq)
3996.25-4029.25	ARB	Geostationary		Arabsat 1A,1B
3997-4033	ITSO	Geostationary		Intelsat 4/4A channel 8
4002-4038	--	Geostationary		Group A satellites channel 16
	INDIA	Geostationary		Insat 1B channel 8
4014.75-4047.75	ARB	Geostationary		Arabsat 1A,1B
4022-4058	--	Geostationary		Group A satellites channel 17
	--	Geostationary		Group B satellites channel 9
4025-4061	MEXICO	Geostationary		Morelos 1 & 2
4032-4068	ISR	Geostationary		AMS 1,2 Ch14(4050 center freq)
4033.25-4066.25	ARB	Geostationary		Arabsat 1A,1B
4033.55	INDIA	Geostationary		Insat 1B TT&C
4037-4073	ITSO	Geostationary		Intelsat 4/4A/5/5A channel 9
4038.100	--	Geostationary		Group A satellites channel 18
	INDIA	Geostationary		Insat 1B channel 9
4042-4118	US	Geostationary		Spacenet 1-3 channel 17

Frequency (MHz)	Country/ Agency	Use	Mode	Notes
4049-4120	MEXICO	Geostationary		Morelos 1 & 2
4051.75- 4084.75	ARB	Geostationary		Arabsat 1A,1B
4057-4133	ITSO	Geostationary		Intelsat 2
4062-4098	--	Geostationary		Group A satellites channel 19
	--	Geostationary		Group B satellites channel 10
4063.000	ESA	Geostationary		Symphonie 2
4065-4101	MEXICO	Geostationary		Morelos 1 & 2
4068-4104	ITSO	Geostationary		Intelsat 1 channel 1
4070.25- 4103.25	ARB	Geostationary		Arabsat 1A,1B
4072-4108	IRS	Geostationary		AMS-1,2 Ch15 (4090 center freq)
4075-4195	FRANCE	Geostationary		Telecom F3,F1 Ch.F4
4075.000	JAPAN	Geostationary		Ayame 2 4.7 watt beacon
4077-4113	ITSO	Geostationary		Intelsat 4/4A/5/5A channel 10
4079.730	US	Geostationary		Telstar 2
4080.000	JAPAN/US	Geostationary		CS 2A/2B/Telstar 1/Ayame 2/ Relay 1,2
4081.000	US	Nongeostationary		Early Bird
4082-4118	--	Geostationary		Group A satellites channel 20
	INDIA	Geostationary		Insat 1B channel 10
4088.75- 4121.75	ARB	Geostationary		Arabsat 1A,1B
4092.000	ESA	Geostationary		Symphonie 2
4095.000	INDIA	Geostationary		Apple 1
4095-4185	ESA	Geostationary		Symphonie 1
4102-4138	--	Geostationary		Group A satellites channel 21
	--	Geostationary		Group B satellites channel 11
4104.000	ITSO	Geostationary		Intelsat 1 TT&C
4104.140	US	Nongeostationary		Early Bird
4105-4141	MEXICO	Geostationary		Morelos 1 & 2
4107.25- 4140.25	ARB	Geostationary		Arabsat 1A,1B
4112-4148	ISR	Geostationary		AMS-1,2 Ch16 (4130 center freq)
4117-4153	ITSO	Geostationary		Intelsat 4/4A/5/5A channel 11
4119.5990	NASA	Geostationary	25F9	ATS satellite series
4120.00	US	Geostationary		ATS-2
4121.00	ITSO	Geostationary		Intelsat II F-1,2,3,4
4122-4158	--	Geostationary		Group A satellites channel 22
	INDIA	Geostationary		Insat 1B channel 11
4122-4158	US	Geostationary		Spacenet 1-3 channel 18
4125.75- 4158.75	ARB	Geostationary		Arabsat 1A,1B
4129-4200	MEXICO	Geostationary		Morelos 1 & 2
4135.172	NASA	Space		Pioneer 7/ATS 3,4
4135.9460	NASA	Geostationary	0A0	ATS satellite series BCN/Pioneer 7
4135.957	NASA	Geostationary		ATS-5
4137.860	US	Nongeostationary		Early Bird
4138.000	ITSO	Geostationary		Intelsat 1 TT&C
4140-4180	INDIA	Geostationary		Apple 1
4142-4178	--	Geostationary		Group A satellites channel 23
	--	Geostationary		Group B satellites channel 12
4144.25- 4177.25	ARB	Geostationary		Arabsat 1A,1B
4145-4181	MEXICO	Geostationary		Morelos 1 & 2
4148-4174	ITSO	Geostationary		Intelsat 1 channel 2
4150.000	US	Geostationary		ATS-6
4152-4188	ISR	Geostationary		AMS-1,2 Ch17 (4170 center freq)
4157-4193	ITSO	Geostationary		Intelsat 4/4A channel 12
4157-4198	ITSO	Geostationary		Intelsat 5/5A channel 12
4160.350	US	Nongeostationary		Early Bird
4162-4198	--	Geostationary		Group A satellites channel 24
	INDIA	Geostationary		Insat 1B channel 12
4162.75-				

Frequency (MHz)	Country/ Agency	Use	Mode	Notes
4195.75	ARB	Geostationary		Arabsat 1A,1B
4165.000	US	Geostationary		Telstar 1/Relay 1 & 2
4169.720	US	Geostationary		Telstar 2
4170.000	US	Geostationary		Telstar 1/Relay 1 & 2
4175.000	US	Geostationary		Telstar 1/Relay 1 & 2
4178.5910	NASA	Geostationary	25F9	ATS satellite series
4179.00	US	Geostationary		ATS-2
4188.7000	ESA	Geostationary		Marecs
4192.5-4200	ITSO	Geostationary		Intelsat 5/5A ship-to-shore
4194-4200.5	ESA	Geostationary		Marecs ship-to-shore
4195-4199	ITSO	Geostationary		Marisat ship-to-shore
4195.000	FRANCE	Geostationary	TM	F-Sat 1 beacon
4195.1720	NASA	Geostationary	0A0	ATS satellite series BCN
4197.250	US	Geostationary		SBS 1-4 TT&C
4197.750	US	Geostationary		SBS 1-4 TT&C
4198.000	CANADA	Geostationary		Anik A1-3 TT&C
4198.250	US	Geostationary		Westar 1-3 TT&C
4198.600	INDO	Geostationary		Palapa A1-2
	CANADA	Geostationary		Anik C1-3 TT&C
4198.750	US	Geostationary		Westar 1-3
4199.000	US	Geostationary		Satcom 1-4 TT&C
	US	Geostationary		Westar 1-4 TT&C
	CANADA	Geostationary		Anik A1-3 TT&C
	CANADA	Geostationary		Anik C1-3 TT&C
4199.100	INDO	Geostationary		Palapa A1-2 TT&C
	CANADA	Geostationary		Anik B TT&C
4199.250	US	Geostationary		Westar 1-3 TT&C
4199.400	US	Geostationary		Satcom 5 (Aurora 1) TT&C
4200.000	US	Nongeostationary		Relay 1 & 2
	ISR	Geostationary	TM	AMS-1
5000-5125	ESA	Geostationary		Aerosat uplink
5000.000	US	Nongeostationary		Atlas-Centaur 4 (beacon)
5100.000	URS	Geostationary		Iskra 1
5125-5250	ESA	Geostationary		Aerosat downlink
5690.000	US	Geostationary		Geos 3
	USN	Geostationary		Fleetsatcom
5690-5765	US	Nongeostationary		Mercury-Atlas manned program
5700-6200	URS	Geostationary		Raduga/Ekran/Gorizont
5725-6225	URS	Geostationary		Molniya class 1/3 COMSATs
5745.000	ESA	Nongeostationary		Spot-1
5764.000	ITSO	Geostationary		Intelsat 5/5A telemtry
5765.000	USAF	Geostationary		Scatha 1
	NASA	Geostationary		Geos 3
	USN	Geostationary		Fleetsatcom
	US	Nongeostationary		Atlas-Centaur 4 (beacon)
	NASA	Nongeostationary		Gemini caps/Gemini Agena Target Veh/ Gemini 9/Target B-ATDA/Apollo CM,LM/ Explorer 36
5850.000	ISR	Ground Station		AMS-1,2 TT&C
5862-5898	ISR	Ground Station		AMS-1,2 Ch1 (5880 center freq)
5902-5938	ISR	Ground Station		AMS-1,2 Ch2 (5920 center freq)
5924-6425	Various	Geostationary		(See following)

C-Band Uplink

Telecom 1A/1B	ASC 1-5	CCC 1-2
Digisat 1B	Digisat 2	Equastar 1-2
Ford 1-3	Spotnet C1-2	AMS 1-2
Arabsat 1-2	Morelos 1-2	Pacific Area Satellite System A/B
CS(Sakura)	Westar 1-5	Anik A1-3
Spacenet 1-3	Anik B	Palapa A1-2
Comstar D1-4	Galaxy 1-4	Satcom 1-5
Anik D1-2	Telstar 3	Aurora 2-3

Frequency (MHz)	Country/ Agency	Use	Mode	Notes
5925-5965	FRANCE	Ground Station		Telecom F1,3 Ch.F1
5927-5963	FRANCE	Ground Station		F-Sat 1, Ch.R1
5930-6155	ITSO	Geostationary		Intelsat 3 channel 1 uplink
5942-5978	ISR	Ground Station		AMS-1,2 Ch3 (5960 center freq)
5967-6003	FRANCE	Ground Station		F-Sat 1, Ch.R2
5980-6100	FRANCE	Ground Station		Telecom F1,3 Ch.F2
5982-6018	ISR	Ground Station		AMS-1,2 Ch4 (6000 center freq)
6000.000	URS	Ground Station		Mars 1
6007-6043	FRANCE	Ground Station		F-Sat 1 Ch.R3
6022-6058	ISR	Ground Station		AMS-1,2 Ch5 (6040 center freq)
6047-6083	FRANCE	Ground Station		F-Sat 1 Ch.R4
6050.348	DOE	Geostationary	45F9	
6050-438	DOE	Geostationary	45F9	
6050.483	DOE	Geostationary	45F9	
6050.538	DOE	Geostationary	45F9	
6050.573	DOE	Geostationary	45F9	
6050.618	DOE	Geostationary	1F0	
6050.633	DOE	Geostationary	45F9	
6062-6098	ISR	Ground Station		AMS-1,2 Ch6 (6080 center freq)
6087-6123	FRANCE	Ground Station		F-Sat 1 Ch.R5
6102-6138	ISR	Ground Station		AMS-1,2 Ch7 (6120 center freq)
6115-6155	FRANCE	Ground Station		Telecom F1,3 Ch.F3
6127-6163	FRANCE	Ground Station		F-Sat 1 ChR.6
6142-6178	ISR	Ground Station		AMS-1,2 Ch8 (6160 center freq)
6167-6203	FRANCE	Ground Station		F-Sat 1 Ch.R7
6175.000	ITSO	Geostationary		Intelsat 3/4/4A/5/5A Command link
6182-6218	ISR	Ground Station		AMS-1,2 Ch9 (6200 center freq)
6195-6420	ITSO	Geostationary		Intelsat 3 uplink
6207-6243	FRANCE	Ground Station		F-Sat 1 Ch.R8
6210.000	FRANCE	Ground Station		Telecom F1,2,3 TT&C
6215.000	FRANCE	Ground Station		Telecom F1,2,3 TT&C
6222-6258	ISR	Ground Station		AMS-1,2 Ch10 (6240 center freq)
6247-6283	FRANCE	Ground Station		F-Sat 1 Ch.9
6262-6298	ISR	Ground Station		AMS-1,2 Ch11 (6280 center freq)
6282-6408	ITSO	Geostationary		Intelsat 2 uplink
6287-6323	FRANCE	Ground Station		F-Sat 1 Ch.R10
6288-6314	ITSO	Geostationary		Intelsat 1 channel 1 uplink
6300-6420	FRANCE	Ground Station		Telecom F1,3 Ch.F4
6301.020	US	Ground Station		Early Bird
6302-6338	ISR	Ground Station		AMS-1,2 Ch12 (6320 center freq)
6327-6363	FRANCE	Ground Station		F-Sat 1 Ch.R11
6342-6378	ISR	Ground Station		AMS-1,2 Ch13 (6360 center freq)
6346.000	US	Ground Station		ATS satellite series
	ITSO	Geostationary		Intelsat II F-1,2,3,4
6367-6403	FRANCE	Ground Station		F-Sat 1 Ch.R12
6377-6403	ITSO	Geostationary		Intelsat 1 channel 2 uplink
6382-6418	ISR	Ground Station		AMS-1,2, Ch14 (6400 center freq)
6389.970	US	Ground Station		Early Bird
6417.5-6425	ITSO	Ground Station		Intelsat V MCS shore-to-ship
6420-6424	ITSO	Ground Station		Marisat shore-to-ship
6420-6425	ESA	Ground Station		Marecs A&B shore-to-ship
6422.000	FRANCE	Ground Station		F-Sat 1 TT&C
6422-6458	ISR	Ground Station		AMS-1,2 Ch15 (6440 center freq)
6462-6498	ISR	Ground Station		AMS-1,2 Ch16 (6480 center freq)
6502-6538	ISR	Ground Station		AMS-1,2 Ch17 (6520 center freq)
6550.000	ISR	Ground Station		AMS-1,2 TT&C
7040-7080	FRANCE	Geostationary		Telecom military
7085-7125	FRANCE	Geostationary		Telecom military
7241.2880	USN	Geostationary	5F9/4.8F9	
7245-7275	NATO	Geostationary		NATO Phase III channel 1
7250.1000	USAF	Geostationary	5F9	DSCS Phase II Earth Coverage Beacon
	USN	Geostationary	5F9	
7250.4400	USN	Geostationary	5F9/4.8F9	

Frequency (MHz)	Country/ Agency	Use	Mode	Notes
7250-7375	USAF	Geostationary		DSCS Phase II channel 1
7250-7310	USAF	Geostationary		DSCS Phase III downlink
7250-7750	URS	Geostationary		Raduga/GALS downlink
7252.5-7262	USAF	Geostationary		TACSAT downlink
7252.7000	USN	Geostationary	5.13F9	
7254.6500	USN	Geostationary	200F9	
7255.0000	USN	Geostationary	500F9	
7255-7370	FRANCE	Geostationary		Telecom 1A-1B
7255.2050	USN	Geostationary	8.21F9	
7255.6280	US	Geostationary	3F9	
7257.3-7259.3	UK	Geostationary	170F9	Skynet (2 MHz channel)
7260.0000	NATO	Geostationary	170F9	NATO 3 series satellites
	USN	Geostationary	170F9	
7261.6090	US	Geostationary	5F98F9	
7263.1000	US	Geostationary	3.88F9	
	USN	Geostationary	8.21F9	
7263.1730	US	Geostationary	1.9F9	
7265.000	USN	Geostationary	500F9	
7266.4-7286.4	UK	Geostationary		Skynet COMSAT (20 MHz channel)
7267.0250	US	Geostationary		DSCS Phase I
7267.5870		Geostationary	1.9F9	
7271.7125	US	Geostationary		DSCS Phase I
7272-7332	NATO	Geostationary		NATO Phase III
7275.0000	USAF	Geostationary	50F9	
	US	Geostationary	50F9	
	USN	Geostationary	50F9	
7276.6000	USAF	Geostationary	M38.4F9/F0	
7276.8000	USAF	Geostationary	M38.4F9/F0	
7277.7000	USAF	Geostationary	M10F9	
	US	Geostationary	M10F9	
	USN	Geostationary	M10F9	
7277.9625	US	Geostationary		DSCS Phase I
7282.1850	USN	Geostationary	910F9/200F9	
7285.7350	US	Geostationary		DSCS Phase I
7287.0000	US	Geostationary	M3F9	
7290.1880	US	Geostationary	M1.9F9	
7290.5800	USN	Geostationary	M1.9F9	
7291.7710	US	Geostationary	M1.9F9	
7294.2530	USN	Geostationary	M1.8F9	
7298.5000	USAF	Geostationary		TACSAT Beacon
7299.1000	UK	Geostationary		Skynet tracking beacon
7299.5000	USAF	Geostationary		DSCS Phase 1 Beacon
7300.0000	USAF	Geostationary	0A0 M2/0A3 M2/0F1 M2/0F3 M2/0F9 M2/P/0 M2/0P9	
7300.8000	USN	Geostationary	M25F9	
7301.1000	USN	Geostationary	M1/0F9	
7301.3000	USN	Geostationary	200F9	
7303.1000	USAF	Geostationary	M1.8F9	
7303.1460	USN	Geostationary	M1.7F9	
7303.5270	USN	Geostationary	420F9	
7304.1230	USN	Geostationary	420F9	
7304.7600	USN	Geostationary	465F9	
7308.6100	USAF	Geostationary	M5F9	
	US	Geostationary	M3.88F9	
	USN	Geostationary	100F9	
7311.8600	US	Geostationary	M4F9	
7312.3290	USN	Geostationary	M5.13F9	
7313.3330	US	Geostationary	M1.9F9	
7315.5260	US	Geostationary	M1.4F9/M4F9/M5.ADF9	
7319.0000	USN	Geostationary	M4.8F9	
7319.5000	USN	Geostationary	M4.8F9	
7322.9000	US	Geostationary	M5.13F9	
	USN	Geostationary	M5.13F9	

Frequency (MHz)	Country/ Agency	Use	Mode	Notes
7323.8900	USAF	Geostationary	M1.94F9	
7325.0500	USN	Geostationary	M50F9	
7327.1230	USAF	Geostationary	M4F9	
	US	Geostationary	M4F9	
7327.7660	USN	Geostationary	500F9	
7329.4260	USN	Geostationary	M5.13F9	
7330-7390	US	Geostationary		DSCS Phase 3 satellites
7331.6710	US	Geostationary	M1F9	
7333.2990	USN	Geostationary	M4F9	
7338.4540	US	Geostationary	M3.09F9	
7339.2000	NATO	Geostationary		NATO Phase 3 satellites BCN
7339.5000	USN	Geostationary	275F9	
7341.5000	USN	Geostationary	M3.88F9	
7344.5020	USN	Geostationary	M4F9	
7344.5820	USAF	Geostationary	M8F9	
7347-7449	NATO	Geostationary		NATO Phase 3 satellites
7349.7700	USAF	Geostationary	M1.94F9	
	USN	Geostationary	M3.88F9	
7349.8910	USN	Geostationary	M3F9	
	USN	Geostationary	690F9	
7350.8870	USN	Geostationary	M3.88F9	
7352.0000	USN	Geostationary	M50F9	
7354.0000	USN	Geostationary	M1.9F9	
7358.0000	USN	Geostationary	M8F9	
7367.0000	USN	Geostationary	M4F9	
7367.5870	USN	Geostationary	M4F9	
7369.6500	US	Geostationary	420F9	
7371.7670	US	Geostationary	M1.29F9	
7371.9160	USAF	Geostationary	M.15F9	
7374.3400	US	Geostationary	420F9	
7400-7450	USAF	Geostationary	0A0 M20A3 M20F1 M20F3 M20F9 M20P0 M20P9	
	US	Geostationary		DSCS Phase 2 satellites
7402.6600	USN	Geostationary	M1.9F9/5.13F9	
7403.7660	USN	Geostationary	M2.4F9	
7405.6500	USN	Geostationary	M1.17F9	
7406.3850	USN	Geostationary	M1.17F9	
7408.2980	US	Geostationary	M2.58F9	
	USN	Geostationary	M3F9	
7410-7495	US	Geostationary		DSCS Phase 3 satellites
7410.2000	US	Geostationary	M1.94F9	
	USN	Geostationary	M1.94F9	
7410.4430	USN	Geostationary	M1.5F9	
7415.2620	USN	Geostationary	M2.58F9	
7420.0000	USN	Geostationary	M10.26F9	
7423.8190	USN	Geostationary	M8F9/4F9	
7425.0000	USAF	Geostationary	M50F9	
7428.1470	USN	Geostationary	647F9	
7429.0720	USN	Geostationary	M3.88F9	
7433.4470	USAF	Geostationary	M1.8F9	
7435.5000	US	Geostationary	M3.88F9	
	USN	Geostationary	M3.88F9	
7441.0660	USN	Geostationary	M1.9F9	
7442.8660	USN	Geostationary	M5F9	
7446.5790	USN	Geostationary	M5.5F9	
7446.9250	USN	Geostationary	M2.58F9	
7490-7675	US	Geostationary		DSCS Phase 2 channel 3
7490.0000	USAF	Geostationary	M10F9 1.2F9 0A0 20A3 20F1 20F3 20F9 20P0 20P9	
	US	Geostationary	M1.1F9 16F9 12F9	
	USN	Geostationary	M1.1F9	
7515-7575	US	Geostationary		DSCS Phase 3 satellites
7515.0000	USAF	Geostationary	M50F9	
7543.8690	USAF	Geostationary	M1F9	
7548.0540	USAF	Geostationary	M1F9	
7549.2210	USAF	Geostationary	M42F9	

Frequency (MHz)	Country/ Agency	Use	Mode	Notes
7556.6390	USAF	Geostationary	M1.5F9	
7560.0000	USAF	Geostationary	M20F9/K32F9	
7595-7655	US	Geostationary		DSCS Phase 3 satellites
7664.0000	USAF	Geostationary	M1.2F9	
	US	Geostationary	M1.1F9	
7668.3000	US	Geostationary	M3.88F9	
7670.1200	US	Geostationary	M1.18F9	
7672.0000	US	Geostationary	K5F9	
7675-7725	US	Geostationary		DSCS Phase 3 satellites
7675.1000	USAF	Geostationary	K5F9	DSCS Phase 2 Beacon
	USN	Geostationary	K5F9	
7700-7750	US	Geostationary		DSCS Phase 2 channel 4
7703.1110	USN	Geostationary	M2.6F9	
7704.4260	USN	Geostationary	M2.6F9	
7710.4530	US	Geostationary	M3.88F9	
7714.4530	US	Geostationary	M4F9	
7727.1620	USAF	Geostationary	M3.88F9	
7734.6700	USAF	Geostationary	M5F9	
7734.6730	US	Geostationary	M3.88F9	
7736.1190	US	Geostationary	420F9	
7738.4290	US	Geostationary	M3.88F9	
7739.9290	USAF	Geostationary	M2.58F9	
	US	Geostationary	M1.18F9	
7743.8450	USAF	Geostationary	M5.13F9	
7744.0000	US	Geostationary	M3.38F9	
7900-8400	USN	Ground Station		FLEETSATCOM satellite to shore link
	US	Ground Station		DSCS Phase 2 channel 4 uplink
7900.0000	USAF	Ground Station	0A0	
7900.2760	USN	Ground Station	500F9	
7902.1280	USN	Ground Station	M3.88F9	
7904.4260	USN	Ground Station	M2.6F9	
7907.3120	USAF	Ground Station	M3F9	
7908.6330	USN	Ground Station	M8.21F9	
7910.4530	USN	Ground Station	M3.88F9	
7913.3330	USN	Ground Station	M3.88F9	
7914.4530	USN	Ground Station	M4F9	
7914.8780	US	Ground Station	M3.88F9	
7921.6900	USN	Ground Station	200F9	
7921.7000	US	Ground Station	M3.88F9	
	USN	Ground Station	M3.88F9	
7928.6680	USN	Ground Station	M5.13F9	
7928.7000	USN	Ground Station	M1.18F9	
7934.6700	USAF	Ground Station	M5F9	
7934.9230	US	Ground Station	M1.94F9	
7940.3000	USN	Ground Station	M3F9	
7948.0000	USN	Ground Station	M2.8F9	
7949.4460	USAF	Ground Station	M1F9	
7975-8162	NATO	Ground Station		NATO satellite uplink
7975.4400	USN	Ground Station	K5F9	
7976.02-7978.02	UK	Ground Station		Skynet uplink (2 MHz channel)
7977.5-7987.5	US	Ground Station		TACSAT uplink
7978.2000	USN	Ground Station	K200F9	
7980-8095	FRANCE	Ground Station		Telecom 1A/1B uplink
7980.0000	USN	Ground Station	800F9	
7980.0230	USN	Ground Station	200F9	
7980.6280	US	Ground Station	420F9	
7981.2000	USN	Ground Station	200F9	
7983.9280	USN	Ground Station	M1.9F9	
7985.12-8005.12	UK	Ground Station		Skynet (20 MHz channel) uplink
7985.745	US	Ground Station		DSCS Phase 1 satellites
7990.4325	US	Ground Station		DSCS Phase 1 satellites

Frequency (MHz)	Country/ Agency	Use	Mode	Notes
7992.2000	USAF	Ground Station	38.4F9	
7992.5870	USN	Ground Station	5F9	
7994.0000	USN	Ground Station	910F9	
7995.0000	USN	Ground Station	M15F9	
7996.6825	USN	Ground Station		DSCS Phase 1 satellites
8000.0000	USAF	Ground Station	50F9/M30F9	
	US	Ground Station	M50F9	
	USN	Ground Station	M50F9	
	USAF	Geostationary		LES 1,2,4
8001.8000	USAF	Ground Station	38F9/38P0	
8002.7000	USAF	Ground Station	M1F9	
	US	Ground Station	M10F9	
	USN	Ground Station	M10F9	
8004.4954	US	Ground Station		DSCS Phase 1 satellites
8012.7000	US	Ground Station	420F9	
8019.1850	US	Ground Station	500F9	
8020.3000	USAF	Ground Station	700F9	
8025.0000	USAF	Ground Station	0A0 M20A3 M20F1 MS0F3 M20F9 M20P0 M20P9	
8025.8000	USN	Ground Station	25F9	
	USA	Ground Station	.10F9	
8026.1000	USN	Ground Station	10F9	
8026.3000	USN	Ground Station	200F9	
8028.1000	USAF	Ground Station	M1.8F9	
8029.4000	USN	Ground Station	M3.88F9	
8030.3600	USN	Ground Station	50F9	
8033.1190	US	Ground Station	M3.88F9	
8033.6100	USAF	Ground Station	M5F9	
8036.5300	USN	Ground Station	429F9	
8036.8600	USN	Ground Station	M3.88F9	
8038.4910	USN	Ground Station	200F9	
8038.5200	USN	Ground Station	910F9	
8042.8080	USN	Ground Station	M5.13F9	
8044.5000	USN	Ground Station	4.80F9	
8044.5060	USAF	Ground Station	M2.40F9	
8051.4520	USAF	Ground Station	M1F9	
8052.1230	USAF	Ground Station	M5.13F9	
8058.2990	US	Ground Station	M4F9	
8060.4430	US	Ground Station	M1.94F9	
8062.0990	USN	Ground Station	M3.88F9	
8069.5020	USN	Ground Station	M4F9	
8076.5000	USAF	Ground Station	M1.29F9	
8077.2410	USN	Ground Station	M3.88F9	
8078.3370	US	Ground Station	200F9	
8083.0000	USAF	Ground Station	M8.21F9	
8083.2630	USN	Ground Station	420F9	
8084.2680	USN	Ground Station	420F9	
8088.2120	US	Ground Station	M2.4F9	
8092.0000	USN	Ground Station	M4F9	
8094.1500	US	Ground Station	420F9	
8094.8760	USN	Ground Station	200F9	
8096.9160	USN	Ground Station	M1.5F9	
8098.4020	USN	Ground Station	200F9	
8098.9200	US	Ground Station	420F9	
8125-8176	US	Ground Station		DSCS Phase 2 channel 2 uplink
8125.0000	USAF	Ground Station	0A0/M20A3/M20F1/M20F3/M20F9/M20P0/M20P9	
8125.2760	USN	Ground Station	200F9	
8128.7660	USAF	Ground Station	M3.88F9	
8133.2980	US	Ground Station	M3F9	
	USN	Ground Station	2.58F9	
8133.6738	US	Ground Station	M2F9	
8133.6740	USAF	Ground Station	M3.88F9	
8135.4430	USAF	Ground Station	M3F9	
8137.2510	USN	Ground Station	200F9	
8145.0000	US	Ground Station	M3.88F9	

Frequency (MHz)	Country/ Agency	Use	Mode	Notes
8146.8820	US	Ground Station	M3F9	
8149.8810	US	Ground Station	200F9	
8150.0000	USAF	Ground Station	M3.88F9	
	JAPAN	Nongeostationary		MOS-1
8153.1470	USN	Ground Station	M3.88F9	
8156.2620	USN	Ground Station	M5.14F9	
8158.4470	USAF	Ground Station	M50F9	
8161.8600	USN	Ground Station	647F9	
8164.2600	US	Ground Station	420F9	
8169.6160	US	Ground Station	M1.8F9	
8171.3650	USN	Ground Station	M3F9	
8171.9250	US	Ground Station	M3.8F9	
8174.4470	USAF	Ground Station	M3.8F9	
8174.7600	USN	Ground Station	250F9	
8212.5000	NASA	Nongeostationary	M170F9	Landsat
8215-8400	US	Ground Station		DSCS Phase 2 channel 3 uplink
8215.0000	US	Ground Station	M1.1F9/M12F9/M16F9	
	USAF	Ground Station	M1.1F9	
	USN	Ground Station	M1.1F9	
	USAF	Ground Station	0A0/M20A3/M20F1/M20F3/M20F9/M20O0/M20P9	
8221.0000	US	Ground Station	M1.1F9	
8230.0000	US	Ground Station	M1.1F9	
8240.0000	US	Ground Station	M1.1F9	
	USAF	Ground Station	M50F9	
8245.6560	US	Ground Station	M1.1F9	
8253.1	ESA	Nongeostationary		Spot-1
8258.0000	US	Ground Station	M1.1F9	
8268.8690	USAF	Ground Station	M1F9	
8273.0540	USAF	Ground Station	M1.29F9	
8276.3210	US	Ground Station	M1.8F9	
8278.9990	US	Ground Station	M3.38F9	
8281.6390	USAF	Ground Station	M1.5F9	
8285.0000	USN	Ground Station	32F9/M20F9	
8292.1460	USAF	Ground Station	M2F9	
8307.1	ESA	Nongeostationary		Spot-1
8348.0000	USAF	Ground Station	M1.2F9	
8349.8350	US	Ground Station	M1.94F9	
8350.000	JAPAN	Nongeostationary		MOS-1
8352.3210	US	Ground Station	M1.8F9	
8379.5120	USN	Ground Station	500F9	
8387.6000	US	Ground Station	M3.38F9	
8388.5210	US	Ground Station	M5.13F9	
8389.0000	US	Ground Station	M1.1F9	
	USAF	Ground Station	M1.2F9/M20F9	
8393.3260	US	Ground Station	200F9	
8400.0000	NASA	Nongeostationary	M10A9/M10F9/M10P9	
8410.9259	NASA	Space	170F9	
8412.2840	NASA	Space	170F9	Voyager 2
8415.0000	NASA	Space	M3.3F9	Voyager 2, 11-12 or 21.3W TT&C/ Mariner 10
8420.430	NASA	Space		Voyager 2
8428.5802	ESA	Space		Giotto Comet Halley satellite
10450-10700		Geostationary		Orion 1-3 satellites
10470.000	ARS	Nongeostationary		OSCAR 9 X-band beacon
10700-10950		Geostationary		ISI 1-2 satellites
10711-10765	US	Geostationary		USA-Sat 1,2 TR1A-East (10738 center) USA-Sat 1,2 TR9B-East (10738 center)
10771-10825	US	Geostationary		USA-Sat 1,2 TR2A-East (10798 center USA-Sat 1,2 TR10B-East (10798 center)
10831-10885	US	Geostationary		USA-Sat 1,2 TR3A-East (10858 center) USA-Sat TR11B-East (10858 center)
10891-10945	US	Geostationary		USA-Sat 1,2 TR4A-West (11865 center) USA-Sat 1,2 TR12B-East (10918 center)
10950-11200	ESA	Geostationary		OTS 2(ESA)

Frequency (MHz)	Country/ Agency	Use	Mode	Notes
	US			Orion 1-3 satellites
	G			Unisat 1-3 satellites
10950-11033	ETSO	Geostationary		Eutelsat-I F1,2,3 channels 1X/1Y
10950-11200	US	Geostationary		Cygnus 1,2; European downlink
10952-10988	US	Geostationary		Orion 1,2,3 Ch1(10970 center freq)
10954-11030	ITSO	Geostationary		Intelsat 5/5A
10958-10992	URS	Geostationary		Loutch channel 1
10972-11008	US	Geostationary		Orion 1,2,3 Ch2 (10990 center freq)
10992-11028	US	Geostationary		Orion 1,2,3 Ch3 (11010 center freq)
11008-11042	URS	Geostationary		Loutch channel 2
11012-11048	US	Geostationary		Orion 1,2,3 Ch4 (11030 center freq)
11032-11068	US	Geostationary		Orion 1,2,3 Ch5 (11050 center freq)
11033.333-11116.667	ETSO	Geostationary		Eutelsat-I F1,2,3 channel 2X/2Y
11052-11088	US	Geostationary		Orion 1,2,3 Ch5 (11070 center freq)
11058-11092	URS	Geostationary		Loutch channel 3
11072-11108	US	Geostationary		Orion 1,2,3 Ch7 (11090 center freq)
11092-11128	US	Geostationary		Orion 1,2,3 Ch8 (11110 center freq)
11108-11142	URS	Geostationary		Loutch channel 4
11112-11148	US	Geostationary		Orion 1,2,3 Ch9 (11130 center freq)
11116.666-11200.000	ETSO	Geostationary		Eutelsat-I F1,2,3 channels 3X/3Y
11119-11191	ITSO	Geostationary		Intelsat 5/51
11132-11168	US	Geostationary		Orion 1,2,3 Ch10 (11150 center freq)
11152-11188	US	Geostationary		Orion 1,2,3 Ch11 (11170 center freq)
11158-11192	URS	Geostationary		Loutch channel 5
11198.000	ITSO	Geostationary		Intelsat 5/5A TT&C
11299-11459	ISR	Geostationary		AMS-1,2
11325.000	ITALY	Geostationary		Sirio COMSAT
11331-11862	ITALY	Geostationary		Sirio COMSAT
11450.0-11453.5	D	Geostationary	TM BCN	DFS 1,2
11450.0-11533.3	ETSO	Geostationary		Eutelsat-I F1,2,3 channel 4X/4Y
11450-11470	ESA	Geostationary		OTS 2
11450-11700	D	Geostationary		DFS 1,2
11450.350	ETSO	Geostationary		Eutelsat-I F2,F3
11451.091	ETSO	Geostationary	TM BCN	Eutelsat-I F1
11452.000	ITSO	Geostationary		Intelsat 5/5A TT&C
11457-11698	ITSO	Geostationary		Intelsat 5/5A
11458-11492	URS	Geostationary		Loutch channel 6
11462-11498	US	Geostationary		Orion 1,2,3 Ch12 (11480 center freq)
11470.000	ITALY	Geostationary		Sirio
11482-11518	US	Geostationary		Orion 1,2,3 Ch13 (11500 center freq)
11490-11530	ESA	Geostationary		OTS 2
11502-11538	US	Geostationary		Orion 1,2,3 Ch14 (11520 center freq)
11508-11542	URS	Geostationary		Loutch channel 7
11508.7500	ESA	Geostationary		ETS 2 TT&C
11510.000	ESA	Geostationary		OTS 2 TT&C
11520.000	ITALY	Geostationary		Sirio
11522-11558	US	Geostationary		Orion 1,2,3 Ch15 (11540 center freq)
11533.333-11616.667	ETSO	Geostationary		Eutelsat-I F1,2,3 channels 5X/5Y
11542-11578	US	Geostationary		Orion 1,2,3 Ch16 (11560 center freq)
11558-11592	URS	Geostationary		Loutch channel 8
11562-11598	US	Geostationary		Orion 1,2,3 Ch17 (11580 center freq)
11575.000	ESA	Geostationary		OTS-2TT&C
11580-11700	ESA	Geostationary		OTS 2
11582-11618	US	Geostationary		Orion 1,2,3 Ch18 (11600 center freq)
11602-11638	US	Geostationary		Orion 1,2,3 Ch19 (11620 center freq)
11608-11642	URS	Geostationary		Loutch channel 9
11616.666-11700.000	ETSO	Geostationary		Eutelsat-I F1,2,3 channel 6X/6Y
11622-11658	US	Geostationary		Orion 1,2,3 Ch20 (11640 center)

Frequency (MHz)	Country/ Agency	Use	Mode	Notes
11640.000	ESA	Geostationary		OTS 2 TT&C
11642-11678	US	Geostationary		Orion 1,2,3 Ch21 (11660 center)
11658-11692	URS	Geostationary		Loutch channel 10
11662-11698	US	Geostationary		Orion 1,2,3 Ch22 (11680 center)
11690-11700	ESA	Geostationary		Marots satellite to shore link
11700.000	CANADA	Geostationary		CTS-1
11700-1220	Various	Geostationary		See below

Domestic Ku-Band Satellites

ABC 1-2	ASC 1-5	CCC 1-2
Comstar K1-3	Digisat 1A	Digisat 2
Ford 1-3	RSI 1-4	Spotnet K1-4
Usat 1-4	Galaxy H5-7	Satcom K1-3
Westar A-F	AMS 1-2	ISI 1-2
Unisat 1-3	Pacific Area Satellite System A & B	

Frequency (MHz)	Country/ Agency	Use	Mode	Notes
11700-11717	US	Geostationary		Gstar A1-3 TT&C
11700-12200	US	Geostationary		Cygnus 1,2 American downlink
11702.000	US	Geostationary		Gstar A1-3 Beacon
11703-11757	CANADA	Geostationary		Anik C1-3 channel 1
	US	Geostationary		Gstar A1-3 channel 1
11703.5-11746.5	US	Geostationary		SBS 1-4 channel 1
11703.5-11795.5	MEXICO	Geostationary		Morelos 1 & 2
11704-11776	CANADA	Geostationary		Anik B channel 13
11706.72	JAPAN	Geostationary		BS-2B
11706-11715	US	Geostationary	TM BCN	USA-Sat 1,2
11706-11774	US	Geostationary		Spacenet 1-3 channel 19
11716-11770	CANADA	Geostationary		Anik C1-3 channel 2
11717-11771	US	Geostationary		Gstar A1-3 channel 9
11718-11772	US	Geostationary		USA-Sat 1,2 TR9B-West (11745 centr)
				USA-Sat 1,2 TR1A-West (11745 center)
11727.000	--	Geostationary		Europe DBS channel 1(Video)
11746.000	--	Geostationary		Europe DBS channel 2 (Video)
	D	Geostationary		TV-Sat A3,A5 DBS channel 2
11752.5-11795.5	US	Geostationary		SBS 1-4 channel 2
11764-11818	CANADA	Geostationary		Anik C1-3 channel 3
	US	Geostationary		Gstar A1-3 channel 2
11765.000	--	Geostationary		Europe DBS channel 3(Video)
11777-11831	CANADA	Geostationary		Anik C1-3 channel 4
11785.02	G	Geostationary		Unisat-1 DBS channel 4
11778-11832	US	Geostationary		Gstar A1-3 channel 10
				USA-Sat 1,2 TR2A-West (11805 center)
				USA-SAT 1,2 TR10B-West (11805 center)
11784-11856	CANADA	Geostationary		Anik B channel 14
11785.000	--	Geostationary		Europe DBS channel 4(Video)
11786-11854	US	Geostationary		Spacenet 1-3 channel 20
11792.5-11797.5	ESA	Geostationary		OTS 2
11801.5-11844.5	US	Geostationary		SBS 1-4 channel 3
11801.5-11893.5	MEXICO	Geostationary		Morelos 1 & 2
11804.000	--	Geostationary		Europe DBS channel 5(Video)
11823.000	--	Geostationary		Europe DBS channel 6(Video)
	D	Geostationary		TV-Sat A3,A5 DBS channel 6
11825-11879	CANADA	Geostationary		Anik C1-3 channel 5
	US	Geostationary		Gstar A1-3 channel 3
11838-11892	CANADA	Geostationary		Anik C1-3 channel 6
	US	Geostationary		USA-Sat 1,2 TR11B-West (11865 center)
				USA-Sat 1,2 TR3A-West (11865 center)

Frequency (MHz)	Country/ Agency	Use	Mode	Notes
11839-11893	US	Geostationary		Gstar A1-3 channel 11
11842.000	--	Geostationary		Europe DBS channel 7(Video)
11843-11928	CANADA	Geostationary		CTS-1
11850.5-11893.5	US	Geostationary		SBS 1-4 channel 4
11855.000	D	Geostationary		Sirio
11861.000	--	Geostationary		Europe DBS channel 8(Video)
	G	Geostationary		Unisat-1 DBS channel 8
11864.11936	CANADA	Geostationary		Anik B channel 15
11880.000	--	Geostationary		Europe DBS channel 9(Video)
11886-11940	CANADA	Geostationary		Anik C1-3 channel 7
	US	Geostationary		Gstar A1-3 channel 4
11886-11904	CANADA	Geostationary		Spacenet 1-3 channel 21
11898-11952	US	Geostationary		USA-Sat 1,2 TR12B-West (11925 center) USA-Sat 1,2 TR4A-West (11925 center)
11899-11953	CANADA	Geostationary		Anik C1-3 channel 8
11899.5-11942.5	US	Geostationary		SBS 1-4 channel 5
11899.5-11991.5	MEXICO	Geostationary		Morelos 1 & 2
11900.000	--	Geostationary		Europe DBS channel 10 (Video)
	D	Geostationary		TV-Sat A3,A5 DBS channel 10
11900-11954	US	Geostationary		Gstar A1-3 channel 12
11919.000	--	Geostationary		Europe DBS channel 11 (Video)
11919.28	JAPAN	Geostationary		BS-2B
11938.000	--	Geostationary		Europe DBS channel 12 (Video)
	G	Geostationary		Unisat-1 DBS channel 12
11944-12016	CANADA	Geostationary		Anik B channel 16
11946-12014	US	Geostationary		Spacenet 1-3 channel 22
11947-12001	CANADA	Geostationary		Anik C1-3 channel 9
	US	Geostationary		Gstar A1-3 channel 6
11948.5-11991.5	US	Geostationary		SBS 1-4 channel 6
11957.000	--	Geostationary		Europe DBS channel 13 (Video)
11958-12012	US	Geostationary		USA-Sat 1,2 TR5A-West (11985 center) USA-Sat 1,2 TR13B-West (11985 center)
11960-12014	CANADA	Geostationary		Anik C1-3 channel 10
11961-12015	US	Geostationary		Gstar A1-3 channel 13
11976.000	--	Geostationary		Europe DBS channel 14 (Video)
	D	Geostationary		TV-Sat A3,A5 DBS channel 14
11996.000	--	Geostationary		Europe DBS channel 15 (Video)
	JAPAN	Geostationary		BS-2B
11997.5-12040.5	US	Geostationary		SBS 1-4 channel 7
11997.5-12089.5	MEXICO	Geostationary		Morelos 1 & 2
12000.000	Japan	Geostationary		CTS
12008-12062	CANADA	Geostationary		Anik C1-3 channel 11
	US	Geostationary		Gstar A1-3 channel 6
12015.000	--	Geostationary		Europe DBS channel 16 (Video)
	G	Geostationary		Unisat-1 channel 16
12018-12072	US	Geostationary		USA-Sat 1,2 TR6A-West(12045 center) USA-Sat 1,2 TR14B-West (12045 center)
12021-12075	CANADA	Geostationary		Anik C1-3 channel 12
12022-12076	US	Geostationary		Gstar A1-3 channel 14
12024-12096	CANADA	Geostationary		Anik B channel 17
12026-12094	US	Geostationary		Spacenet 1-3 channel 23
12034.000	--	Geostationary		Europe DBS channel 17 (Video)
12038-12123	CANADA	Geostationary		CTS-1
12046.5-12089.5	US	Geostationary		SBS 1-4 channel 8
12053.000	--	Geostationary		Europe DBS channel 18 (Video)
	D	Geostationary		TV-Sat A3,A5 DBS channel 18
12069-12123	CANADA	Geostationary		Anik C1-3 channel 13

Frequency (MHz)	Country/ Agency	Use	Mode	Notes
	US	Geostationary		Gstar A1-3 channel 7
12072.000	--	Geostationary		Europe DBS channel 19 (Video)
12078-12132	US	Geostationary		USA-Sat 1,2 TR7A-West (12105 center) USA-Sat 1,2 TR15B-West (12105 center)
12082-12136	CANADA	Geostationary		Anik C1-3 channel 14
12083-12137	US	Geostationary		Gstar A1-3 channel 15
12091.000	--	Geostationary		Europe DBS channel 20 (Video)
	ESA	Geostationary		Olympus DBS channel 20
12091.900	G	Geostationary		Unisat-1 DBS channel 20
12095.5-12138.5	US	Geostationary		SBS 1-4 channel 9
12104-12176	CANADA	Geostationary		Anik B channel 18
12106-12174	US	Geostationary		Spacenet 1-3 channel 24
12111.000	--	Geostationary		Europe DBS channel 21 (Video)
12130.000	--	Geostationary		Europe DBS channel 22 (Video)
12130-12184	CANADA	Geostationary		Anik C1-3 channel 15
	US	Geostationary		Gstar A1-3 channel 8
12138-12192	US	Geostationary		USA-Sat 1,2 TR8A-West (12165 center) USA-Sat 1,2 TR16B-West (12165 center)
12143-12197	CANADA	Geostationary		Anik C1-3 channel 16
12144-12198	US	Geostationary		Gstar A1-3 channel 16
12144.5-12187.5	US	Geostationary		SBS 1-4 channel 10
12145.34	ESA	Geostationary		Olympus DBS channel 28
12149.000	--	Geostationary		Europe DBS channel 23 (Video)
12168.000	--	Geostationary		Europe DBS channel 24 (Video)
12168.62	ITALY	Geostationary		Sarit 1,2 DBS channel 24
12168.62	ESA	Geostationary		Olympus DBS channel 24
12184-12200	US	Geostationary		Gstar A1-3 TT&C
12187.000	--	Geostationary		Europe DBS channel 25 (Video)
12198.000	US	Geostationary		Gstar A1-3 beacon
12200-12700	US	Geostationary		US DBS satellite band CBS 1-2 DBSC 1-2 GSC 1-2 RCA (BSS) 1-4 STC 1-4 USSB 1-2 VSS 1-2 WU (BSS) 1-4
12206	--	Geostationary		Europe DBS channel 26 (Video)
12206.98	S/NOR/FIN	Geostationary		Tele-X DBS channel 26
12224.000	--	Geostationary		America DBS channel 1 (Video)
12226	--	Geostationary		Europe DBS channel 27 (Video)
12238.800	--	Geostationary		America DBS channel 2 (Video)
12245	--	Geostationary		Europe DBS channel 28 (Video)
	ITALY	Geostationary		Sarit 1,2 DBS channel 28
12253.16	--	Geostationary		America DBS channel 3 (Video)
12264	--	Geostationary		Europe DBS channel 29 (Video)
12267.74	--	Geostationary		America DBS channel 4 (Video)
12282.32	--	Geostationary		America DBS channel 5 (Video)
12283	--	Geostationary		Europe DBS channel 30 (Video)
12296.90	--	Geostationary		America DBS channel 6 (Video)
12302	--	Geostationary		Europe DBS channel 31 (Video)
12311.48	--	Geostationary		America DBS channel 7 (Video)
12322	--	Geostationary		Europe DBS channel 32 (Video)
	ITALY	Geostationary		Sarit 1,2 DBS channel 32
	S/NOR/FIN	Geostationary		Tele-X DBS channel 32
12326.06	--	Geostationary		America DBS channel 8 (Video)
12340.64	--	Geostationary		America DBS channel 9 (Video)
12341	--	Geostationary		Europe DBS channel 33 (Video)
12355.22	--	Geostationary		America DBS channel 10 (Video)
12360	--	Geostationary		Europe DBS channel 34 (Video)
12369.80	--	Geostationary		America DBS channel 11 (Video)
12379	--	Geostationary		Europe DBS channel 35 (Video)
12384.38	--	Geostationary		America DBS channel 12 (Video)
12398	--	Geostationary		Europe DBS channel 36 (Video)
	ITALY	Geostationary		Sarit 1,2 DBS channel 36

Frequency (MHz)	Country/ Agency	Use	Mode	Notes
12398.96	--	Geostationary		America DBS channel 13 (Video)
12413.54	--	Geostationary		America DBS channel 14 (Video)
12417	--	Geostationary		Europe DBS channel 37 (Video)
12428.12	--	Geostationary		America DBS channel 15 (Video)
12437	--	Geostationary		Europe DBS channel 38 (Video)
12442.70	--	Geostationary		America DBS channel 16 (Video)
12456	--	Geostationary		Europe DBS channel 39 (Video)
12457.28	--	Geostationary		America DBS channel 17 (Video)
12471.86	--	Geostationary		America DBS channel 18 (Video)
12475	--	Geostationary		Europe DBS channel 40 (Video)
	S/NOR/FIN	Geostationary		Tele-X DBS channel 40
	ITALY	Geostationary		Sarit 1,2 DBS channel 40
12489-12500	D	Geostationary	TM BCN	TV-Sat A3,A5
12500-12750	FRANCE	Geostationary		Telecom 1A/1B
		Geostationary		ISI 1-2
		Geostationary		Unisat 1-3
		Geostationary		Videosat 1-2
	US	Geostationary		Cygnus 1,2 European downlink
	L	Geostationary		GDL-5,6
12500-14500	D	Geostationary		DFS 1,2
12500-12583.3	ETSO	Geostationary		Eutelsat-I F1,2,3 ESS channel
12504-12540	FRANCE	Geostationary		Videosat 1,2 TR-R1/ F-Sat 2/ Telecom F1,2,3 Ch R1
12507-12561	US	Geostationary		USA-Sat 1,2 TR13B-East (12534 center)
12546-12582	FRANCE	Geostationary		Videosat 1,2 TR-R2/ F-Sat 2/ Telecom F1,2,3 Ch R2
12559.340	--	Geostationary		America DBS channel 24 (Video)
12567-12621	US	Geostationary		USA-Sat 1,2 TR6A-East (12596 center) USA-Sat 1,2 TR14B-East (12594 center)
12573.920	--	Geostationary		America DBS channel 25 (Video)
12588-12624	FRANCE	Geostationary		Videosat 1,2 TR-R3/ F-Sat 2 Telecom F1,2,3 Ch R3
12588.500	--	Geostationary		America DBS channel 26 (Video)
12595-12685	S/NOR/FIN	Geostationary		Tele-X telecom channel
12603.080	--	Geostationary		America DBS channel 27 (Video)
12617.660	--	Geostationary		America DBS channel 28 (Video)
12627-12681	US	Geostationary		USA-Sat 1,2 TR7A-East (12654 center) USA-Sat 1,2 TR15B-East (12654 center)
12630-12666	FRANCE	Geostationary		Videosat 1,2 TR-R4/ F-Sat 2/ Telecom F1,2,3 Ch R4
12632.240	--	Geostationary		America DBS channel 29 (Video)
12646.820	--	Geostationary		America DBS channel 30 (Video)
12661.400	--	Geostationary		America DBS channel 31 (Video)
12672-12708	FRANCE	Geostationary		Videosat 1,2 TR-R5/ F-Sat 2/ Telecom F1,3 Ch R5
12675.980	--	Geostationary		America DBS channel 32 (Video)
12687-12741	US	Geostationary		USA-Sat 1,2 TR8A-East (12714 center) USA-Sat 1,2 TR16B-East (12714 center)
12705-12745	S/NOR/FIN	Geostationary		Tele-X Telecom channel
12714-12750	FRANCE	Geostationary		Videosat 1,2 TR-R6 F-Sat 2/ Telecom F1,3 Ch R6
12744-12750	US	Geostationary	TM BCN	USA Sat 1,2
12750-13250	ISR	Ground Station		AMS 1,2
13724.2500	NASA	Geostationary	M650F9	
13766.0000	NASA	Nongeostationary	M6F9	Space Shuttle Ku-band/TDRSS/Landsat
13775.0000	NASA	Geostationary	M6F9	Space Shuttle Ku-band/TDRSS/Landsat
13778.0000	NASA	Geostationary	M26F9	
13900.0000	NASA	Geostationary	M200F9	Geos 3
14000.0000	US	Geostationary	M43F9/M72F9	
	Japan	Geostationary		CTS uplink
14000-14017	US	Geostationary		Gstar A1-3 TT&C
14000-14250	D	Ground Station		DFS 1,2
14000-14500	Various	Ground Station		See below

Ku-Band Uplink

Telecom 1A/1B	ABC 1-2	ASC 1-5
CCC 1-2	Comstar K1-3	Digisat 1A
Digisat 2	Ford 1-3	RSI 1-4
Spotnet K1-4	Usat 1-4	AMS 1-2
ISI 1-2	Morelos 1-2	Unisat 1-3
Galaxy H5-7	Satcom K1-3	Westar A-F
Pacific Area Sat. System A/B	Orion 1-3	Videosat 1-2
	SBS 1-4	USA-Sat 1-2
Cygnus 1-2		

Frequency (MHz)	Country/ Agency	Use	Mode	Notes
14001.0	D	Ground Station		DFS 1,2 command channel
14001.5	US	Ground Station		Gstar A1-3 command channel
14002-14038	US	Ground Station		Orion 1,2,3 Ch1 (14020 center)
14003-14057	US	Ground Station		Gstar A1-3 channel 1
14004-14040	FRANCE	Ground Station		Videosat 1,2 TR-R1/ F-Sat 2/ Telecom F1,2,3 Ch R1
14008-14042	URS	Ground Station		Loutch channel 1
14017-14071	US	Ground Station		Gstar A1-3 channel 9
14022-14058	US	Ground Station		Orion 1,2,3 Ch2 (14040 center freq)
14042-14078	US	Ground Station		Orion 1,2,3 Ch3 (14060 center freq)
14046-14082	FRANCE	Ground Station		Videosat 1,2 TR-R2/ F-Sat 2/ Telecom F1,2,3 Ch R2
14058-14092	URS	Ground Station		Loutch channel 2
14062-14098	US	Ground Station		Orion 1,2,3 Ch4 (14080 center freq)
14064-14418	US	Ground Station		Gstar A1-3 channel 2
14078-14132	US	Ground Station		Gstar A1-3 channel 10
14082-14118	US	Ground Station		Orion 1,2,3 Ch5 (14100 center freq)
14088-14124	FRANCE	Ground Station		Videosat 1,2 TR-R3/ F-Sat 2/ Telecom F1,2,3 Ch R3
14095-14185	S/NOR/FIN	Ground Station		Tele-X Telecom channel
14102-14138	US	Ground Station		Orion 1,2,3 Ch6 (14120 center freq)
14108-14142	URS	Ground Station		Loutch channel 3
14112.5	US	Ground Station	M22F9	
14122-14158	US	Ground Station		Orion 1,2,3 Ch7 (14140 center freq)
14125-14179	US	Ground Station		Gstar A1-3 channel 3
14250-14500	D	Ground Station		DFS 1,2
14130-14166	FRANCE	Ground Station		Videosat 1,2 TR-R4/ F-Sat 2/ Telecom F1,F2,F3 Ch R4
14139-14193	US	Ground Station		Gstar A1-3 channel 11
14142-14178	US	Ground Station		Orion 1,2,3 Ch8 (14160 center)
14152.5-14192.5	ESA	Ground Station		ECS/OTS
14158-14192	URS	Ground Station		Loutch channel 4
14162-14198	US	Ground Station		Orion 1,2,3 Ch9 (14180 center)
14172-14208	FRANCE	Ground Station		Videosat 1,2 TR-R5 F-Sat 2/ Telecom F1/F3 Ch R5
14182-14218	US	Ground Station		Orion 1,2,3 Ch10 (14200 center)
14186-14240	US	Ground Station		Gstar A1-3 channel 4
14200-14254	US	Ground Station		Gstar A1-3 channel 12
14202-14238	US	Ground Station		Orion 1,2,3 Ch11 (14220 center)
14205-14245	S/NOR/FIN	Ground Station		Tele-X Telecom channel
14208-14242	URS	Ground Station		Loutch channel 5
14214-14250	FRANCE	Ground Station		Videosat 1,2 TR-R6/ F-Sat 2/ Telecom F1,F3 Ch R6
14261-14315	US	Ground Station		Gstar A1-3 channel 13
14262-14298	US	Ground Station		Orion 1,2,3 Ch12 (14280 center)
14236.9625	US	Ground Station	M10F9	
14242.5-14362.5	ESA	Ground Station		ECS/OTS
14247-14301	US	Ground Station		Gstar A1-3 channel 5
14258-14292	URS	Ground Station		Loutch channel 6

Frequency (MHz)	Country/ Agency	Use	Mode	Notes
14282-14318	US	Ground Station		Orion 1,2,3 Ch13 (14300 center)
14302-14338	US	Ground Station		Orion 1,2,3 Ch14 (14320 center)
14308-14342	URS	Ground Station		Loutch channel 7
14308-14362	US	Ground Station		Gstar A1-3 channel 6
14322-14358	US	Ground Station		Orion 1,2,3 Ch15 (14340 center)
14322-14376	US	Ground Station		Gstar A1-3 channel 14
14342-14378	US	Ground Station		Orion 1,2,3 Ch16 (14360 center)
14358-14392	URS	Ground Station		Loutch channel 8
14362-14398	US	Ground Station		Orion 1,2,3 Ch17 (14380 center)
14369-14423	US	Ground Station		Gstar A1-3 channel 7
14382-14418	US	Ground Station		Orion 1,2,3 Ch18 (14400 center)
14383-14437	US	Ground Station		Gstar A1-3 channel 15
14387.5000	US	Ground Station	M225F9	
14402-14438	US	Ground Station		Orion 1,2,3 Ch19 (14420 center)
14408-14442	URS	Ground Station		Loutch channel 9
14422-14458	US	Ground Station		Orion 1,2,3 Ch20 (14440 center)
14430-14484	US	Ground Station		Gstar A1-3 channel 8
14442-14478	US	Ground Station		Orion 1,2,3 Ch21 (14460 center)
14444-14498	US	Ground Station		Gstar A1-3 channel 16
14455-14460	ESA	Ground Station		ECS/OTS
14458-14492	URS	Ground Station		Loutch channel 10
14462-14498	US	Ground Station		Orion 1,2,3 Ch22 (14480 center)
14484-14500	US	Ground Station		Gstar A1-3 TT&C
14490-14500	ESA	Ground Station		Marots
14497.000	US	Ground Station	M5F9/M1F9	
14499.000	D	Ground Station		DFS 1,2 command channel
14912.500	NASA	Ground Station	M6F9	
15000.85000	NASA	Ground Station		STS Ku-band
15000.85000	NASA	Ground Station	M225F9/ M170F9	Landsat
17100-17850	ITALY	Geostationary		Sarit 1,2 Telecom transponder
17300-17800	US	Ground Station		US DBS uplink:
				CBS 1-2 DBSC 1-2
				GSC 1-2 RCA (BSS) 1-4
				STC 1-4 USSB 1-2
				VSS 1-2 WU (BSS) 1-4
	G	Ground Station		Unisat-1 telecom uplink
17307.000	D	Ground Station		TV-Sat A3,A5 TT&C
17324.300	--	Ground Station		America DBS channel 1
17338.580	--	Ground Station		America DBS channel 2
17353.160	--	Ground Station		America DBS channel 3
17367.740	--	Ground Station		America DBS channel 4
17382.320	--	Ground Station		America DBS channel 5
17396.900	--	Ground Station		America DBS channel 6
17441.480	--	Ground Station		America DBS channel 7
17426.060	--	Ground Station		America DBS channel 8
17440.640	--	Ground Station		America DBS channel 9
17455.220	--	Ground Station		America DBS channel 10
17469.800	--	Ground Station		America DBS channel 11
17484.380	--	Ground Station		America DBS channel 12
17498.960	--	Ground Station		America DBS channel 13
17513.540	--	Ground Station		America DBS channel 14
17528.120	--	Ground Station		America DBS channel 15
17542.700	--	Ground Station		America DBS channel 16
17557.280	--	Ground Station		America DBS channel 17
17571.860	--	Ground Station		America DBS channel 18
17586.44	--	Ground Station		America DBS channel 19
17601.020	--	Ground Station		America DBS channel 20
17615.600	--	Ground Station		America DBS channel 21
17630.180	--	Ground Station		America DBS channel 22
17644.760	--	Ground Station		America DBS channel 23
17659.320	--	Ground Station		America DBS channel 24
17673.920	--	Ground Station		America DBS channel 25
17688.500	--	Ground Station		America DBS channel 26

Frequency (MHz)	Country/ Agency	Use	Mode	Notes
17700-21200	JAPAN	Geostationary		CS-1
17700-11714	D	Ground Station		TV-Sat A3,A5 TT&C
17703.080	--	Ground Station		America DBS channel 27
17717.660	--	Ground Station		America DBS channel 28
17732.240	--	Ground Station		America DBS channel 29
17746.820	--	Ground Station		America DBS channel 30
17761.400	--	Ground Station		America DBS channel 31
17775.980	--	Ground Station		America DBS channel 32
17806.980	S/NOR/FIN	Ground Station		Tele-X DBS channel 26
17810-17940	JAPAN	Geostationary		CS-2A
17922.060	S/NOR/FIN	Ground Station		Tele-X DBS channel 32
17990-17782	ITALY	Ground Station		Sirio
18075.50	S/NOR/FIN	Ground Station		Tele-X DBS channel 40
18090-18220	JAPAN	Geostationary		CS-2A
18100-20200	ITALY	Geostationary		Italsat
18370-18500	JAPAN	Geostationary		CS-2A
18650-18780	JAPAN	Geostationary		CS-2A
18685.000	ITALY	Geostationary	TM BCN	Italsat
18850-19550	ESA	Geostationary		Olympus Ka-Telecom Ch2
18905-18945	ESA	Geostationary		Olympus Ka-Telecom Ch1
18930-19060	JAPAN	Geostationary		CS-2A
19000-21000	US	Geostationary		ATS-6
19040.000	US	Geostationary		Comstar 2-4 satellites
19210-19340	JAPAN	Geostationary		CS-2A
19450.000	JAPAN	Geostationary		CS-2A TT&C
19455-19495	ESA	Geostationary		Olympus Ka-Telecom Ch3
19700-20200	US	Geostationary		ASC 4-5
	D	Geostationary		DFS 1,2
20000.000	JAPAN	Geostationary		Japanese COMSAT
27500-30000	ITALY	Geostationary		Italsat
27550-30050	JAPAN	Ground Station		CS (Sakura) uplink
27800.000	US	Ground Station		Geos 3
27997.255-28697.255	ESA	Ground Station		Olympus Ka-Telecom Ch2
28050-28650	ITALY	Ground Station		Sarit 1,2 Telecom transponder
28052.255-28092.255		Ground Station		Olympus Ka-Telecom Ch1
28560.000	US	Ground Station		Comstar 2-4
28602.255-28642.255	ESA	Ground Station		Olympus Ka-Telecom Ch3
29000-31000	US	Geostationary		ATS-6
29500-29600	FRANCE	Ground Station		F-Sat 1,2
29500-30000	US	Ground Station		ASC 4-5
	D	Ground Station		DFS 1,2
29665-29765	FRANCE	Ground Station		F-Sat 1,2
29830-29930	FRANCE	Ground Station		F-Sat 1,2
29997.000	FRANCE	Ground Station		F-Sat 1,2 TT&C
31650.000	JAPAN	Geostationary		Ayame-2 3.2 watt beacon
33400.000	USAF	Geostationary		M500F1
	USAF	Space		M500F1
34526.2000	JAPAN	Geostationary		ETS-2
36786.000	USAF	Geostationary	M500F1	
36787.600	USAF	Ground Station		LES 8 uplink
36833.300	USAF	Ground Station		LES 9 uplink
36840.000	USAF	Ground Station	M500F9	LES 9 downlink
	USAF	Nongeostationary	M500F9	LES 9 downlink
	USAF	Nongeostationary	M500F9	LES 9 downlink
36900-38000	USAF	Nongeostationary		LES 8/9 downlink
36900.000	USAF	Nongeostationary	M500F9/ M200F9	LES 9 downlink
37980.000	USAF	Nongeostationary	M500F9/M200F9	
38040.000	USAF	Nongeostationary	M500F9	
	USAF	Nongeos; Up-Ngeos	M500F9	
38047.400	USAF	Ground Station		LES 9 uplink

Frequency (MHz)	Country/ Agency	Use	Mode	Notes
38092.400	USAF	Ground Station		LES 8 uplink
38094.000	USAF	Ground Station	M500F9	
39087.000	ITALY	Geostationary		Italsat experimental beacons
39592.000	ITALY	Geostationary		Italsat experimental beacons
40097.000	ITALY	Geostationary		Italsat experimental beacons
41700.000	NASA	Ground Station		Geos 3
49490.000	ITALY	Geostationary		Italsat experimental beacons
217000- 219000	URS	Nongeostationary		Used by Soviets for precise orbital determination of spacecraft
281700	USAF	Ground Station	6PO/20PO	
563300	USAF	Ground Station	1PO	
	USAF	Nongeostationary	6PO	

Author's Note:

Not all satellites covered in the frequency list are covered in other sections and chapters of this book. However, future editions will cover these satellites as more information is added to the author's files. All frequencies listed above with a mode indicated are extracted from the federal government's IRAC microfiche files available from Grove Enterprises.

Emission Designators

The abbreviations used above were replaced at WARC '79 by a new system. We have used the old set here because of familiarity to many readers.

The first number is the bandwidth in kilohertz (or megahertz is preceded by "M");
the next letter indicates the type of modulation (Frequency, Amplitude of Pulse);
the suffix number indicates the type of intelligence carried (0 = unmodulated; 1 = on-off keying; 3 = voice; 9 = non-standard, usually multiplex).

Other mode abbreviations are:

- AM Amplitude Modulation
- BCN Beacon
- CW Morse code
- FM Frequency Modulation
- LSB Lower Side Band
- PDM Pulse Data Modulation
- SS Spread Spectrum
- TM Telemetry
- WB Wide Band

Appendix F

GLOSSARY OF TERMS AND ABBREVIATIONS

ABM -- Anti-ballistic missile
AFRTS -- American Forces Radio and Television Service
AFSCF -- Air Force Satellite Control Facility
AFTAC -- Air Force Technical Applications Center
A/G -- Air to ground
AGRI -- Department of Agriculture
AGRISTARS - Agriculture and Resources Inventory Surveys Through Aerospace Remote Sensing
Altitude -- The distance between a satellite and the point on the surface of the earth directly below it.
AM -- Amplitude Modulation
AMSAT -- Amateur Satellite (Radio Amateur Satellite Corp)
AMSU -- Advanced Microwave Sounder Unit
AOS -- Acquisition of Signal: The point in time when the signal from a satellite is first received.
APT -- Automatic Picture Transmission
Apogee -- The highest point in a satellite's orbit above the earth.
ARB -- Arab League
ARG - Argentina
Argument of Perigee -- The position in the orbital plane of the satellite's perigee, measured in degrees and measured counterclockwise from the equator back to where the perigee occurs. Abbreviated by some as ArgP.
ARO -- Amateur Radio Operator
ARPA -- Advanced Research Projects Agency
ARS -- Amateur Radio Satellites
ASC -- American Satellite Corp
Ascending Node -- The point at which a satellite crosses the equator headed in a northerly direction.
ASW -- Anti-submarine warfare
AT & T -- American Telephone & Telegraph
AUS -- Australia
AVHRR -- Advanced Very High Resolution Radiometer
AVCS -- Advanced Vidicon Camera System
Azimuth -- Compass bearing; the angle in the horizontal plane measured clockwise with respect to north (0 degrees)
B -- Brazil
Ballistic Coefficient -- See Drag
BEL -- Belgium
BMD -- Ballistic Missile Defense
BMEWS -- Ballistic Missile Early Warning
C - Department of Commerce
CAN -- Canda
CCD -- Charge Coupled Device
CCR -- Corner Cube Reflectors
CDA -- Command and Data Acquisition
CHN -- China
CIA -- Central Intelligence Agency
Clarke Belt -- The equatorial corridor in space in which geosynchronous satellites appear stationary above the earth. Named after science fiction author Arthur C. Clarke.
CLM -- Colombia
CNES -- Centre National d'Etudes Spatiales
COMINT -- Communications Intelligence
COMIREX -- Committee on Imagery Requirements and Exploitation
COMOR -- Committee on Overhead Reconnaissance
COPUOS -- Committee on the Peaceful Uses of Outer Space
CPA -- Compressed Pulse Altimeter
CSOC -- Consolidated Space Operations Center
CW -- Continuous Wave (Morse)
CZCS -- Coastal Zone Color Scanner
Czech/TCH -- Czechoslovakia
D -- Federal Republic of Germany (West)
Decay -- A decrease in a satellite's orbit, leading to disintegration
DARPA -- Defense Advanced Research Projects Agency
DCA -- Defense Communications Agency
DCL -- Direct Communications Link (Washington-Moscow Hotline)
DCP -- Data Collection Platform
DCS -- Data Collection System
DEFSMAC -- Defense Spacial Missile and Astronautics Center
Descending Node -- The point at which a satellite crosses the equator headed in a southerly direction.
DIA -- Defense Intelligence Agency
DIP -- Digital Image Processing
DISCOS -- Disturbance Compensation System
DOC -- Canadian Department of Communications
DOD -- Department of Defense

DOE -- Department of Energy
Downlink -- Frequency or signal transmitted from the satellite to ground stations
Drag -- The forces that act on a satellite to bring it back towards the earth surface (gravity, solar wind, etc.)
DSB -- Direct Sounder Broadcast
Eccentricity -- The figure used to describe mathematically the elongation of a circle into an ellipse. A figure of zero equates to a circle and maximum elongation equates to a figure of one.
ECM -- Electronic Countermeasures
Elevation -- The angle above the horizon measured in degrees.
ELINT -- Electronic Intelligence
ELT -- Emergency Locator Transmitter
EME -- Environmental Measuring Experiment
EMP -- Electronic Pulse
EOPAP -- Earth and Ocean Physical Applications Program
EORSAT -- Elint ocean recon satellite
EPIRB -- Emergency Indicating Radio Beacon
Epoch -- This is the Julian day and fractional parts of that day that orbital elements are referenced to (i.e. 87 365.015636 equates to year day.fractional part of a day).
Equatorial plane -- The plane containing the earth's equator
EQX -- The equator crossing point (longitude) measured in degrees.
ERBE -- Earth Radiation Budget Experiment
ESA -- European Space Agency
ESOC -- European Space Operations Center
ESRO -- European Space Research Organization (forerunner to ESA)
ETR -- Eastern Test Range off Cape Canaveral and Cape Kennedy
EVA -- Extra Vehicular Activity (spacewalk)
EVM -- Earth Viewing Module
F -- France
FAX - Facsimile
FIN -- Finland
First Derivative of Mean Motion -- See Drag
FM -- Frequency Modulation
FOBS -- Fractional Orbital Bomb System
FOC -- Faint Object Camera
FOV -- Field of View
FPR -- Flat Plate Radiometer
FSK -- Frequency Shift Keying
G -- Great Britain (United Kingdom)
GARP -- Global Atmospheric Research Program
GEODSS -- Ground-based Electro-Optical Deep Space Surveillance System
Geos -- Geostationary Orbit
Geostationary satellite -- a satellite that is placed approximately 36,000 km (22,300 miles) above the earth will have a speed matching that of the earth's rotation, and the satellite appears to remain stationary above the equator.
GHz -- Gigahertz
Gnd Stn -- Ground Station
Ground track (subsatellite point) -- A point on the earth's surface that is directly below a satellite. Also known as SSP.
GOT -- Geostationary Transfer Orbit
HCMR -- Heat Capacity Mapping Radiometer
HEPAD -- High Energy Proton and Electron Detector
HF -- High Frequency (3-30 MHz)
HIRS/2 -- High Resolution Infrared Radiation Scanning
HRIR -- High Resolution Infrared Scanning Radiometer
HRPT -- High Resolution Picture Transmission
HRS -- High Resolution Spectograph
HRV -- High Resolution Visible
HUMINT -- Human intelligence
Hz -- Hertz
IAF -- International Astronautical Federation
ICBM -- Intercontinental Ballistic Missile
IK -- Intercosmos
IMS -- International Magnetic Study
Inclination -- The angle between the plane of the satellite orbit and the earth's equatorial plane.
Increment (longitudinal) -- The change in longitude between two successive ascending node passes of a specific satellite; measured in degrees west per orbit.
IND -- India
INS/INDO -- Indonesia
Interior -- Department of Interior
International Satellite Designation -- The International Committee of Space Research (COSPAR) registers all satellites, fragments of launches and the number of successful launches duirng the year.

The designation for Salyut 7 is 1982-033A. Usually the letter A is given to the payload (instrumented satellite). B is given to the rocket body and C, D, E, etc. are assigned to fragments and debris. The letters I and O are never used in these designations.

Many of the Soviet eight-payload launched satellites that are used for military communications are assigned designations A-H with J being the rocket body. When more than 24 pieces are assigned to a particular launch, as can happen after an explosion, the sequence continues after Z with AA, AB, AC, etc. The greatest number of fragments recorded from one launch is 462, resulting from the explosion of OV2/LCS-2, a USAF payload numbered 1965-82A.

In the early years (1957-1962) a different system was used. Satellites were designated using the 24 Greek letters; thus the first launch of 1959 was 1959 alpha, the second was 1959 beta, and so on. After 24 launches, double letters were used (i.e. -- 1959 alpha alpha, 1959 alpha beta, etc.) Following the Greek letters a number would be used to indicate each piece of debris, payload or rocket body associated with that launch.

IOC -- Initial Operational Capability

IR -- Infrared
IRBM -- Intermediate-Range Ballistic Missile
ISAS -- Institue of Space and Aeronautical Science
ISRO -- Indian Space Research Organization
ITSO -- International Telecommunications Satellite Organization (Intelsat)
ITT -- International Telephone & Telegraph
ITU -- International Telecommunications Union; Geneva, Switzerland
J -- Japan
JCS -- Joint Chiefs of Staff
JPL -- Jet Propulsion Laboratory
JRC -- Joint Reconnaissance Center
KeV -- Thousands Electron Volts
KH -- Keyhole
kHz -- kilohertz
km -- kilometer
KY -- Kapustin Yar
Lant -- Atlantic Ocean
LASINT -- Laser intelligence
LIMS -- Limb Infrared Monitor of the Stratosphere
LOS (Loss of Signal) -- That point in time when the signal from a satellite is no longer received.
LSB -- Lower sideband
LUT - Local User Terminal
MAP -- Middle Atmospheric Program
Mean Anomaly -- An angle in degrees that increases uniformly with time, used to indicate where the satellite is along its orbit.
Mean Motion -- The number of complete orbits a satellite makes in one day.
MEPED -- Medium Energy Proton and Electron Detector
MESSR -- Multispectrum Electronic Self-Scanning
MeV -- Million Electron Volts
MHz -- Megahertz
Molniya orbit -- A unique eccentric orbit characterized by the Russian Molniya Comsats. Orbit inclined between 60-70 degrees, with a perigee point low over the south pole (200-300 miles) and an apogee over the northern hemisphere (22,000-23,000 miles).
MRBM -- Medium-range ballistic missile
MSI -- Multispectral Imaging
MSR -- Microwave Scanning Radiometer
MSS -- Multispectural Scanner System
MUX -- Multiplex
NAL -- National Aerospace Laboratory
NASA - National Aeronautics and Space Administration
NASDA -- National Space Development Agency
NATO -- North Atlantic Treaty Organization
NAVSPASUR -- Naval Space Surveillance System
NBFM -- Narrow-band frequency modulation
NCA -- National Command Authority
NCMC -- NORAD Cheyenne Mountain Center
NEC -- Nippon Electric Company
NESC -- National Satellite Center
NESS -- National Environmental Satellite Service
NG -- Non-government
NHK -- Japanese Broadcasting Corp
NMC -- National Meteorological Centers
NOAA -- National Oceanographic and Atmospheric Administration
Node -- The point where a satellite ground track crosses the equator
NOIC -- Navy Operational Intelligence Center
Nongeos -- Non-geostationary orbit
NORAD -- North American Aerospace Defense (Command)
NRL -- Naval Research Laboratory
NRO -- National Reconnaissance Office
NSA -- National Security Agency
NSF -- National Science Foundation
NSP -- Navy Space Project
NTM -- National Technical Means
OBC -- Optical Bar Camera
Oblique -- Recon imagery taken at an angle
OCC -- Operations Control Center
PAM -- Payload Assist Module; also pulse-train amplitude modulation
PCA (point of closest approach) -- The point in a satellite's ground track where the satellite passes closest to a particular ground station.
PDM -- Pulse data modulation
Perigee -- The lowest point in a satellite's orbit above the earth.
Period -- The amount of time it takes a satellite to make one complete revolution of the earth. **Anomalistic period** represents the elapsed time between two successive perigees of a satellite. **Nodal period** represents the period of time between two ascending nodes of a satellite's orbit.
PHOTINT -- Photographic intelligence
PL -- Plesetsk
Polar orbit -- An orbit that takes a satellite over, or nearly over, the earth's poles, making every spot on earth accessible to the satellite every day.
RA -- Radar Altimeter
RAAN -- Right Ascension of Ascending Node
RADINT -- Radar intelligence
RAF -- Royal Air Force
RBV -- Return Beam Vidicon
Reference Epoch -- See Epoch
Reference orbit -- The orbit that follows the first ascending node of the UTC date.
RORSAT -- Radar Ocean Recon Satellite
RTG -- Radioisotope Thermal Generator
RTS -- Remote Tracking Site
RTTY -- Radio teletype
S -- Sweden
SAC -- Strategic Air Command
SAM II -- Stratospheric Aerosol Measurement experiment
SAMS -- Stratospheric and Mesospheric Sounder
SAR -- Search and Rescue
SAR -- Synthetic Aperture Radar
SASS -- Seasat-A Satellite Scatterometer
SCF -- Satellite Control Facility
SCI -- Sensitive Compartmented Information
SCPC -- Single carrier per channel
SDI -- Strategic Defense Initiative
SEM -- Space Environmental Monitor
SIGINT -- Signals intelligence

SIR -- Shuttle Imaging Radar
SMMR -- Scanning Multichannel Microwave Radiometer
Spacecom -- Space Communications Company
SPADATS -- Space Tracking and Detection System
SPADOC -- Space Defense Operations Center
SR -- Scanning Radiometer
SR -- Strategic Reconnaissance
SRUV/2 -- Solar Backscatter Ultraviolet Radiometer/2
SSC -- Special Service Channel
SSP -- Subsatellite point; see ground track
SSTS -- Space Surveillance and Tracking System
SSU -- Stratospheric Sounding Unit
STC -- Satellite Television Corp
Sub-satellite point (SSP) -- See ground track
Sun-synchronous -- An orbit, generally used by weather and imaging recon satellites, which follows an inclination of about 98 degrees and therefore keeps the satellite's orbital plane the same relative to the sun. This makes it easier to spot change by noting whether the length or position of shadows varies from one day to another.
SWL -- Shortwave listener
TAC -- Tactical Air Command; also tactical
TAV -- Trans-Atmospheric Vehicle
TBUS -- APT Prediction message
TCA -- Time of Closest Approach. The time that a satellite makes its closest approach to a ground station during a specific orbit.
TDMA -- Time Division Multiple Access
TECHINT -- Technical Intelligence
TED -- Total Energy Detector
Telemetry -- Radio emission from a satellite (or other remote transmitter) which conveys information on the performance or status of onboard subsystems; also refers to the information itself.
TELINT -- Telemetry intelligence
THIR -- Temperature-Humidity Infrared Radiometer
TK -- Talent Keyhole
TLM -- See telemetry
TM -- Thermatic Mapper; also see telemetry
TOMS -- Total Ozone Mapping Spectrometer
TOVS -- Tiros OperationalVertical Sounder
Transponder (Transp) -- This repeater-like device receives the uplink signal from ground stations, amplifies the signal and retransmits the signal back to ground stations on another freuqency range.
TT -- Tyuratam
TT & C -- Tracking, telemetry, and command
UK -- United Kingdom (Great Britain)
Uplink -- The frequency signal that is transmitted from the ground station to the satellite.
URS/USSR -- Soviet Union
US -- United States
USA -- United States Army
USAF -- United States Air Force
USB -- Upper sideband
USCG -- United States Coast Guard
US-DOD -- United States Department of Defense
USN -- United States Navy
VAS -- Visible Infrared Spin-Scan Radiometer Atmospheric Sounder
VHRR -- Very High Resolution Radiometer
VIRGS -- VISSR Image Registration and Gridding System
VIRR -- Visible-Infrared Radiometer
VIS -- Visible
VISSR -- Visible Infrared Spin-Scan Radiometer
VNIR -- Visible and Near-Infrared Radiometer
VTIR -- Visible andThermal Infrared Radiometer
VTPR -- Vertical Temperature Profile Radiometer
WBFM -- Wide-band frequency modulation
WEFAX -- Weather Facsimile
WMO -- World Meteorological Organization
WPM -- Words per minute
WTR -- Western Test Range

Mathematical Conversions

Readers will notice that the use of kilometers is found extensively in this book as well as in most archival information on satellites. To be complete, I have included some metric conversion to aid the monitor in using available satellite information.

Symbol	When You Know	Multi- By	To Find	Symbol
Length:				
in	inches	2.54	centimeters	cm
ft	feet	30.48	centimeters	cm
mi	miles	1.6	kilometers	km
mm	millimeters	0.04	inches	in
cm	centimeters	0.4	inches	in
m	meters	3.3	feet	ft
km	kilometers	0.6	miles	mi
Mass (weight):				
oz	ounces	28.0	grams	g
lb	pounds	0.45	kilograms	kg
g	grams	0.035	ounces	oz
kg	kilograms	2.2	pounds	lb

Constants

Mass of Earth (M) = 5.98×10^{21} kg
Mean equatorial radius of earth = 3963 statute miles
Mean radius of earth = 3959 statute mile
Siderial day = 1436.07 minutes
Solar day = 1440 minutes

Frequency Conversion

Kilohertz (kHz), megahertz (MHz), and gigahertz (GHz), are all related by x1000 (three decimal place). Thus, 2.2 GHz = 2200 MHz = 2,200,000 kHz.

Frequency Band Designators

Band	Frequency Range
HF	3-30 MHz
VHF	30-300 MHz
UHF	300-3000 MHz
L	500-1549 MHz
S	2000-2300 MHz
C	3700-6500 MHz
X	8000-11000 MHz
Ku	10.9-17.15 GHz